Thirty-Fourth Hanford Symposium
on Health and the Environment

Biomacromolecules: From 3-D to Applications

October 23–26, 1995
Pasco, Washington, U.S.A.

Edited by
Rick L. Ornstein

BATTELLE PRESS
Columbus • Richland

DISCLAIMER

This report was prepared as an account of work sponsored by an agency of the United States Government. Neither the United States Government nor any agency thereof, nor Battelle Memorial Institute, nor any of their employees, makes any warranty, expressed or implied, or assumes any legal liability or responsibility for the accuracy, completeness, or usefulness of any information, apparatus, product, or process disclosed, or represents that its use would not infringe privately owned rights. Reference herein to any specific commercial product, process, or service by trade name, trademark, manufacturer, or otherwise does not necessarily constitute or imply its endorsement, recommendation, or favoring by the United States Government or any agency thereof, or Battelle Memorial Institute. The views and opinions of authors expressed herein do not necessarily state or reflect those of the United States Government or any agency thereof.

PACIFIC NORTHWEST NATIONAL LABORATORY
operated by
BATTELLE
for the
UNITED STATES DEPARTMENT OF ENERGY
under Contract DE-AC06-76RLO 1830

Library of Congress Cataloging-in-Publication Data

Hanford Symposium on Health and the Environment (34th : 1995 : Pasco, Wash.)
Biomacromolecules : from 3-D structure to applications / thirty-fourth Hanford Symposium on Health and the Environment, October 23-26, 1995, Pasco, Washington, U.S.A. / edited by Rick L. Ornstein.
p. cm.
Includes bibliographical references and index.
ISBN 1-57477-019-5
1. Structure-activity relationships (Biochemistry)—Congresses.
I. Ornstein, Rick L., 1950– . II. Title.
QP517.S85H36 1995 96-27519
574.8′8—dc20 CIP

Battelle Press
505 King Avenue
Columbus, OH 43201-2693
614–424–6393
Toll Free: 1–800–451–3543
Fax: 614–424–3819

CONTENTS

PREFACE

The understanding of biomacromolecular structure and dynamics has broad implications for health and environmental science. The Thirty-Fourth Hanford Symposium on Health and the Environment, held in Pasco, Washington, October 23-26, 1995, examined the current state of information and future prospects for determining and applying structure-function-dynamic relationships in proteins, DNA, and their complexes.

The first major topical area in which papers were presented dealt with the current and future potential of experimental and theoretical methods for obtaining three-dimensional structural information on biomacromolecules. The first full day of the symposium and numerous papers in this volume address this topic. The second topical area addressed at the symposium was focused on several specific examples of structure-function-dynamic relationships in biomacromolecules, with an emphasis on nucleic acids and enzymes. This area of discussion lasted for one and one-half days during the symposium. The third topical area, addressed during the final day of the symposium, involved a discussion of several health and environmental areas, such as bioremediation, in which current and future progress could benefit from a greater understanding of the underlying structure and dynamics of biomacromolecules.

A benefit of the breadth of the three topical areas discussed at the 34th Hanford Symposium was the exposure of scientists and bioengineers to ideas and concepts beyond those that they normally encounter in their personal research activities. One objective in publishing the proceedings of this symposium has been to capture in a single book the broad range of new information and ideas on macromolecular structure and dynamics that were presented, together with fresh perspectives on the relevance of this information to practical applications. Many of the complex health, environmental, and industrial problems that we face today will require a greater understanding of macromolecular structure and dynamics if timely solutions are to be forthcoming.

As Chairman of the 34th Hanford Symposium, I would like to extend my appreciation to members of the Technical Program Committee: T. H. Dunning, Jr., J. H. Miller, M. D. Paulsen, M. F. Romine, D. L. Springer, T. S. Tenforde, and R. A. Walters. My thanks also go to B. J. Abbott for assistance in the text processing, and to D. L. Felton for editing the proceedings.

Rick L. Ornstein
Editor

BIOMACROMOLECULAR STRUCTURE AND FUNCTION: PROGRESS AND FUTURE OUTLOOK

STRUCTURE AND DYNAMICS OF T4 LYSOZYME AS SEEN IN MULTIPLE CRYSTAL FORMS

X.-J. Zhang[1] and B. W. Matthews[2]

Institute of Molecular Biology, Howard Hughes Medical Institute and Department of Physics

University of Oregon, Eugene, Oregon

Present address: [1]Oklahoma Medical Research Foundation, Oklahoma City, Oklahoma
[2]Author to whom correspondence should be addressed

Key Words: *Lysozyme, dynamics, crystals, α-helices, hydrogen bonding*

ABSTRACT

The structures of various mutants of T4 lysozyme have been determined in 25 different crystal forms. This provides an unusually diverse data base to compare the structures and dynamics of a closely related set of proteins in different crystal packing environments.

The molecules display a wide range of "hinge-bending" angles which may be relevant to catalysis.

Comparison of the different structures indicates that crystal contacts perturb the backbone structure of the protein by 0.2 to 0.5 Å. Helices and β-sheet strands seem to be affected equally, suggesting that protein structures can be defined and maintained equally well by polar or nonpolar interactions. The discrepancies between the lysozyme structures in different crystallographic environments are in line with other comparisons of independently determined protein crystal structures. They suggest that protein structures in general are subject to low energy changes in conformation of 0.2 to 0.5 Å. The thermal factors of the side chains in the individual crystal structures agree well with the average values (overall correlation, 0.74), suggesting that side-chain mobility in the crystals is representative of side-chain mobility in solution. The thermal factors of the main-chain atoms, however, do not agree so well (correlations with the average value ranging from 0.16 to 0.76). This suggests that in extreme cases the main-chain motion in some crystal forms may be constrained by intermolecular crystal contacts to such a degree that it is no longer representative of dynamic behavior in solution.

INTRODUCTION

One of the questions in macromolecular crystallography is the degree to which crystal packing influences structure, conformation, and dynamics.

In the course of characterizing mutants of T4 lysozyme, a large number of crystal structures, both in isomorphous and nonisomorphous forms, have been determined (Matthews, 1993; Wozniak et al., 1994; Zhang et al., 1995). These provide an unusually diverse sample to compare the structures of a closely related set of proteins under different crystallization conditions in different crystal packing environments (Zhang et al., 1995; Zhang and Matthews, 1995).

The crystal structure of wild-type lysozyme, as well as mutants Met 6 Ile (M6I) and Ile 3 Pro (I3P) were determined by multiple isomorphous replacement (Matthews and Remington, 1974; Bell et al., 1991; Faber and Matthews, 1990; Dixon et al., 1992). The other mutants were determined by molecular replacement. In some cases, however, the structure determination was complicated by substantial "hinge-bending" within the lysozyme molecule, by the presence of multiple copies of the molecule in the asymmetric unit, or by a combination of both. In order to deal with such situations, methods were developed to enhance the method of molecule replacement by the incorporation of known structural information (Zhang and Matthews, 1994).

The Influence of Crystal Packing on Structure

In order to investigate the lysozyme structures for the possible influence of crystal packing it is necessary to take hinge-bending into account. It is also necessary to differentiate structural changes due to crystal packing from those due to the introduction of mutations. Therefore, the lysozyme structures were divided into three groups: (1) lysozymes with mutations in the hinge-bending region (residues 1-14) and with disulfide bridges across the mouth of the active site; (2) lysozymes with mutations in the amino-terminal domain (residues 15-59); (3) lysozymes with mutations in the carboxy-terminal domain (residues 80-162).

Figure 1A summarizes the overall correspondence between the α-carbon atoms of each mutant and those of wild-type lysozyme. The discrepancies can be in excess of 3 Å, but are dominated by the change in hinge-bending angle of the mutant relative to the wild type.

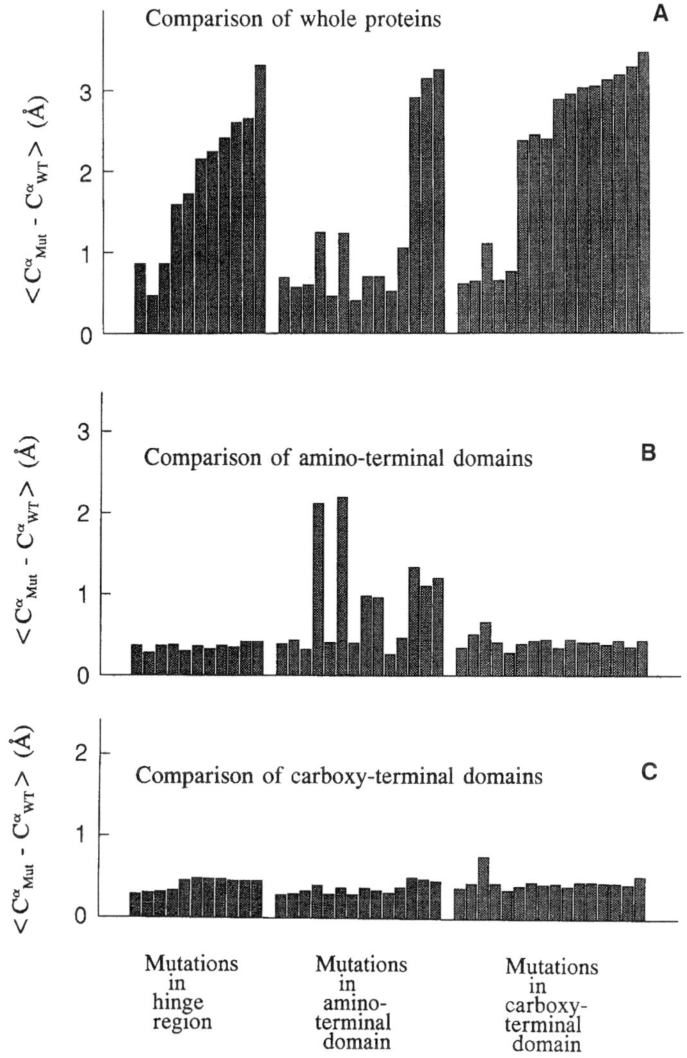

Figure 1. A, Comparison of root-mean-square discrepancy between the -carbon atoms of each mutant lysozyme (residues 1-162) with wild type. Lysozyme molecules are separated into three groups: (1) includes 11 variants with mutations in the hinge region arranged, in sequence from the left, with increasing hinge-bending angle; (2) 14 variants with mutations in the amino-terminal domain, again arranged in sequence with increasing hinge-bending angle; (3) 16 variants with mutations in the carboxy-terminal domain. **B**, Comparisons of amino-terminal domains (α-carbons 15-59) of mutant lysozymes relative to wild type (mutants as in Figure 1A). **C**, Comparisons of carboxy-terminal domains (α-carbons 80-162) of mutant lysozymes relative to wild type.

A more meaningful comparison is to compare the amino- and carboxy-terminal domains separately. These comparisons (Figures 1B and 1C) allow one to differentiate the effects of the mutations, *per se*, from other effects. Looking first at the amino-terminal domain (Figure 1B) some combinations of mutants can lead to root-mean-square backbone changes of 1 to 2 Å. These are, however, due either to insertions (S44-[AA]) or to multiple replacements (R2, 4008A and J002A). (Mutant lysozymes are identified in detail in the publication by Zhang et al., 1995.) When the mutations are restricted to the carboxy-terminal domain, the α-carbon shifts in the amino-terminal domain are in the range 0.28 to 0.66 Å and average 0.42 Å. (The 0.66-Å value corresponds to the disulfide mutant linking *Cys* 127 and *Cys* 154.) Conversely, when the carboxy-terminal domains are compared for lysozymes with mutations in the amino-terminal domain (Figure 1C) the range is 0.29 to 0.51 Å, and the average discrepancy is 0.37 Å. The differences do not appear to depend on the quality or accuracy of the x-ray structure determination. Also, they do not depend on the hinge-bending angle (Figures 1B, C). The discrepancies exceed the estimated coordinate error, which, for most of the lysozymes shown in Figure 1, is in the range 0.1 to 0.2 Å. In contrast, the backbone structures of mutant lysozymes crystallized in the *same* (i.e., wild-type) crystal form typically have discrepancies of 0.1 to 0.2 Å, i.e., approximately equal to the estimated error (e.g., Eriksson et al., 1993; Blaber *et al.*, 1994). The discrepancies between the nonisomorphous lysozymes are comparable with those between protein structures determined independently in differing crystalline environments (e.g., Chothia and Lesk, 1986; Wagner et al., 1992; Hohenester and Jansonius, 1994; Kishan et al., 1994). They clearly suggest that crystal packing does influence the structure of the protein to some degree.

The Influence of Crystal Packing on Dynamics

Refined models of crystal structures give not only the coordinates of each atom but also the thermal factor ("B-factor"). The latter provides a measure of the mean square amplitude of displacement or "mobility" of each atom. Similar molecules in different packing environments can retain similar thermal factors (Artymiuk et al., 1979). This suggests that mobility as estimated by crystallographic thermal factors is related to mobility in solution, and that crystal contacts influence thermal factors to only a limited degree. The degree to which the thermal factors for the individual residues vary from structure to structure, is illustrated in Figure 2. There is substantial fluctuation from structure to structure although some regions remain consistently more mobile than others. Not surprisingly, the side-chain thermal factors show greater variability than the main-chain factors. Consistent patterns are, however, apparent: for example, the alternation between high mobility and low mobility within the long interdomain helix (residues 60-79), corresponding to residues that are alternatively on the solvent-exposed and the buried sides of the helix.

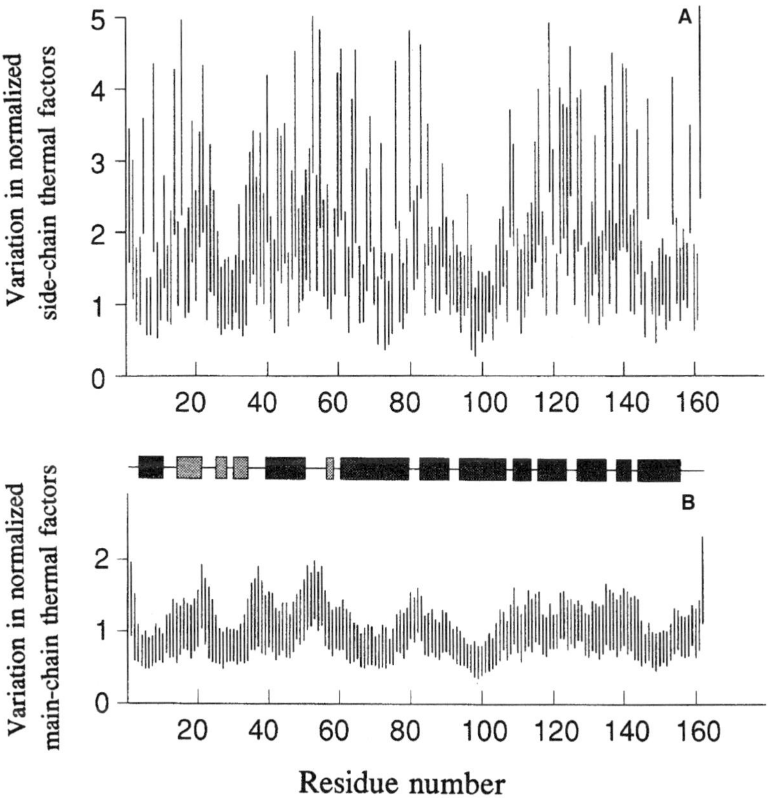

Figure 2. Variability of thermal factors for lysozyme structures in different crystalline environments. Overall thermal factor of each structure was first normalized so that average value for backbone atoms was equal to unity. On this same scale, the average thermal factor $\bar{B}$ and the standard deviation, $\sigma(B)$, were then determined for each residue in all structures. Vertical bars correspond to $\bar{B} \pm \sigma(B)$. Darker rectangles show locations of α-helices, lighter rectangles indicate β-sheet. **A**, Side-chain atoms; **B**, main-chain atoms.

The consistency of the thermal factors from structure to structure was evaluated by determining the correlation between the thermal factors of all pairs of molecules in the set. The side-chain thermal factors in the various structures (Figure 3) agree substantially better than do those for the main-chain atoms (Figure 3). Presumably this is in part because the thermal factors of the side chains have greater variability from residue to residue than do the main-chain atoms. Also, although crystal contacts may restrict the motion of some side chains, most side chains clearly retain similar thermal factors throughout all the structures. Unexpectedly, there are a number of pairs of molecules for which the correlation coefficient between the main-chain thermal factors is essentially zero, or even negative in some cases. In the most extreme case (I3P$_B$ and the double alanine insertion mutant S44-[AA]$_B$), the correlation coefficient between the main-chain thermal factors is -0.32.

	146C	3SS	4008A	6004A_A	6004A_B	6004A_C	6004A_D	6004A_E	9001A_A	9001A_B	9001A_C	9001A_D	9001A_E	BOBSR2_A	BOBSR2_B	DD39_A	DD39_B	DDC	E45A	I3C	I3L	I3P_A	I3P_B	J002A_A	J002A_B	M6I_A	M6I_B	M6I_C	M6I_D	QUAD	R96A_A	R96A_B	S44E	S44F_A	S44F_B	S44_AA_A	S44_AA_B	SS	SS_B	T26EX	T26E_NAG_A	T26E_NAG_B	T4165RF
146C	100	82	32	33	57	32	-17	30	3	14	-1	8	-15	56	58	28	10	74	50	48	56	49	-14	55	43	24	17	54	27	63	36	55	54	37	11	42	75	63	54	65	30	62	62
3SS	72	100	42	34	52	31	-23	25	4	15	-4	6	-14	48	45	19	22	57	55	51	37	32	-16	45	38	30	26	52	31	68	26	44	47	37	0	30	69	44	40	74	33	64	52
4008A	44	52	100	17	38	37	25	21	31	25	22	30	3	40	34	30	23	19	31	40	5	23	3	30	37	35	54	43	36	50	39	49	36	34	33	30	44	19	44	54	51	52	35
6004A_A	46	54	28	100	38	40	18	58	26	26	-2	30	28	-5	3	27	22	16	11	23	-1	11	1	20	41	38	20	36	29	47	30	32	43	36	11	34	38	14	45	30	43	43	27
6004A_B	69	63	44	54	100	58	32	53	33	27	34	29	16	48	45	43	15	41	36	54	25	38	11	48	31	38	43	57	47	56	56	59	53	48	33	51	50	34	51	50	57	66	54
6004A_C	52	57	46	54	76	100	43	56	43	19	52	40	21	35	30	31	27	18	26	49	18	13	13	32	37	32	54	46	45	55	46	54	39	31	45	31	30	13	52	38	65	45	39
6004A_D	27	32	35	39	57	62	100	46	50	29	51	48	36	11	5	51	12	-29	-1	17	-26	9	51	15	-3	50	54	15	61	-1	65	35	28	39	56	45	-17	-13	32	-21	41	17	7
6004A_E	55	60	36	61	75	70	62	100	42	32	29	34	37	10	15	37	33	15	13	35	12	14	30	29	36	53	36	35	50	49	49	38	45	44	34	33	17	11	50	20	51	47	30
9001A_A	44	48	42	49	53	55	54	60	100	18	38	34	16	7	1	19	23	-10	18	20	-14	7	17	10	17	41	44	33	37	26	33	22	20	26	26	23	12	-16	35	10	43	25	8
9001A_B	38	40	28	55	45	38	32	50	36	100	20	19	18	-1	0	25	14	-9	12	11	-18	4	14	-1	1	32	27	25	32	24	34	24	24	25	16	29	12	-3	23	18	18	25	11
9001A_C	39	39	33	30	59	70	53	52	47	36	100	34	20	19	3	18	7	-15	10	26	-7	-6	29	9	3	18	49	20	36	14	33	27	16	13	37	17	-1	-11	22	5	41	11	5
9001A_D	48	46	36	53	56	56	52	60	50	38	40	100	34	4	2	33	19	-5	6	31	-8	14	23	27	16	53	43	34	50	30	45	28	26	40	29	34	5	5	39	16	41	27	27
9001A_E	35	37	24	51	54	58	54	57	48	38	50	52	100	-20	-14	21	1	-19	-13	11	-18	-1	19	8	10	27	25	6	36	7	20	11	11	19	17	20	-13	-7	26	-9	24	13	3
BOBSR2_A	68	57	40	32	67	59	41	47	41	26	46	40	43	100	89	30	13	70	57	51	62	49	-5	58	34	1	34	29	19	44	45	55	32	32	34	40	50	65	36	44	40	39	57
BOBSR2_B	72	60	41	31	67	56	37	48	40	28	43	41	42	93	100	33	14	80	49	49	72	51	-11	63	39	-4	23	25	16	46	44	52	32	31	26	38	53	78	35	44	32	38	64
DD39_A	58	66	47	52	62	59	57	63	56	48	42	52	48	61	59	100	2	23	19	26	7	32	34	41	16	41	36	26	53	26	64	44	54	50	36	53	15	34	37	11	41	42	51
DD39_B	46	60	40	47	55	59	46	63	56	35	37	44	50	50	50	74	100	3	9	45	19	8	10	7	36	28	37	21	18	36	14	19	14	11	21	-4	11	1	21	23	33	27	20
DDC	80	59	32	40	62	45	18	52	35	29	26	43	34	73	78	57	47	100	42	39	77	51	-24	57	47	-10	-1	30	2	56	24	38	33	29	3	28	63	83	35	53	23	40	66
E45A	63	67	39	44	58	53	33	49	38	36	39	40	33	72	69	66	52	63	100	41	35	32	-4	54	29	18	23	36	24	39	33	33	28	32	5	28	39	35	32	49	31	39	50
I3C	66	69	49	50	69	64	46	63	49	41	50	50	57	65	62	67	64	59	61	100	38	43	6	56	44	31	56	57	40	55	39	43	40	35	24	35	48	42	46	64	57	54	48
I3L	72	62	33	37	63	53	32	57	40	31	36	42	42	74	79	62	61	78	67	66	100	43	-13	56	49	-14	0	16	-6	40	13	37	16	13	18	9	37	68	35	30	17	27	50
I3P_A	52	50	38	34	58	53	40	54	40	24	38	46	38	59	65	52	45	59	52	53	59	100	5	51	35	25	22	40	25	28	38	41	41	39	25	47	49	55	43	32	29	36	44
I3P_B	27	34	30	25	41	40	53	48	36	28	33	35	33	27	28	54	47	20	30	41	39	33	100	5	-4	48	25	13	39	-16	48	21	31	32	42	29	-32	-6	20	-29	16	13	10
J002A_A	61	60	39	50	61	58	51	62	44	34	40	53	52	66	66	71	65	59	64	72	71	58	47	100	63	20	36	41	37	40	45	55	42	40	28	48	49	58	50	43	45	48	55
J002A_B	57	56	42	51	48	53	37	53	45	28	33	46	43	53	55	61	61	53	56	67	66	46	42	84	100	14	38	43	17	54	21	49	35	34	21	21	47	41	49	50	51	48	45
M6I_A	55	62	48	52	58	62	57	62	55	45	45	51	52	50	48	71	65	42	54	69	51	45	52	59	58	100	50	44	68	27	58	39	49	55	35	44	10	2	48	17	36	44	20
M6I_B	52	59	64	34	55	63	57	54	51	32	54	41	46	59	58	62	56	38	50	68	54	50	47	58	57	64	100	52	58	42	51	55	37	36	53	35	28	7	46	39	63	43	26
M6I_C	65	69	56	48	60	57	44	58	54	34	47	47	49	59	59	62	55	54	56	78	65	47	46	63	65	71	74	100	42	58	41	57	50	41	28	42	54	27	52	59	47	59	46
M6I_D	54	61	45	46	55	57	67	66	58	34	46	52	54	49	50	64	56	41	47	68	53	46	54	59	51	75	73	67	100	40	67	48	52	52	46	55	14	14	43	22	54	44	42
QUAD	66	77	52	49	61	60	35	66	50	32	38	50	33	60	60	63	62	62	60	67	65	53	38	60	57	59	62	65	62	100	33	51	40	42	16	23	57	39	49	71	62	62	61
R96A_A	63	62	48	48	66	61	63	66	58	42	51	58	61	67	67	76	64	58	55	66	62	50	51	66	56	75	69	67	73	63	100	61	55	65	53	64	21	40	53	20	51	43	56
R96A_B	67	63	53	46	65	62	58	60	52	36	52	48	52	65	67	74	62	58	60	71	67	56	45	76	69	65	75	76	64	60	80	100	56	38	62	58	47	43	66	40	54	61	50
S44E	70	67	44	65	62	57	51	66	54	52	43	54	51	54	56	74	57	60	52	61	56	50	44	66	65	67	56	62	65	59	72	72	100	55	38	58	47	37	51	35	46	63	58
S44F_A	64	69	43	59	65	58	53	66	51	45	43	53	51	58	58	81	65	58	68	72	65	53	51	72	65	73	57	67	69	63	76	74	74	100	25	48	29	37	41	36	47	46	53
S44F_B	49	53	44	43	55	60	64	63	53	36	49	51	46	56	56	70	58	46	55	56	62	57	51	69	63	60	68	60	65	55	72	78	63	69	100	44	11	16	43	-1	51	29	26
S44_AA_A	68	60	49	48	65	54	59	55	53	42	43	48	49	64	62	67	51	57	51	58	56	50	41	58	49	65	62	63	63	51	76	70	70	67	61	100	42	42	58	19	39	44	36
S44_AA_B	78	67	40	47	59	45	19	46	38	31	30	34	29	64	63	53	41	70	60	60	59	46	20	55	56	46	49	61	38	55	54	59	65	58	48	63	100	52	43	73	40	55	47
SS	72	59	38	45	60	51	33	58	40	36	33	54	45	67	70	66	53	85	65	61	72	59	29	71	58	47	45	56	52	59	65	64	62	66	60	59	60	100	39	45	22	33	65
SS_B	70	62	48	57	73	73	54	70	56	43	52	59	59	64	64	76	67	67	62	70	68	56	38	73	65	65	64	65	62	67	74	73	68	71	69	68	56	73	100	33	39	51	37
T26EX	69	68	55	47	55	45	23	43	46	40	39	49	36	57	61	58	53	60	58	71	57	46	18	58	60	52	55	66	47	59	55	62	62	61	47	52	66	59	62	100	50	58	53
T26E_NAG_A	57	67	43	57	63	60	52	65	53	42	48	55	51	56	58	73	71	51	56	69	65	50	45	69	65	63	61	64	64	64	68	66	67	73	65	58	52	61	67	63	100	61	51
T26E_NAG_B	64	72	48	59	64	58	43	60	55	44	44	55	49	55	59	73	67	57	60	66	66	50	38	68	64	62	58	64	58	64	67	68	68	72	58	60	56	65	72	70	89	100	56
T4165RF	68	66	41	58	64	54	40	65	48	46	43	56	52	62	65	71	58	70	64	70	67	62	38	73	65	63	51	67	61	59	70	71	71	74	61	59	59	77	66	67	69	71	100

Figure 3. Matrix showing correlation coefficients between thermal factors of each mutant lysozyme with all other molecules in the set. Mutants are identified in detail in Zhang et al., 1995. Upper diagonal half of matrix shows correlations for main-chain thermal factors. Lower diagonal shows correlations for side-chain thermal factors.

Overall, the generally high correlations between the side-chain thermal factors (Figure 3 is reassuring. This tends to confirm that these thermal factors are a reliable indication of side-chain mobility in solution. On the other hand, the low or negative correlations in some cases for main-chain thermal factors (Figure 3) show that these must be used with caution. Indeed, in some cases, main-chain motion, as judged by thermal factors, appears to be constrained by intermolecular contacts to such a degree that it is no longer indicative of motion in solution. Tightly packed crystal structures tend to diffract to high resolution and so give more accurate thermal factors. At the same time, these are the very structures where the main-chain thermal factors may be constrained by crystal contacts and so are less representative of dynamic behavior in solution.

ACKNOWLEDGMENTS

We are especially grateful to Sheila Snow and Joan Wozniak for purifying and crystallizing many of the mutant lysozymes described here. This work was supported in part by NIH grant GM21967 to B.W.M.

REFERENCES

Artymiuk, PJ, CCF Blake, DEP Grace, SJ Oatley, DC Phillips, and MJE Sternberg. **1979**. Crystllographic studies of the dynamic properties of lysozyme. Nature 280:563-568.

Bell, JA, KP Wilson, X-J Zhang, HR Faber, H Nicholson, and BW Matthews. **1991**. Comparison of the crystal structure of bacteriophage T4 lysozyme at low, medium and high ionic strengths. Proteins 10:10-21.

Blaber, M, X-J Zhang, JD Lindstrom, SD Pepiot, WA Baase, and BW Matthews. **1994**. Determination of α-helix propensity within the context of a folded protein: Sites 44 and 131 in bacteriophage T4 lysozyme. J. Mol. Biol. 235:600-624.

Chothia, C, and AM Lesk. **1986**. The relation between the divergence of sequence and structure in proteins. EMBO J. 5:823-826.

Dixon, MM, H Nicholson, L Shewchuk, WA Baase, and BW Matthews. **1992**. Structure of a hinge-bending bacteriophage T4 lysozyme mutant, Ile 3 Pro. J. Mol. Biol. 227:917-933.

Eriksson, AE, WA Baase, and BW Matthews. **1993**. Similar hydrophobic replacements of Leu 99 and Phe 153 within the core of T4 lysozyme have different structural and thermodynamic consequences. J. Mol. Biol. 229:747-769.

Faber, HR, and BW Matthews. **1990**. A mutant T4 lysozyme displays five different crystal conformations. Nature 348:263-266.

Hohenester, E, and JN Jansonius. **1994**. Crystalline mitochondrial aspartate aminotransferase exists in only two conformations. J. Mol. Biol. 236:963-968.

Kishan, KVR, JP Zeelen, MEM Noble, TV Borchert, and RK Wierenga. **1994**. Comparison of the structures and the crystal contacts of trypanosomal triosephosphate isomerase in four different crystal forms. Protein Sci. 3:779-787.

Matthews, BW. **1993**. Structural and genetic analysis of protein stability. Ann. Rev. Biochem. 62:139-160.

Matthews, BW, and SJ Remington. **1974**. The three-dimensional structure of the lysozyme from bacteriophage T4. Proc. Natl. Acad. Sci. USA 71:4178-4182.

Wagner, G, SG Hyberts, and TF Havel. **1992**. NMR structure determination in solution: a critique and comparison with X-ray crystallography. Ann. Rev. Biophys. Biomol. Struct. 21:167-198.

Wozniak, JA, X-J Zhang, LH Weaver, and BW Matthews. **1994**. Structural and genetic analysis of the stability and function of T4 lysozyme. In: *Molecular Biology of Bacteriophage T4*, pp. 332-339. American Society for Microbiology, Washington, DC.

Zhang, X-J, and BW Matthews. **1994**. Enhancement of the method of molecular replacement by incorporation of known structural information. Acta Crystallogr. D50:675-686.

Zhang, X-J, and BW Matthews. **1995**. Determination, dynamics and deviations of the structure of T4 lysozyme in 25 crystal forms. In: *Proceedings of the CCP4 Study Weekend, "Making the Most of Your Model,"* pp. 129-136. Daresbury Laboratory, Chester, UK.

Zhang, X-J, JA Wozniak, and BW Matthews. **1995**. Protein flexibility and adaptability seen in 25 crystals forms of T4 lysozyme. J. Mol. Biol. 250:527-552.

NMR STUDIES OF THE CALCIUM-INDUCED STRUCTURAL CHANGES THAT TRIGGER MUSCLE CONTRACTION

B. D. Sykes,[1] G. Audette,[2] S. M. Gagné,[2] M. X. Li,[2] C. M. Slupsky,[3] and S. Tsuda[4]

[1]Department of Biochemistry and MRC Group in Protein Structure and Function, University of Alberta, Edmonton, Alberta, Canada

[2]Department of Biochemistry, University of Saskatchewan, Saskatoon, Saskatchewan, Canada

[3]Department of Biochemistry and Molecular Biology, Faculty of Medicine, University of British Columbia, Vancouver, B.C., Canada

[4]Bioscience and Chemistry Division, Hokkaido National Industrial Research Institute, Toyohira, Sapporo, Japan

Key Words: *NMR, protein structure, calcium, muscle, conformational change*

ABSTRACT

Skeletal and cardiac muscle contraction is triggered by the binding of calcium ions to the muscle thin-filament protein troponin C (TnC). The resulting conformational change alters the interaction of TnC with other proteins in the thin filament leading to muscle contraction. We have used multinuclear, multidimensional, high-resolution nuclear magnetic resonance (NMR) spectroscopic techniques to study this calcium-induced conformational change in TnC.

This paper discusses the problems that exist with respect to NMR structural approaches in complex biological systems, i.e., the size limit and the aggregation of the protein, and the strategies we have taken to deal with these problems. These include the expression of only the regulatory domain of the protein and finding unique solution conditions for the intact protein. TnC undergoes a calcium-induced dimerization at neutral pH. Our studies indicate that the mode of dimerization involves the N-domain of one monomer interacting with the N-domain of another monomer. Addition of the solvent trifluoroethanol breaks the dimer effectively, making TnC predominantly a monomer.

Under this condition, we were able to study the ^{13}C, ^{15}N-labeled TnC protein using NMR and to determine its structure in solution. The structure shows that the two domains are connected by a flexible linker. We have cloned and expressed the regulatory N-domain of the protein TnC and solved the complete three-dimensional structure of this protein in both the apo- and calcium-saturated forms using the ^{13}C- and ^{15}N-labeled proteins. These structures provide a clear description of the calcium-induced structural change. A simple method for characterizing the NMR properties of a given new protein to determine its chances of success in subsequent NMR experiments is also discussed.

INTRODUCTION

Muscle contraction involves a large number of protein molecules assembled into complex structures. In particular, the two main ultrastructural components of striated muscle are the thick and thin filaments. The thick filament is formed from the protein myosin, which contains two heavy chains, two molecules of the regulatory light chain, and two molecules of the essential light chain. The thin filament is formed from the proteins actin, tropomyosin, and the troponin complex (itself comprising troponin-C, troponin-I, and troponin-T), present in a stoichiometry of 7:1:1. These proteins range in molecular weight from 18,000 to approximately 500,000.

There are two main questions that are often posed relative to muscle contraction. The first is to understand how the chemical energy derived from the cleavage of ATP is converted into the mechanical force used during contraction. This force is generated by the actomyosin complex, wherein the globular head of the thick-filament myosin molecule, or so-called cross-bridge, is the ATPase. The second major question is how muscle contraction is regulated. This occurs via the binding of calcium ions to the thin-filament protein troponin-C and the subsequent transmission of induced structural change to other components of the thin filament which changes the rate of the actomyosin ATPase reaction. The question addressed in this paper is the understanding of these processes on a molecular level. In particular, can we determine the structures of these complex macromolecular assemblies and/or their components and use this information to help answer the questions posed above?

Considerable information about the overall structure and organization of the muscle filament has been obtained from techniques such as electron microscopy, low-angle x-ray diffraction experiments, and various spectroscopic studies involving fluorescence or electron spin resonance methods as examples. In addition, there has been recent tremendous success in the determination of the three-dimensional molecular structures of the various protein components, using x-ray crystallography. This includes the structure of troponin-C (Herzberg and James, 1988; Satyshur et al., 1988), tropomyosin (Phillips et al., 1979, 1986), actin

(Kabsch et al., 1990), the S-1 fragment or cross-bridge portion of myosin (Rayment et al., 1993), and the regulatory domain of myosin (Xie et al., 1994). These studies have shed considerable light on the mechanism/function of these proteins.

What, then, about nuclear magnetic resonance? The NMR spectroscopic techniques have developed into a versatile tool for the study of the structures of biological macromolecules in solution. As such, NMR provides a complementary approach to x-ray crystallography and, in 1995, approximately 100 new NMR structures were reported in comparison with approximately 350 new x-ray structures! Can NMR then be used to study muscle proteins? We have recently reviewed the previous applications of NMR to muscle proteins; they are not extensive (Slupsky and Sykes, 1996). With respect to NMR structural approaches two problems exist: first, these proteins are too large for present NMR methods; and second, the isolated proteins are often not that well behaved in solution, possibly because they are normally found as a part of larger structures.

Nuclear magnetic resonance has been successful, however, in one important area: determination of the calcium-induced structural change in troponin-C, which is the first step in the triggering of muscle contraction. A recent issue of *Nature Structural Biology*, for example, contains the NMR-derived solution structures for calmodulin, troponin-C, and calcyclin in the apo- and calcium-saturated states submitted by five research groups, all using NMR (Finn et al., 1995; Gagné et al., 1995; Kuboniwa et al., 1995; Potts et al., 1995; Zhang et al., 1995). In addition, we have recently completed the solution structure of troponin-C in the calcium-saturated state (Slupsky and Sykes, 1995). These studies represent several different strategies for the study of these proteins, including the expression of important domains and the use of unique conditions so that the proteins are well behaved.

A protein such as troponin-C (162 amino acids; molecular weight, 18,257) is near the upper molecular weight range for structure determination by modern multinuclear, multidimensional NMR techniques if it is well behaved. However, TnC has been shown to aggregate in solution when it is in the calcium-saturated state (Slupsky et al., 1995). It is the actual effective molecular weight of a protein and its corresponding hydrodynamic properties that are important to the success of NMR structural studies, not the formal molecular weight. This determines the rotational correlation time for the protein, which is proportional to the molecular weight:

$$\tau_{ROT} = \left[\frac{4}{3}\pi r^3\right]\frac{\eta}{kT} \quad . \tag{1}$$

This influences the success of multidimensional NMR experiments, because the value of the transverse relaxation time, T_2, is inversely proportional to the line width of the NMR resonance, which is, in turn, proportional to the rotational correlation time:

$$\frac{1}{T_2} = \pi\Delta\nu = A\tau_{ROT} \quad . \tag{2}$$

The shorter the T_2, the less time the magnetization remains in multidimensional NMR experiments and the less is the sensitivity of the NMR experiment. Thus, a large MW and corresponding long $_{ROT}$ lead to a short T_2 and poor NMR data.

In this manuscript we would like to describe two strategies we used for studying the structure of calcium-saturated TnC which relate to the effective molecular weight issue discussed above. We will also report a simple method for characterizing the NMR properties of a given new protein to determine its chances of success in subsequent NMR experiments.

For TnC we found, as had others, that the protein dimerized in solution in the calcium-saturated state (Slupsky et al., 1995). Using multinuclear, multidimensional NMR we were able to characterize this dimer and determine that the aggregation occurred via the regulatory N-domain of the protein, and that the linkage to the C-domains of each molecule in the dimer was flexible. This led to differential line widths in the NMR spectrum, as shown in Figure 1A. We also discovered that the dimerization could be reversed by the addition of small molar concentrations of the co-solvent trifluoroethanol (TFE). This led to a removal of the differential line widths in the NMR spectrum by sharpening the resonances from the N-domain portion of the troponin-C, as shown in Figure 1B. Subsequently we were able to study the ^{13}C, ^{15}N expressed TnC protein using NMR and to determine its structure in solution (Slupsky and Sykes, 1995). The average structure of the N- and C-domains of troponin-C are shown in Figure 2. The linker region between the two domains was flexible in solution (Slupsky and Sykes, 1995). This structure allows one to visualize the calcium-induced structural change by comparison with the x-ray structure (Herzberg and James, 1988; Satyshur et al., 1988) in which the two calcium sites in the regulatory domain are not occupied. A question remains about the relevance of the structure in TFE/H_2O. We have argued from NMR chemical shift measurements that the structural changes caused by TFE are small (Slupsky et al., 1995), and the structure is similar to that of the other proteins which have been determined, e.g., calmodulin (Babu et al., 1988) and that of the isolated N-domain of TnC described below.

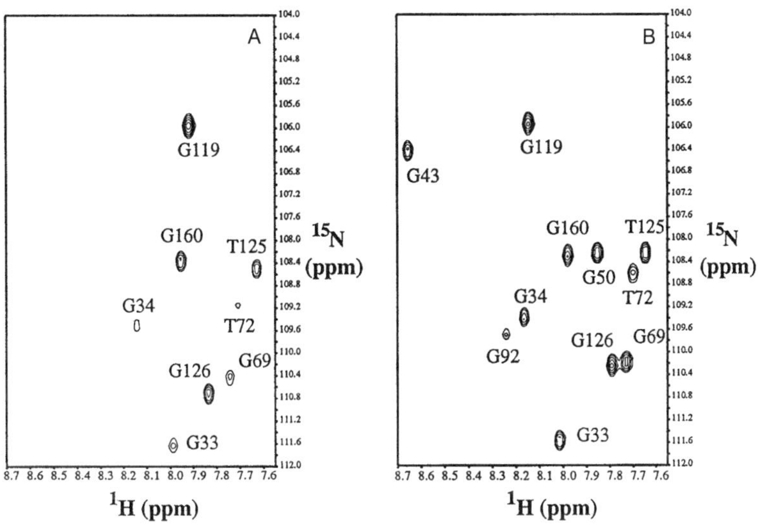

Figure 1. **A**, Portion of the 2D ^{1}H-^{15}N hsqc NMR spectrum showing the existence of differential line widths indicative of the N-domain N-domain dimerization state for calcium-saturated troponin-C. **B**, Same portion of spectrum for same sample of troponin-C in the presence of 15% (v/v) of trifluoroethanol, showing similar line widths for all resonances, indicating monomeric state of protein.

The second approach we used to study the TnC protein was to clone and express the regulatory N-domain of the protein (NTnC). We have shown that the isolated NTnC has the same properties as the same region in the intact protein (Li et al., 1994). NTnC in the apo form is completely monomeric; NTnC in the calcium-saturated form is in an equilibrium between the monomeric and dimeric form. We have determined the solution structure of NTnC in both apo and calcium-saturated forms using the ^{13}C- and ^{15}N-labeled proteins, and these structures are shown in Figure 3. These structures provide a clear

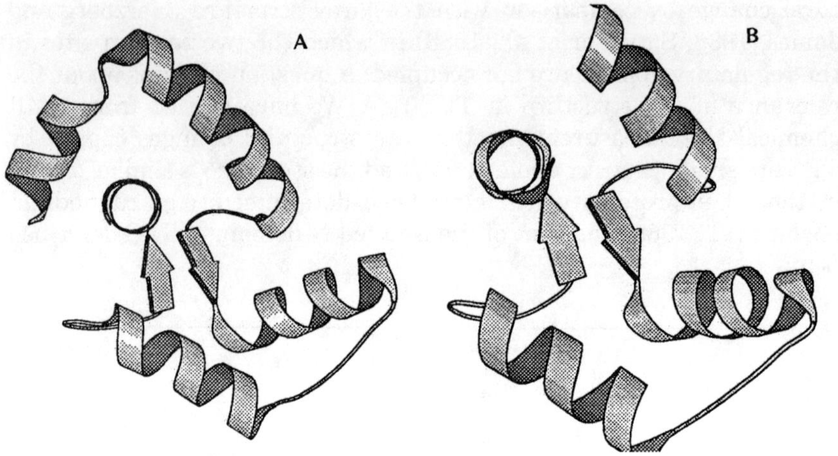

Figure 2. NMR-derived solution structures for N-domain (A) and C-domain (B) regions of calcium-saturated troponin-C.

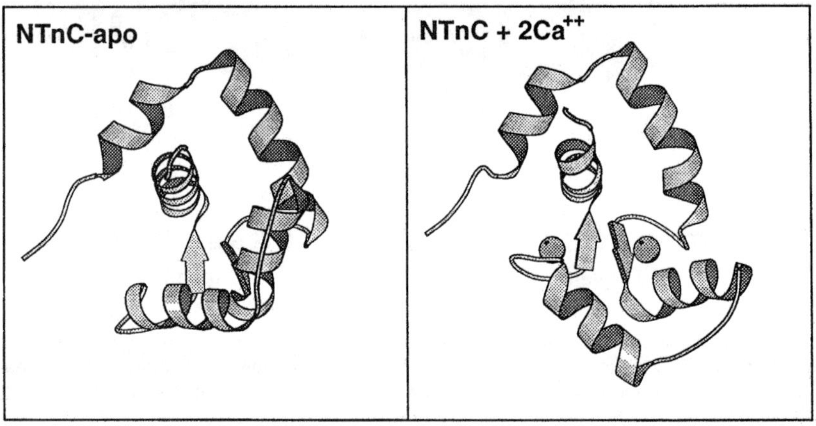

Figure 3. NMR-derived solution structures of N-domain of troponin-C in the apo- and calcium-saturated states.

description of the calcium-induced structural change. Again, one may question the influence of dimerization on the structure of calcium-saturated NTnC. It is, however, the same as that observed for intact TnC.

From the discussion above it is obvious that it is important to know the state of the protein in solution for NMR studies. We have discovered this in our initial studies of new systems such as troponin-C complexed with troponin-I, or the regulatory light chains of myosin complexed with a heavy-chain peptide. We have had considerable success using a simple NMR approach, described by Bax and co-workers (Anglister et al., 1993), to determine the effective rotational correlation time of protein in solution.

Normally, the T_2 for an NMR resonance is determined using a spin echo pulse sequence,

$$90° — \text{delay} — 180° — \text{delay} — \text{echo},$$

where the NMR signal intensity for a given resonance in the Fourier transform of the echo is followed as a function of the delay time. The T_2 is measured for the envelope of the NH resonances, and τ_{ROT} calculated from $\tau_{ROT} = 1/(5*T_2)$ (Anglister et al., 1993). Two problems remain: the suppression of the water resonance in the NMR spectrum and resulting partial transfer of saturation to NH resonances, and the modulation of the T_2 delay of the NH resonances caused by spin-spin coupling to the attached αCH protons. Both problems are removed by using the jump and return echo pulse sequence,

$$(90_x - D - 90_{-x}) \text{ delay } (90_x - 2D - 90_{-x}) \text{ delay - echo}.$$

Using a series of known proteins, we have been able to construct a plot of measured τ_{ROT} versus actual molecular weight, as shown in Figure 4. This, then, allows the rapid determination of the aggregation and/or hydrodynamic state of the protein understudy under NMR conditions.

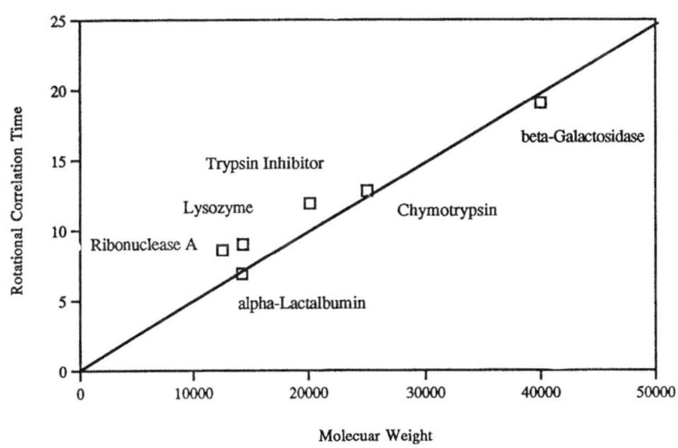

Figure 4. Plot of effective rotation correlation time of protein standards versus protein molecular weight determined from ^{1}H jump and return echo measurements of NH proton transverse relaxation.

We hope to have demonstrated some of the issues involved, problems encountered, techniques used, and recent successes of NMR in the study of proteins involved in complex biological systems such as muscle. In our opinion, a large number of important questions in complex systems can be reduced to binary or ternary complexes of domains of the proteins involved. Far more difficult to understand and/or predict in advance is whether a given protein or protein domain will be well behaved in solution, whether it will exist as a monomer, and whether it will have a unique structure representative of the physiological state.

REFERENCES

Anglister, J, H Ren, CB Klee, and A Bax. **1993**. Isotope-edited multidimensional NMR of calcineurin B in the presence of the non-deuterated detergent CHAPS. J. Biol. NMR 3:121-126.

Babu, YS, CE Bugg, and WJ Cook. **1988**. Structure of calmodulin refined at 2.2Å resolution. J. Molec. Biol. 203:191-204.

Finn, B, J Evenas, T Drakenberg, JP Waltho, E Thulin, and S Forsén. **1995**. Calcium-induced structural changes and domain autonomy in calmodulin. Nat. Struct. Biol. 2:777-783.

Gagné, SM, S Tsuda, MX Li, LB Smillie, and BD Sykes. **1995**. Structures of the troponin C regulatory domain in the apo and calcium-saturated states. Nat. Struct. Biol. 2:784-789.

Herzberg, O, and MNG James. **1988**. Refined crystal structure of troponin C from turkey skeletal muscle at 2.0Å resolution. J. Mol. Biol. 203:761-779.

Kabsch, W, HG Mannherz, D Suck, EF Pai, and KC Holmes. **1990**. Atomic structure of actin:DNase I complex. Nature 347:37-44.

Kuboniwa, H, N Tjandra, S Grsesiek, H Ren, CB Klee, and A Bax. **1995**. Solution structure of calcium-free calmodulin. Nat. Struct. Biol. 2:768-776.

Li, MX, M Chandra, JR Pearlstone, KI Racher, G Trigo-Gonzalez, T Borgford, CM Kay, and LB Smillie. **1994**. Properties of isolated recombinant N and C domains of chicken troponin C. Biochemistry 33:917-925.

Phillips, GN, Jr, JP Fillers, and C Cohen. **1986**. Tropomyosin crystal structure and muscle regulation. J. Mol. Biol. 192:111-131.

Phillips, GN, Jr, EE Lattman, P Cummins, KY Lee, and C Cohen. **1979**. Crystal structure and molecular interactions of tropomyosin. Nature 278:413-417.

Potts, BCM, J Smith, M Akke, TJ Macke, O Katsuo, H Hiroyoshi, DA Case, and WJ Chazin. **1995**. The sturcture of calcyclin reveals a novel homodimeric fold for S100 Ca^{+2}-binding proteins, Nat. Struct. Biol. 2:790-796.

Rayment, I, WR Rypniewski, K Schmidt-Base, R Smith, DR Tomchick, MM Benning, DA Winkelmann, G Wesenberg, and HM Holden. **1993**. Three-dimensional structure of myosin subfragment-1: a molecular motor. Science 261:50-58.

Satyshur, KA, ST Rao, D Pyzalska, W Drendel, M Greaser, and M Sundaralingham. **1988**. Refined structure of chicken skeletal muscle troponin C in the two-calcium state at 2Å resolution. J. Biol. Chem. 263:1628-1647.

Slupsky, CM and BD Sykes. **1995**. The NMR solution structure of calcium-saturated skeletal muscle troponin C. Biochemistry 34:15753-15764.

Slupsky, CM and BD Sykes. **1996**. Muscle Proteins, pp. 3188-3200. In: *Encyclopedia of Nuclear Magnetic Resonance*. John Wiley and Sons, New York.

Slupsky, CM, CM Kay, FC Reinach, LB Smillie, and BD Sykes. **1995**. Calcium-induced dimerization of troponin C: mode of interaction and use of trifluoroethanol as a denaturant of quaternary structure. Biochemistry 34:7365-7375.

Xie, X, DH Harrison, I Schlichting, RM Sweet, UN Kalaboks, AG Szent-Gyorgyi, and C Cohen. **1994**. Structure of the regulatory domain of scallop myosin at 2.8Å resolution. Nature 368:306-312.

Zhang, M, T Tanaka, and M Ikura. **1995**. Calcium-induced conformational transition revealed by the solution structure of apo calmodulin. Nat. Struct. Biol. 2:759-767.

APPLICATIONS OF MOLECULAR MECHANICS AND MOLECULAR DYNAMICS TO COMPLEX MOLECULAR SYSTEMS

P. Kollman

Department of Pharmaceutical Chemistry, University of California, San Francisco, San Francisco, California

Key Words: *Nucleic acids, proteins, ionophores, molecular dynamics, free energy calculations*

ABSTRACT

We will review our recent results using molecular dynamics (MD) to study proteins, nucleic acids, ionophores, and model systems. The emphasis will be on evaluating, with appropriate experiments, force fields, long-range electrostatics, and conformational sampling as they relate to making meaningful contact.

INTRODUCTION

What are the most important issues in molecular mechanics/dynamics applications to biological molecules? They can be grouped into four areas. The first is the development of accurate force-field representation. The second is to describe all the low-energy conformations of a molecule, ideally in solution as well as *in vacuo*. The third is to describe the free energies of solvation, interaction, and stability of a complex molecule in solution. Finally, if one wishes to study chemical reactions in condensed phases, one can combine the use of quantum mechanical calculations on the atoms in which electronic structure/bonding changes are occurring with molecular mechanics/dynamics for the rest of the system in order to calculate structural and free-energy changes occurring in a complex reacting system in condensed phases. Below, we summarize some of the current state of the art in these areas, focusing on our own work.

FORCE FIELDS/REPRESENTATION

In the last year, we completed a new force field (Cornell et al., 1995) and applied it to a number of interesting problems (Howard et al.,

1995). This force field appears to be a clear improvement over our previous one (Weiner et al., 1986). In collaboration with Tom Darden, we have also incorporated particle-mesh Ewald long-range electrostatics into AMBER 4.1 and showed it to be very effective in molecular dynamics simulations of complex molecules (Cheatham et al., 1995). In addition, we have begun simulations of a fully nonadditive force field and have applied it to a number of prototype systems (Dang et al., 1991; Caldwell and Kollman 1995a,b; Sun et al., 1995; Meng et al., 1994).

THE SAMPLING ISSUE

The multiple copy, locally enhanced sampling (LES) method developed by Elber and coworkers (Roitberg and Elber, 1991) and further developed in our laboratory by Simmerling (unpublished), is particularly promising. This method uses multiple copies of pieces of peptide chains; they do not interact with each other, and each interacts with the surrounding atoms with a strength of 1/n (n = number of copies). After using high-temperature dynamics to randomize the copies, one anneals the system back to room temperature. The amount of convergence of these multiple copies, perhaps to a few, gives one a sense of the lowest energy conformations of the system. Such a model has been shown to have the same global minimum with a single copy as with multiple copies; it is only that the force-field parameters have been reduced to 1/n of their value for the single copy (n = number of copies) and, thus, conformational transitions take place with much greater facility in the multiple-copy case. Elber and Simmerling estimate an increased sampling on the order of 10-100 for small peptides (Simmerling and Elber, 1994). One envisages very exciting applications of such an approach in understanding the conformational variability of antibody binding loops, saccharide and nucleic acid polymers with multiple OH groups, and many other systems.

There have been two exciting recent studies from our laboratory which show that the capability to "routinely" perform simulations for nanoseconds rather than hundreds of picoseconds is a major step forward in the ability to simulate complex molecular assemblies. The first study, by Debolt (DeBolt and Kollman, 1995), was a definitive study of the properties of liquid octanol and saturated water/octanol. Simply stated, the relaxation times of water are in the range of a few picoseconds, whereas more complex solvents/solutions, such as octanol and octanol/water, require hundreds of picoseconds for their slowest relaxation times.

The structural, thermodynamic, and dynamic properties of this solvent and solution have been studied using molecular dynamics. First, the molecular mechanical parameters were carefully evaluated and slightly adjusted from Jorgensen's values (Jorgensen, 1986), based on results of simulations of simpler liquid hydrocarbons and liquid alcohols. Trajectories of liquid octanol at two different temperatures showed a very interesting segregation of the OH and hydrocarbon parts of the molecules, with a radius of 6 to 10 Å for the hydrophobic and hydrophilic clusters. Encouragingly, the percent of free O-H groups at both 40 and 75 °C were calculated in excellent agreement with experiments. The calculated dielectric constants were in reasonable agreement with the experiment, and the dielectric relaxation properties were fitted to three relaxation times. In this fit, the amplitudes of the three relaxation times were in excellent agreement with the experiment, despite the fact that all three calculated relaxation times were about three times faster than in the experiment. The most logical reason for this was the approximations inherent in using a united atom model for the CH_n groups. An analysis of the slowest relaxation time suggested that it might correspond to the formation/breakup of the "hydrophilic" clusters. In the water-saturated octanol, there were about 1/3 as many water molecules as octanol and, not surprisingly, the water molecules occupied the same area of space as the O-H groups. Thus, as in the pure liquid, the solution could be thought of as a minimicelle, with segregation of hydrophobic and polar groups. Finally, using parallel computing, the relative free energy of solvation of benzene and phenol in octanol was calculated. Such a simulation took a long time to converge due, as computer graphics visualization showed, to the long time it took the solute position to change. Phenol sits at the interface between the hydrophilic region, into which its O-H group protrudes, and the hydrophobic region, which is the location of the rest of its aromatic ring. As the O-H group is mutated to a H, it takes considerable time for the phenol/benzene to relax into a location within the hydrophobic region of the octanol/water solution. The calculated relative solvation free energy of phenol and benzene in octanol/water of 3.8 ± 0.4, when combined with the value found in pure water of 4.9 ± 0.2 kcal/mole, leads to a relative partition coefficient for phenol and benzene in excellent agreement with the experiment. Thus, in summary, current computer capabilities have allowed exciting insight into the properties of this model system for biological partitioning and membranes, something that would not have been possible 5 years ago.

Until this last year, simulations of nucleic acids with explicit representation of a solvent were very problematic, with large distortions from x-ray- or NMR-derived structures occurring within tens or hundreds of picoseconds. Darden's implementation of particle-mesh Ewald methods (York et al., 1993), a particularly efficient form of describing the long-range electrostatic effects in systems with a high density of ions and related work by Pettitt (Weerasinghe et al., 1995) and Zichi (1995), have changed this situation. One can now carry out stable trajectories for nucleic acid systems that are close to those inferred from experiments.

Recently, Cheatham (Cheatham and Kollman, in press) has taken this one step further. He has run separate 1-nanosecond trajectories of d(CCAACGTTGG)2, two beginning from the A form of DNA and one beginning from the B form. For this system, the A and B forms differ by approximately 6 Å. All three structures converge to the same average structure, which is similar to B-DNA. Furthermore, the central six residues of this structure are approximately 1.5 Å rms from the crystal structure of this sequence, with the deviation increasing to about 3 Å when the entire decamer is considered. This is encouraging, because NMR studies indicate that TpG steps, which occur between base pairs 2 and 3 and 8 and 9 in the sequence above appear to have nonstandard stacking properties, in contrast to the crystal structure. Since our simulation corresponds to a full representation of water and counterions and should correspond to full aqueous solution, it should be more directly relatable to the solution than the crystal structure.

Cheatham is continuing these studies (unpublished) with further analysis of DNA-RNA hybrids as well as other DNA sequences. Miller is studying the properties of an RNA hairpin loop with either DNA or RNA sugars in the loop (unpublished), and Cieplak is analyzing phosphoramide-phosphate helices (unpublished) in which one strand of the DNA has N-H at the 3' position, replacing the oxygen normally there. This modified backbone is one of the few that is not less stable (Gryaznov and Chen, 1994) than the canonical backbone and it is apparently more stable in an A conformation than in B. Thus, there appear to be many exciting MD studies possible on nucleic acid systems, given that one can achieve long-time stable trajectories.

FREE ENERGY CALCULATIONS

Although calculation of the absolute free energy of any complex molecular system in solution is difficult, it is not so difficult, at least in principle, to calculate relative free energies of closely related systems. Application of free-energy perturbation or thermodynamic integration approaches have led to many exciting applications of relative free-energy calculations on molecular systems. Such applications include relative solvation free energies, relative binding free energies of different ligands to a macromolecule, or relative binding free energies of a given ligand to two closely related macromolecules (e.g., site-specific mutants of a native protein), sequence-dependent macromolecular stability, and relative transition-state stabilization of a protein and site-specific mutants. The studies described have featured an "alchemical approach" first proposed by McCammon (Tembe and McCammon, 1984). Additional computational effort is required to calculate free-energy reaction coordinates for molecular association or conformational changes, but these involve an approach analogous to the alchemical studies. In each case, one evaluates the free energy as a function of a reaction coordinate, which may correspond to a change in molecular mechanical topology (alchemical change) or geometrical parameter (conformational change).

Of course, free-energy calculations suffer from the same limitation of sampling conformational space as regular MD trajectories, and the accuracy of such calculations is limited by this sampling issue. How, then, can one hope to relate such results to experiments. This is feasible because, for many small systems (e.g., simple molecule solvation and rigid host guest systems), due to degeneracy or rigidity, one can adequately sample the relevant conformations. When this is possible, the free-energy approaches often become an exquisitely sensitive test of force-field parameters. This offers useful guidance in force-field development. Nonetheless, one must always be aware of the sampling difficulties. In this context, one should stress that more sampling is not necessarily an improvement if, as has often happened in complex macromolecular systems, the structure drifts away from the correct experimental structure during the MD trajectory. In our view, it is often more useful in these cases to put structural restraints on the macromolecule to ensure that it remains near the correct structure during the trajectory.

There are many examples of excellent agreement, but there are also some areas of disagreement in such small-molecule solvation calculations, most notably the methyl effect on the solvation free energies of amines and amides (Morgantini and Kollman, 1995). It is important to try to resolve this discrepancy. We would like, however, to note another calculation on the relative solvation free energy of ortho, meta, and para hydroxymethyl phenols by Sun and Kollman (in press). These calculations were used to challenge the interpretation from the experimental free energies for toluene to water transfer of these compounds. In brief, Ben-Naim (1990) has used the unusually more favorable free energy of toluene to water transfer of the meta than the para isomer to support his view that the meta isomer is better solvated in water. This is because the separation of the two oxygen atoms in this molecule fits much better into water structure, where second-nearest neighbors in the liquid are approximately 4.5 Å apart. Our calculations, which correspond to the free energies for gas phase to water transfer, do not find special meta stabilization, suggesting that there are some interesting solvent-dependent effects in toluene which might preferentially destabilize the meta isomer in that solvent.

Given these difficulties, it is important to prove, at least in principle, that if sampling is not an issue, one can reproduce the observed relative free energies of association of host/guest systems. Below we present applications to four host-guest systems: anisole spher-and-alkali ion, thermolysin-phosphorus inhibitors, streptavidin-biotin analogs, and DNA-distamycin analogs.

Bayly, Thomas and Kollman (Bayly and Kollman, 1994; Thomas and Kollman, 1994) have simulated the relative binding free energy of alkali ion guests (Li^+,Na^+,K^+,Rb^+ and Cs^+) binding to an octaspherand host. The first set of simulations were done in water, and the experimental binding free energies came from studies in mixed $CHCl_3/H_2O$ solutions. Nonetheless, both experiment and theory agree that there is an unprecedented "double minimum" in the free energy as a function of ion size curve, with the lowest free energy minimum at the large-ion Cs^+ and a secondary minimum for Na^+. It was encouraging that the simulations reproduced the experiment in that regard, and they gave further structural insight into why Na^+ was a secondary minimum. The cavity in the spher-and is almost the perfect size for Cs^+ and also for $Na^+(H_2O)_2$, in which the Na^+ ligates to four ether oxygens and the two water oxygens, and each of the water hydrogens hydrogen-bonds to one of the remaining four ether oxygens. Li^+ is too small and K^+ too large to satisfy all of these coordinations. To test that the free-energy minimum was due to this perfect fit, the simulations were repeated with methanol solvent. Because methanol cannot satisfy the ionic/hydrogen bonding potential for all eight ether oxygens, we expected (and found) (Thomas and Kollman, 1994) only one local minimum for Cs^+ in the free energy as a function of ion size for this system in methanol. This prediction has not yet been addressed by experiment.

Although with thermolysin and a large protein, when we applied free-energy calculations to phosphoramidate and phosphate ester inhibitors, we had the benefit that the x-ray structures of these inhibitor complexes were available and showed that these two inhibitors bound almost identically. Our initial free-energy calculation on the relative free energy of binding of these two inhibitors found, consistent with experiments (Bash et al., 1987), that the phosphoramidate should be more tightly bound by about 10**3. However, we found substantial contributions from both relative interactions with the protein and with the solvent, corresponding to a different interpretation of the relative binding free energies than proposed by our collaborator Bartlett (Bartlett and Marlowe, 1987). The latter suggested that the relative free energy corresponded to the loss of hydrogen bonding in mutating the phosphoramidate, with its N-H forming a hydrogen bond to a backbone carbonyl of the protein, to the phosphate ester, which replaced the N-H with a CH_2 and, thus, could not form such a hydrogen bond. In our view, because the rest of the inhibitor locks the phosphate ester into almost identical positions in the active site as the amidate, as confirmed by the crystal structures, the phosphate oxygen will experience "forced repulsion" with the carbonyl oxygen that could hydrogen-bond with the amidate N-H. Thus, this difference between the two ligands would have contributions both DG_{solv} and DG_{bind}, including this forced repulsion, and would not correspond to a simple hydrogen bond energy. Together with Bartlett, we decided to study these interpretations further by studying the phosphonates, in which the NH (or O) is replaced by a CH_2 group. This CH_2 group should not experience any forced repulsion with the carbonyl and should be easier to desolvate than the NH or O, but it should also be incapable of forming a hydrogen bond with the protein carbonyl. We completed the calculations and submitted them for publication (Merz and Kollman, 1989), since the results were quite intriguing and consistent with our view of the competition between solvation and binding. We predicted that the CH_2 would bind nearly as tightly as the NH, with G_{solv} favoring the CH_2 and

G_{bind} favoring the NH, both by ~2 kcal/mole. The prediction of the relative binding free energy was confirmed (Morgan et al., 1991). It was exciting for us to determine a somewhat counterintuitive, or at least not obvious, result by simulation and then have it confirmed experimentally.

The appearance of the x-ray structure of the streptavidin interaction was particularly exciting to us, since this is the strongest known noncovalent binding interaction involving a small guest and a protein host. Weber et al. (1989) speculated, based on their x-ray structure, that the resonance of the ureido group of biotin, leading to an unusually negative carbonyl oxygen, was the reason for the unusually tight binding of biotin to streptavidin. We (Miyamoto and Kollman, 1993a,b) came to a very different conclusion. First, we studied the relative free energies of binding of biotin, imino biotin(replacing the C=O by C=NH), and sulfur biotin (replacing the C=O by C=S) and were able to reproduce the experimental relative binding free energies, finding that imino biotin was less favorably bound than biotin by 6 kcal/mole, and that thiobiotin was less favorably bound by 4 kcal/mole. This was another beautiful example of the competition between solvation and binding, in that imino binding interacted with the protein only 1 kcal/mole more weakly but was 5 kcal/mole harder to desolvate. Thiobiotin, on the other hand, was 8 kcal/mole easier to desolvate but interacted 12 kcal/mole less favorably with the protein. These results suggested that any particular "ureido resonance" could only be responsible for 4-6 kcal/mole of binding out of the known free energy of binding of biotin to streptavidin of 18.3 kcal/mole. We then attempted a more ambitious calculation, in which we mutated biotin to nothing, first "disappearing" the charges and then the Van der Waals parameters, both in solvent and in the protein binding site. Consistent with our relative free energies, we found that electrostatic effects contributed 3-6 kcal/mole to the binding free energy, and Van der Waals 15-18 kcal/mole, leading to a predicted free energy of association (DG_1) of 21 kcal/mole, qualitatively consistent with the experimental value of 18.3 kcal/mole. For further validation, we applied this same approach to the complex of alpha-chymotrypsin interacting with D-N acetyl tryptophan amide and calculated a DG_1 binding of 9 kcal/mole, compared to the experimental value of 5 kcal/mole.

What is the basis for this large Van der Waals contribution to biotin binding? It results from both hydrophobic interactions in solvent and because of unusually favorable dispersion attractions, featuring four tryptophans in the binding site, between biotin and the protein. We (Dixon R., work in progress) are now assessing the contributions of the individual protein residues by "mutating" the host with a fixed guest, in parallel with experimental studies on such mutations (Chilkoti et al., 1995).

The last system we studied for relative binding affinities was the DNA-distamycin complexes characterized by Wemmer (Geierstanger et al., 1993), which involved two distamycins sandwiched antiparallel in the DNA minor groove. Wemmer had also studied, by NMR and thermodynamics, the properties of complexes that involved replacing either one of the distamycins with an imidazole analog which changed a C-H of distamycin into an N: or replacing both distamycins with such an analog. Such an analog could, in principle, form a hydrogen bond in the minor groove, whereas the C-H from distamycin could not. The NMR studies of Wemmer determined that the two ligands bound near an AAGTT:AACTT part of the DNA duplex. Our calculations (Singh et al., 1994) found, consistent with experiments, that it made a tremendous difference which of the two distamycins was replaced with imidazoles, since the two ligands were in distinct locations in the binding site. Replacing the distamycin closer to the G, which enabled the imidzole N: to hydrogen-bond to the guanine 2-NH_2, leading to a calculated DDG of -0.7 kcal/mole, whereas the experimental value was -1.6 kcal/mole. It should be noted that this -0.7 kcal/mole difference was due to DG_{bind} being -4.2 kcal/mole and compensating for the fact that the imidazole analog is more favorably solvated (DG_{solv} = -3.5 kcal/mole), due to its hydrogen bonding capability. Replacing the other distamycin by imidazole in the binding site costs the solvation free energy and little is gained by interacting with the cytosine. This is consistent with the fact that NMR studies found no evidence for this mixed complex, only for the one with the imidazole near the G. Furthermore, by mutating both distamycins to imidazoles, we calculated that this would bind less favorably to the DNA by 1.8 kcal/mole, compared to the experimental difference of 2.2 kcal/mole. Thus, the unfavorable solvation free energy in this case is greater than any gain in favorable interactions with the DNA minor groove. The agreement with experiment, given that we did not know the thermodynamic data when the calculations were carried out, was most encouraging.

In addition to studying host/guest binding, free-energy calculations have been used extensively to study sequence-dependent stability in macromolecules.

QUANTUM MECHANICAL/MOLECULAR MECHANICAL (QM/MM) CALCULATIONS

In molecular design, one might also want to design new catalytic properties. This would require simulation of chemical reactions, which necessitates the use of quantum mechanical electronic structure methods as well as molecule mechanics. The simplest way to approach this is to carry out a gas-phase quantum mechanical reaction profile, fit this to an analytical potential curve, then calculate the solvation free energy along this potential curve. This was the approach taken in the first such calculation by Jorgensen on the CH_3Cl + Cl^- exchange reaction (Chandresekhar et al., 1985). This reaction has a dramatically different barrier in solution from that *in vacuo* and Jorgensen, by combining *ab initio* calculations on the gas-phase profile with Monte Carlo free-energy calculations, was able to calculate the solvation free energy along the reaction coordinate. The dramatic increase in activation free energy in aqueous solution compared to that *in vacuo* and in nonpolar solvents was clearly shown to be due to the better solvation of the localized Cl^- in the reactant state compared to the delocalized ion $[Cl\text{-}CH_3\text{-}Cl]^-$ in the transition state. Subsequently, Jorgensen successfully applied this approach to many organic reactions (Lim et al., 1994). Other approaches to this problem combine semiempirical molecular orbital calculations with continuum solvation (Cramer and Truhlar, 1992), *ab initio* methods combined with continuum solvation (Coitino et al., 1995) and semiempirical methods combined with explicit solvation (Gao and Xia, 1992).

However, if one wants to simulate reactions of enzymes, the problem is more complicated. For simple reactions in solution, one can study the reactants quantum-mechanically and the solvent with molecular mechanics, so there is no covalent bond connecting the quantum and molecular mechanically described parts of the system. In enzymes, one must have a way to handle such junctions. Warshel (1991) and coworkers have developed the only very useful approach we currently have to the study of enzyme reactions, using an empirical valence bond approach for the quantum mechanical calculations. Using a simple valence bond approach for the quantum mechanical part of the system allows a very efficient evaluation of the energy and gradient, something not possible with *ab initio* calculations. Furthermore, the valence bond method can often be empirically calibrated to reproduce the free energy of the reaction in solution, and the same parameters carried over to the enzyme active site. Another QM/MM method that can and has been applied to enzymes is the combination of semiempirical quantum mechanics with molecular mechanics/dynamics (Field et al., 1990; Stanton et al., 1995). Because semiempirical molecular orbital methods are not as time-consuming as *ab initio* methods, one can feasibly carry out the thousands of energy and gradient evaluation required in long trajectories. In the future, perhaps parallel computing and further computer speedup will allow effective coupling between *ab initio* or density functional approaches with molecular mechanics. In any case, improvement of these combined QM/MM methods is clearly a field for intense study, and the potential rewards for catalysis design are significant. On the other hand, more limited modeling of enzymatic processes has given intriguing insight as well, and we have recently reviewed work in this area (Kollman, 1992).

SUMMARY AND FUTURE DIRECTIONS

The prospects for achieving ever-increasing realism in the simulation of complex organic and biological molecules and using the insights for such simulations in molecular design is great. In the past few years, the molecular mechanical force fields $U(\underline{\mathbf{R}})$ have been improved significantly. We expect future approaches to be even better and to include, as appropriate, nonadditivity in the intermolecular interactions. Methods to accurately represent long-range electrostatics are now efficient and appear essential in the simulation of highly charged systems as well as useful in the representation of any macromolecule. New methods to improve conformational searching are on the horizon, and these methods will likely impact the range of systems to which free-energy calculations can be applied. A final challenge is to further develop the accuracy and applicability of coupled QM/MM methods, which should ultimately be of great significance in modeling enzyme systems. Thus, the future is very bright in the application of molecular and quantum mechanics methods to systems of organic and biochemical interest.

REFERENCES

Bartlett, PA and CK Marlowe. **1987**. Evaluation of intrinsic binding energy from a hydrogen bonding group in an enzyme inhibitor. Science 235:569-571.

Bash, PA, UC Singh, FK Brown, R Langridge, and PA Kollman. **1987**. Calculation of the relative binding free energy of a protein-inhibitor complex. Science 235:574-576.

Bayly, CI and PA Kollman. **1994**. Molecular dynamics and free energy calculations on the peculiar bimodal alkali ion selectivity of an 8-subunit cavitand. J. Am. Chem. Soc. 116:697-703.

Ben-Naim, A. **1990**. Solvent effects on protein association and protein folding. Biopolymers 29:567-596.

Caldwell, JW and PA Kollman. **1995a**. The structure and properties of neat liquids using non-additive molecular dynamics: water, methanol and N-methyl acetamide. J. Phys. Chem. 99:6208-6219.

Caldwell, JW, and PA Kollman. **1995b**. Cation-Pi Interactions: Non-additive effects are critical in their accurate representation. J. Am. Chem. Soc. 117:4177-4178.

Chandresekhar, J, SF Smith, and WL Jorgensen. **1985**. Theoretical examination of the SN2 reaction involving chloride ion and methyl chloride in the has phase and aqueous solution. J. Am. Chem. Soc. 107:154.

Cheatham, TE III and PA Kollman. Observation of the A DNA to B DNA transition during unrestrained molecular dynamics in aqueous solution. J. Mol. Biol. (in press).

Cheatham, TE III, JL Miller, F Fox, TA Darden, and PA Kollman. 1995. Molecular dynamics simulations on solvated biomolecular systems: The particle mesh Ewald method leads to stable trajectories of DNA, RNA and Proteins. J. Am. Chem. Soc. 117:4193-4194.

Chilkoti, A, PH Tan, and PS Stayton. **1995**. Site-directed mutagenesis studies of the high-affinity streptavidin-biotin complex-contributions of tryptophan residues 79, 108, and 120. Proc. Natl. Acad. Sci. USA 92:1754-1758.

Coitino, EL, J Tomasi, and R Cammi. **1995**. On the evaluation of the solvent polarization apparent charges in the polarizable continuum model—a new formulation. J. Comp. Chem. 16:20-30.

Cornell, WD, P Cieplak, DI Bayly, IR Gould, KM Merz, DM Ferguson, DC Spellmeyer, T Fox, JW Caldwell, and PA Kollman. **1995**. A second generation force field for the simulation of proteins and nucleic acids. J. Am. Chem. Soc. 117:5179-5196.

Cramer, CJ and DG Truhlar. **1992**. An Scf solvation model for the hydrophobic effect and absolute free energies of aqueous solvation. Science 256:213-217.

Dang, LX, JE Rice, J Caldwell, and PA Kollman. **1991**. Ion solvation in polarizable water. J Amer. Chem. Soc. 113:2481-2486.

DeBolt, SE and PA Kollman. **1995**. An investigation of the structure, dynamics and solvation in 1-octanol and its water saturated solution: molecular dynamics and free energy perturbation studies. J. Am. Chem. Soc. 117:5316-5340.

Field, MJ, PA Bash, and M Karplus. **1990**. A combined quantum mechanical and molecular mechanical potential for molecular dynamics simulations. J. Comp. Chem. 11:700-733.

Gao, J and X Xia. **1992**. A priori evaluation of aqueous polarization effects through Monte-Carlo QM-MM simulations. Science 258:631-635.

Geierstanger, BH, TJ Dwyer, Y Bathini, JW Lown, and DE Wemmer. **1993**. NMR characterization of a heterocomplex formed by distamycin and its analog 2-IMD with D(CGCAAGTTGGC)-D(GCCAACTTGCG)—preference for the 1/1/1 2-IMD-DST-DNA complex over the 2/1 2-IMD-DNA and the 2/1 DST-DNA complexes. J. Am. Chem. Soc. 115:4474-4482.

Gryaznov, SM and J-K Chen. **1994**. Oligodeoxyribonucleotide N3'-) P5' phosphoramidates—synthesis and hybridization properties. J. Am. Chem. Soc. 116:3143-3144.

Howard, AE, P Cieplak, and PA Kollman. **1995**. A molecular mechanical model that reproduces the relative energy of chair and twist-boat conformations of 1,3 dioxanes. J. Comp. Chem. 16:243-261.

Jorgensen, WL. **1986**. Optimized intermolecular potential functions for liquid alchohols. J. Phys. Chem. 90:1276-1284.

Kollman PA. **1992**. Theory of enzyme mechanisms. Curr. Opin. Struct. Biol. 2:765-771.

Lim, DC, DA Hrovat, WT Borden, and WL Jorgensen. **1994**. Solvent effects on the ring opening of cyclopropanones to oxyallyls—A combined *ab initio* and Monte Carlo study. J. Am. Chem. Soc. 116:3494-3499.

Meng, EC, P Cieplak, JW Caldwell, and PA Kollman. **1994**. Accurate solvation free energies of acetate and methylammonium ions calculated with a polarizable water model. J. Am. Chem. Soc. 116:12061-12062.

Merz, KM, Jr and PA Kollman. **1989**. Free energy perturbation simulations of the inhibition of thermolysin: prediction of the free energy of binding of a new inhibitor. J. Am. Chem. Soc. 111:5649-5658.

Miyamoto, S and PA Kollman. **1993a**. Absolute and relative binding free energy calculations of the interaction of biotin and its analogs with steptavidin using molecular dynamic/free energy perturbation approaches. Proteins Struct. Funct. and Genet. 16:226-245.

Miyamoto, S and PA Kollman. **1993b**. What determines the strength of noncovalent association of ligands to proteins in aqueous solution. Proc. Natl. Acad. Sci. USA 90:8402-8406.

Morgan, BP, JM Scholtz, MD Ballinger, ID Zipkin, and PA Bartlett. **1991**. Differential binding energy—a detailed evaluation of the influence of hydrogen-bonding and hydrophobic groups on the inhibition of thermolysin by phosphorus-containing inhibitors. J. Am. Chem. Soc. 113:297-307.

Morgantini, P-Y and PA Kollman. **1995**. Free energies of amides and amines: disagreement between free energy calculations and experiment. J. Am. Chem. Soc. 117:6057-6063.

Roitberg, A and R Elber. **1991**. Modeling side chains in peptides and proteins—application of the locally enhanced sampling and the simulated annealing methods to find minimum energy conformations. J. Chem. Phys. 95:9277-9287.

Simmerling, C and R Elber. **1994**. Hydrophobic collapse in a cyclic hexapeptide—computer simulations of CHDLFC and CAAAAC in water. J. Am. Chem. Soc. 116:2534-2547.

Singh, SB, Ajay, DE Wemmer, and PA Kollman. **1994**. Relative binding affinities of distamycin and its analog to d(CGCAAGTTGGC)°d (GCCAACTTGCG): Comparison of simulation results with experiment. Proc. Natl. Acad. Sci. USA 91:7673-7677.

Stanton, RW, LR Little, and KM Merz. **1995**. Quantum free energy perturbation study within a Pm3 MM coupled potential. J. Phys. Chem. 99:483-486.

Sun, Y and PA Kollman. Are there water-bridge induced hydrophilic interactions? J. Phys. Chem. (in press).

Sun, Y, Caldwell, JW, and PA Kollman. **1995**. Molecular dynamics and free energy perturbation study of spher-and complexation with metal ions employing additive and non-additive force fields. J. Phys. Chem. 99:10081-10085.

Tembe, B and A McCammon. **1984**. Ligand-receptor interactions. Biochem. Res. Comput. Chem. 8:281-284.

Thomas, BE IV and PA Kollman. **1994**. Free energy perturbation calculations of the relative binding affinities of an 8-subunit cavitand for alkali ions in methanol. J. Am. Chem. Soc. 116:3449-3452.

Warshel, A. **1991**. *Computer Modeling of Chemical Reactions in Enzymes and Solutions.* Wiley-Interscience, New York.

Weber, PC, DH Ohlendorf, JJ Wendoloski, and FR Salemme. **1989**. Structural origins of high-affinity biotin binding to streptavidin. Science 243:85-8.

Weerasinghe, S, PE Smith, V Mohan, YK Cheng, and BM Pettitt. **1995**. Nanosecond dynamics and structure of a model DNA triple helix in saltwater solution. J. Am. Chem. Soc. 117:2147-2158.

Weiner, SJ, PA Kollman, DT Nguyen, and DA Case. **1986**. An all atom force field for simulations of proteins and nucleic acids. J. Comp. Chem. 7:230-252.

York, DM, TA Darden, and LG Pedersen. **1993**. The effect of long-range electrostatic interactions in simulations of macromolecular crystals—a comparison of the Ewald and truncated list methods. J. Chem. Phys. 99:8345-8348.

Zichi, DA. **1995**. Molecular dynamics of RNA with the OPLS force field—aqueous simulation of a hairpin containing a tetranucleotide loop. J. Am. Chem. Soc. 117:2957-2969.

STUDY OF MOLECULAR RECOGNITION USING QM/MM METHODS: THE POTENTIAL OF MEAN FORCE FOR THE ASSOCIATION OF K^+ WITH 1,2-DIMETHOXYETHANE IN AQUEOUS SOLUTION

M. A. Thompson

Environmental Molecular Sciences Laboratory, Pacific Northwest National Laboratory, Richland, Washington

Key Words: *Quantum mechanical/molecular mechanical, molecular recognition*

ABSTRACT

We present results of a molecular dynamics study of the potential of mean force (pmf) free-energy profile for the association of K^+ with the bidentate ligand 1,2-dimethoxyethane (12DME) in H_2O. We use a hybrid quantum mechanical/molecular mechanical (QM/MM) approach that utilizes the QM semiempirical TAM1 method to describe 12DME, the MM parameterization of Dang for K^+, and the SPC/E model for H_2O. The total simulation time required to generate the pmf profile was 1.87 ns. At large K^+/12DME COM separation, 12DME is "preorganized" for cation binding; the central OCCO dihedral angle is predominantly gauche, being stabilized by 1 water having a bifurcated hydrogen bond to both oxygens of 12DME. K^+ begins to interact primarily with one of the oxygens of 12DME, at 4.7 Å separation, which breaks the solvent-induced preorganization of 12DME and causes 12DME to undergo significant conformational changes. The first of two minima in the pmf occurs at 4.15-Å separation. At shorter distances, the second oxygen of 12DME interacts significantly with the cation, giving an inner minimum at 2.65-Å separation. At the inner minimum, the gauche OCCO configuration of 12DME is again observed, due to both oxygens of 12DME interacting with the cation. The integrated free energy of association is calculated to be $+0.5 \pm 0.1$ kcal/mol, indicating that K^+/12DME is not a strongly bound species in aqueous solution. These results are useful for understanding the effects of guest species on solvent-induced preorganization and complexation by multidentate ligands, such as the crown-ethers.

INTRODUCTION

An understanding of host/guest molecular recognition in the condensed phase is of considerable importance to environmental chemistry. One major class of these interactions is that between metal cations and neutral organic ligands (Parker, 1983; Samec, 1990; DeJong and Reinhoudt, 1980). The best known examples are the crown-ethers, which show a remarkable range of specificity for a variety of cations that depends, in part, on the number and type of the donor atoms in the crown as well as the polarity of the solvent (Parker, 1983; Samec, 1990; DeJong and Reinhoudt, 1980). Several studies have addressed the selectivity of 18-crown-6 (18c6) for K^+ in polar solvents using molecular mechanics (MM) force fields (Wipff et al., 1982; Dang et al., 1991; Dang and Kollman, 1995; Marrone et al., 1994; Ha and Chakraborty, 1993; Kowall and Geiger, 1995). While these studies have done much to elucidate the chemistry of ion-macrocycle interaction in solution, it is clear that the charge distribution of the ligand can be a sensitive function of both the instantaneous conformation of the ligand and nearby solvent, as well as the proximity and charge of the cation (Wipff et al., 1982; Marrone et al., 1994). For this reason, we recently began to study cation/crown-ether interactions for gas phase and clusters, using *ab initio* quantum mechanical methods, (Glendening et al., 1994) and for condensed phase, using a hybrid quantum mechanical/molecular mechanical (QM/MM) approach (Thompson et al., 1994; Thompson, 1995a). Hybrid QM/MM methods treat part of the system with QM (typically, the solute(s) of interest) and the remainder of the system with MM (such as the solvent). The two motifs are then coupled by a QM/MM interaction Hamiltonian (Field et al., 1990). In previous QM/MM studies, we have shown the importance of QM polarization in treating cation-neutral interactions for K^+/18c6 in bulk H_2O (Thompson et al., 1994; Thompson, 1995a).

There have been several K^+/18c6 potential of mean force (pmf) studies reported in the literature, each showing qualitatively and quantitatively different results (Dang and Kollman, 1990, 1995; Kowall and Geiger, 1995; van Eerden et al. 1989). These studies have generated disagreement with regard to the placement and relative energetics of the peaks and minima in the pmf. Furthermore, there has been very little discussion on whether cation binding is concerted or nonconcerted with regard to the oxygen donor atoms of 18c6. These previously reported pmf all employ pairwise additive MM potential, and thus ignore polarization effects in the crown-ether.

We assert that a detailed understanding of cation binding by 18c6 must necessarily begin with an understanding of how cations interact with the "functional subunits" of 18c6, dimethylether (Thompson, 1995b) and 1,2-dimethoxyethane (12DME). Thus, we have undertaken a series of studies examining the interaction of K^+ with dimethoxyethane (Thompson, 1995) and 12DME. These results are reported in three separate communications. We previously reported a QM/MM study of the free-energy pmf profile for the interaction of K^+ with dimethylether in aqueous solvent (Thompson, 1995b). We calculated the presence of a solvent-separated ion-dipole pair and a contact ion-dipole pair between the cation and ligand. We discussed the nonadditive effects of the cation and its first solvation shell waters on repolarizing the QM wavefunction of the organic ligand. In this paper we present results of extensive QM/MM molecular dynamics simulations of the free energy of association for K^+ with 12DME in aqueous solvent by calculating the pmf. 1,2-Dimethoxyethane is the first functional subunit of 18c6 that exhibits significant internal flexibility and solvent-induced preorganization for cation binding. Recently, Liu et al. (1995) have reported QM/MM simulation results of free 12DME in H_2O using a MNDO-based approach. Their results indicate that the QM/MM approach is useful for understanding the solvation properties of 12DME.

METHODS

The QM/MM parameterization used here has been described previously (Thompson, 1994, 1995a). Briefly, 12DME is treated by Hartree-Fock molecular orbital theory, while the solvent and K^+ are described with MM. Specifically, we use the AM1 parameterization of the semi-empirical neglect of diatomic differential overlap model Hamiltonian (Pople et al., 1965) to represent the QM solute. We further augment the AM1 method with MM torsion terms (TAM1) to better describe rotamer energetics for 12DME (Thompson, 1995a). Water and K^+ are treated with SPC/E 3-site model for H_2O (Berendsen et al., 1987) and the parameters of Dang (personal communication), respectively. Wavefunction polarization effects in the QM motif arise from inclusion of the MM-atom/QM-electron term in the 1-electron matrix used in the Fock equations. We treat the TAM1 implementation of the QM/MM electrostatic terms in a fashion similar to that used by Field et al. (1990), but we fit the actual values of these parameters to MP2/6-31+G* (CP corrected) binding energies and structures (Thompson et al., 1994).

All calculations used the Argus computer program version 3.0 (Thompson, 1994). MD simulations were carried out in the NVT ensemble in a periodic system, using the minimum image convention (Allen and Tildesley, 1987). The periodic box size was 25 Å, SHAKE constraints

(Ryckaert et al., 1977) were used for rigid SPC/E H_2O and 12DME (C-H bonds; 1.121 Å). Velocities were scaled by the method of Berendsen et al. (1987) to simulate contact with an infinite thermal bath. The MM/MM and QM/MM nonbonded lists were generated with a 12.0-Å cutoff while the actual interactions were truncated at 9.0 Å. The nonbonded lists use neutral group cutoffs and were rebuilt every 10 steps. Initial configurations for solutes embedded in a box of H_2O were obtained using the Insight molecular modeling program 2.2 (Biosym Technologies, San Diego, CA). These files were then converted to input files for subsequent simulations with Argus. The system was first energy-minimized by steepest descents; then annealed to 10 K using a temperature coupling constant of π = 0.01 ps; then slowly warmed up to 300 K. A time step of 2.0 fs was used for all simulations. For the pmf, we used a temperature coupling constant of π = 0.20 ps.

The potential of mean force between the center of mass (COM) of two molecules A and B, in solution, is the reversible work required to assemble the A/B pair at a given COM separation averaging over all other degrees of freedom for the solvent and the other degrees of freedom for A and B that do not affect their COM separation (Dang and Kollman, 1990, 1995; van Gunsteren et al., 1993). The coordinate for our pmf was defined as the distance between K^+ and the COM of 12DME. The pmf profile, consisting of a series of equilibrium molecular dynamics simulations, was generated in the following fashion: (1) The initial K^+/12DME distance was set at 2.157 Å; (2) The system was equilibrated to 300 K as described above. The K^+/12DME COM distance was constrained in a fashion similar to that used by Dang and Kollman (1995). Thus, 12DME was allowed to tumble freely, and all bond angles were free to change during the simulation; (3) The pmf consisted of a series of equilibrium molecular dynamics simulations, each with 4 ps of equilibration, at the new K^+/12DME separation, followed by 100 ps of data collection. The relative free-energy values were calculated with the free-energy perturbation technique (Dang and Kollman, 1990, 1995; van Gunsteren et al., 1993; Zwanzig, 1954) and used a step size of 0.125 Å with double-wide sampling. Configurations were saved every 200 fs to disk for later analysis. The entire pmf profile required a total of 18 simulations each 104 ps in length, for a total of 1.872 ns of simulation time. Simulations were carried out on an IBM model 590 Power2 workstation.

The association constant for M^+/12DME complexation can be calculated by the following equation (Prue, 1969):

$$K_a = N \int_0^{r_c} 4\,\pi\, r^2 \exp\,[-w(R)/kT]dr, \qquad (1)$$

where K_a is the association constant, N is Avogadro's number, r_c is the cutoff distance for the association, w(R) is the value of the pmf, k is Boltzmann's constant, and T is the temperature. Equation (1) is derived by assuming that the associating species are spherical, and thus provides only an approximation to the true association constant for M^+/12DME. The calculated value of K_a can be used to obtain the Helmholtz free energy of binding:

$$\Delta A_{bind} = -RT1nK_\gamma\, K_a\,, \qquad (2)$$

where K_γ is the contribution of the activity coefficients of the interacting solutes to the thermodynamic equilibrium constant. Our simulations are run at infinite dilution, therefore we set all activity coefficients = 1 in equation (2). Our simulations are run with constant NVT, thus providing an estimate of the Helmholtz free energy (ΔA_{bind}) rather than the Gibbs free energy (ΔG_{bind}). For liquid systems such as the one studied here, values of ΔA_{bind} should be comparable to ΔG_{bind}. The uncertainties we report for ΔA_{bind} were obtained by integrating the pmf, using uncertainties obtained by a block averaging procedure over 10 separate subintervals within each simulation (Smith and Dang, 1994).

RESULTS AND DISCUSSION

Energy-Minimized Structure of K^+/1,2-Dimethoxyethane

We first compare the energy-minimized structure for K^+/12DME with our QM/MM model and *ab initio* results of Feller (personal communication). In Figure 1 we show the conformation we minimized: K^+ bound by 12DME, where the central OCCO angle is gauche, and the outer two COCC angles are nearly trans. In Table 1 we give the binding energies and selected geometric elements for this optimized structure. The *ab initio* results are at the MP2/6-31+G*(hybrid) level of theory with counterpoise corrected binding energies, using the Hartree-Fock minimized structure. As shown in Table 1, the agreement between our QM/MM method and the *ab initio* results is generally very good. The binding energies agree to less than 1 kcal/mol, the torsion angles are within 1 to 4, and the K^+/12DME-oxygen (K^+/O_D) distances agree to within ~0.1 Å, with the K^+/O_D distance shorter in the QM/MM method relative to the *ab initio* result. These results, along with our previously reported parameterization studies (Thompson et al., 1994; Thompson, 1995a,b), indicate that our QM/MM potential gives results of sufficient quality to proceed with free-energy calculations.

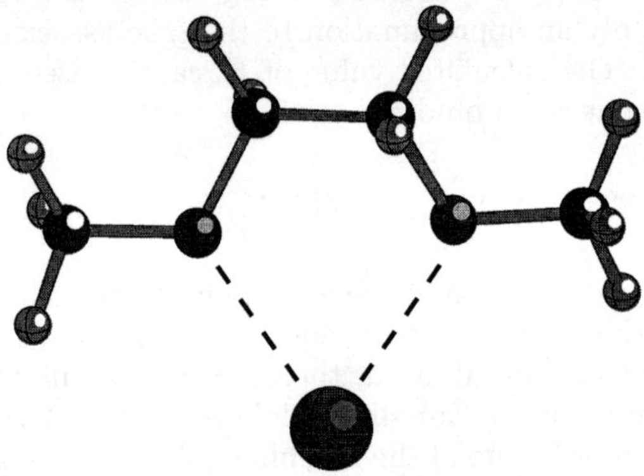

Figure 1. Gas-phase optimized structure for 1,2-dimethoxyethane (12DME) bound to K^+. Comparison of quantum mechanical/molecular mechanical and *ab initio* binding energies and selected geometric elements given in Table 1.

Table 1. K^+/1,2-dimethozyethane gas-phase binding energy and geometry.

	BE (kcal/mol)[(a)]	OCCO (degree)	COCC (degree)	K^+/12DME (oxygen; Å)[(b)]
QM/MM[(c)]	–31.3	60	167	2.57
ab initio[(d)]	–31.4	59	163	2.68

[(a)]Binding energy relative to infinite separation, and 12DME optimized with its OCCO angle gauche.
[(b)]The distance between K^+ and the oxygens of 12DME.
[(c)]Quantum mechanical/molecular mechanical
[(d)]MP2/6-31+G* (counterpoise corrected) as the HF/6-31+G* optimized geometry (Feller, personal communication).

Potential of Mean Force

The K^+/12DME pmf curve is shown in Figure 2. The coordinate for the pmf (abscissa) is the center of mass (COM) separation between K^+ and 12DME. We normalized the pmf to zero at 6.55 Å as it does not change beyond this point. We will refer to the two minima located at 2.65 Å and 4.15Å separation as the inner and outer minimum, respectively. The two barriers separating the inner/outer minima and the outer minima/dissociated limit we will refer to as the inner and outer barriers. They are centered at 3.5 Å and 4.8 Å, respectively. The barrier height going from the inner to the outer minimum is 2.6 kcal/mol, while that going from the outer minimum to the dissociated limit is 1 kcal/mol. The shape of the pmf is suggestive of K^+ interacting with the oxygens of 12DME in a sequential, rather than concerted, fashion (*discussed later*).

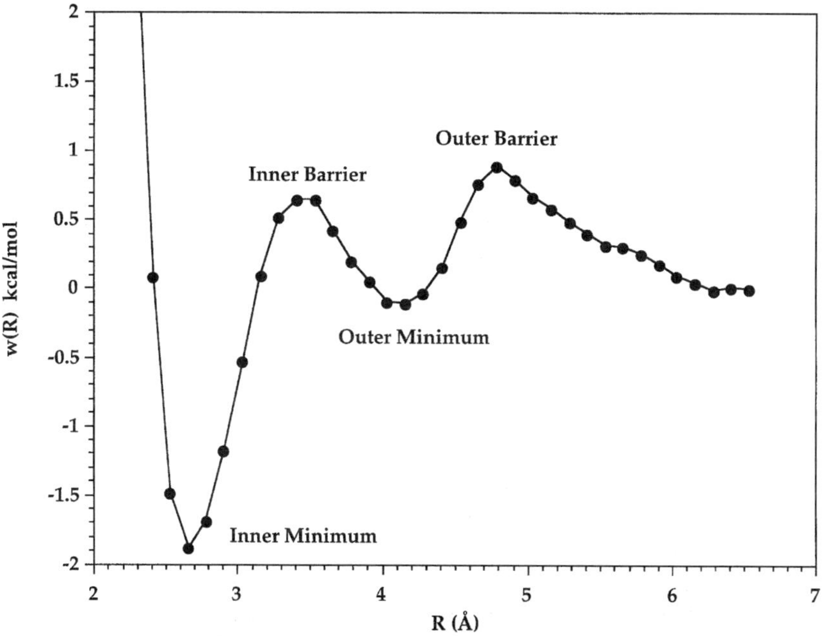

Figure 2. Potential of mean force (pmf) profile for K^+ interacting with 12DME in H_2O; the pmf coordinate is the center of mass separation between K^+ and 12DME.

In Figure 3, we show the simulation average distribution of K^+/12DME-oxygen (O_D) distances, for both oxygens of 12DME, for all of the 18 simulations that constitute the pmf profile. At large K^+/12DME separations, the K^+/O_D distances are broadly distributed, with no general preference for K^+ interacting with a specifice oxygen of the ligand.

At 4.4-Å separation (just inside the outer barrier, Figure 2), the K^+/O_D distribution splits into two distinct peaks, with the inner peak centered at 2.65 Å. This is similar to the K^+/oxygen distance in the pmf of K^+/dimethoxyethane reported previously (Thompson, 1995b). The two K^+/O_D distance peaks begin to coalesce at the inner barrier (~3.4 Å separation) and at the inner minimum they form a single peak, centered at a K^+/O_D distance of 2.65 Å. Note that from 4.4-Å separation inward, the inner K^+/O_D peak stays centered at ~2.65 Å. These results are indicative that K^+ first interacts with 12DME in a fashion similar to K^+/dimethoxyethane and, at closer distances, "picks up" the second oxygen to establish K^+/O_D distances that are only slightly longer than the corresponding gas-phase optimized distances (Figure 1, Table 1). This interpretation is confirmed by visualizing the simulation results (not shown).

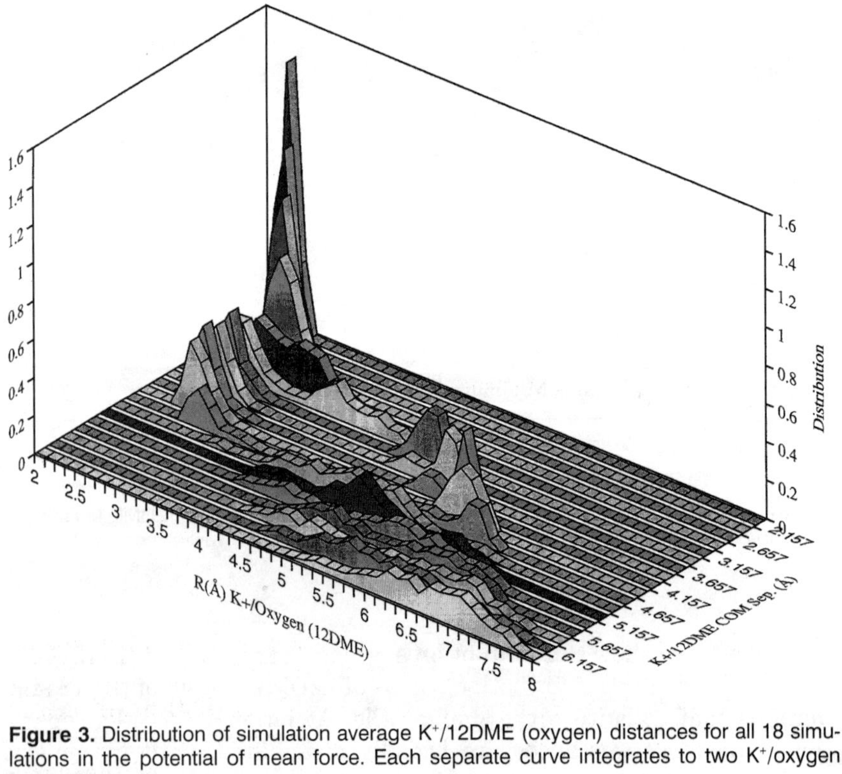

Figure 3. Distribution of simulation average K^+/12DME (oxygen) distances for all 18 simulations in the potential of mean force. Each separate curve integrates to two K^+/oxygen interactions.

1,2-Dimethoxyethane is the smallest functional subunit of 18-crown-6 that exhibits significant internal structural flexibility and the possibility of solvent effects on its conformation. In Figure 4 we show the simulation average distribution of the unsigned OCCO torsion angle of 12DME at selected points along the pmf profile. The points are shown for the following K^+/12DME separations (see Figure 2): (1) separated complex at 6.4 Å; (2) outer barrier; (3) outer minimum; (4) inner barrier; and (5) inner minimum. At large separations the OCCO angle is almost exclusively gauche, with the distribution centered around 60. At the outer barrier, this distribution shows a slight increase in the population of trans OCCO configurations. Just inside the outer barrier, the OCCO distribution changes to trans (not shown). We show the OCCO distribution at the outer minima, where we see mostly trans configurations, with a small population near 60°. The predominantly trans OCCO configuration persists to the maximum of the inner barrier, where it returns to gauche once again as K^+ interacts significantly with both oxygens of 12DME.

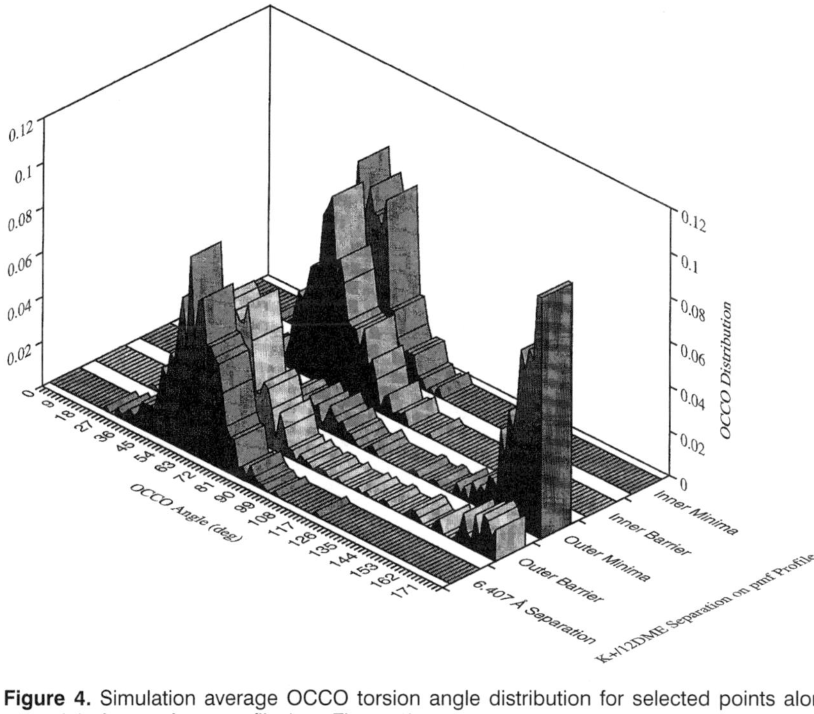

Figure 4. Simulation average OCCO torsion angle distribution for selected points along potential of mean force profile (see Figure 2).

In Figure 5 we show representative frames from the simulations for four K^+/12DME separations: (a) 6.6-Å separation; (2) 5.16-Å separation (just outside the outer barrier); (3) the outer minimum; (4) and the inner minimum (Figure 5D). For all four frames, we show K^+, 12DME, and the first solvation shell of K^+ (all waters having oxygens within 3.5 Å of K^+). In Figures 5A and 5C we also show a water molecule that interacts directly with 12DME. All other solvent molecules from the simulation are removed for clarity.

In Figure 5A, note that 12DME is rotated away from K^+ in this particular frame (our constraint scheme allows for the fragments in the pmf to rotate). There is a single water molecule, with a bifurcated H-bond

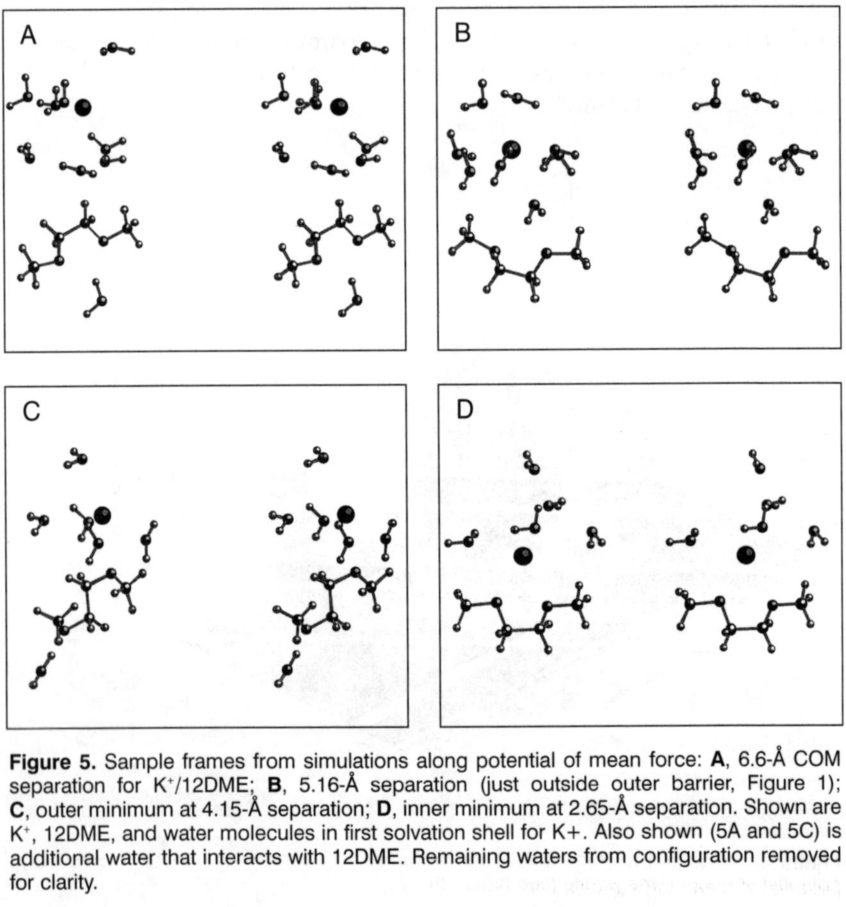

Figure 5. Sample frames from simulations along potential of mean force: **A**, 6.6-Å COM separation for K^+/12DME; **B**, 5.16-Å separation (just outside outer barrier, Figure 1); **C**, outer minimum at 4.15-Å separation; **D**, inner minimum at 2.65-Å separation. Shown are K^+, 12DME, and water molecules in first solvation shell for K+. Also shown (5A and 5C) is additional water that interacts with 12DME. Remaining waters from configuration removed for clarity.

from one of its hydrogens to both oxygens of 12DME. At this separation, 12DME and its bound water are outside the first solvation shell of the cation. This configuration is typical of free 12DME, where a single water acts to stabilize the gauche OCCO conformation. Less frequently observed were structures where a single water had both of its hydrogens separately H-bonded to each oxygen of 12DME (e.g., bridged structure, not shown). We note here a caveat that the SPC/E model is for rigid water with an HOH angle of 109.5°. In reality, water is flexible and may indeed form bridged structures with 12DME. Modeling this would require us to employ a flexible water model, which is beyond the scope of this study. However, we note that in both experimental (Mootz et al., 1994) and theoretical (Thompson, 1995a; Ranghino et al., 1985; Kowall and Geiger, 1994) studies of 18c6, water molecules form bridged structures with crown-ether oxygens separated 1,3, not 1,2 as in 12DME. In the gas phase, the all-trans 12DME (i.e., COCC, OCCO, and CCOC all trans) is the lowest energy configuration (Thompson, 1995a). However, the predominantly gauche OCCO conformation of 12DME, observed at the separated complex limit in this study, is consistent with theoretical and experimental results. Thus, it is known that in the solid state and in polar solvents the 12DME OCCO angle is predominantly gauche (Smith et al., 1995). This can be readily understood because solvent effects tend to stabilize gauche OCCO conformations, which have a sizable dipole moment, while the all-trans configuration of 12DME has a zero dipole moment. In Figure 5B, at a K^+/12DME separation of 5.16 Å, we see that 12DME now uses a water from the first solvation shell of K^+ to stabilize the gauche OCCO conformation. Thus, 12DME is still preorganized for cation binding, but its oxygens are now oriented toward the cation, being influenced by its first solvation shell. At separations slightly shorter than this, the gauche-OCCO-preorganization of 12DME is broken as one of its oxygens begins to penetrate the first solvation shell of the cation (Figures 3, 4). In Figure 5C (the outer minimum) we observe that 1 oxygen of 12DME is interacting primarily with K^+, while the other, which is trans, has a linear H-bond with a solvent molecule. The interaction of this second oxygen with solvent is typical of what we have previously observed in simulations of dimethoxyethane in H_2O (Thompson et al., 1994). At this separation, the K^+/O_D distance with the first oxygen is short enough to prevent both oxygens from simultaneously having a favorable K^+-interaction with 12DME having a gauche OCCO conformation. Note, however (Figure 4), there are some gauche OCCO conformations of 12DME at this outer minimum. Visualization of the simulation shows that these occasional gauche-OCCO structures "rock" back and forth, with one or the other oxygen interacting more closely with K^+, until eventually thermal fluctuations drive this angle to trans (not shown). In Figure 5C, both COCC angles are gauche, tending to maximize the exposure of the oxygen of each -CH_2-O-CH_2- subunit of 12DME to solvent or the cation. In Figure 5D (the inner minimum), we see the final K^+/12DME bound structure, where both oxygens of 12DME now interact with K^+, and the OCCO dihedral angle is once again gauche.

As 12DME interacts with K^+, it displaces solvent from the first solvation shell of the cation. In Figure 6, we summarize the first solvation shell coordination number for K^+/H_2O along the pmf profile. The coordination numbers represent simulation averages that are obtained by integrating the K^+/O_{water} radial distribution function from zero to 3.5-Å separation (not shown) for each of the independent simulations of the pmf. Free K^+ has a coordination number of ~7 (Dang and Kollman, 1995), which we observe in Figure 6 at large K^+/12DME separations. At the outer minimum in the pmf (Figure 1), the K^+/H_2O coordination number has dropped to 6, with the seventh coordinating oxygen coming from 12DME. At the inner minimum, the K^+/H_2O

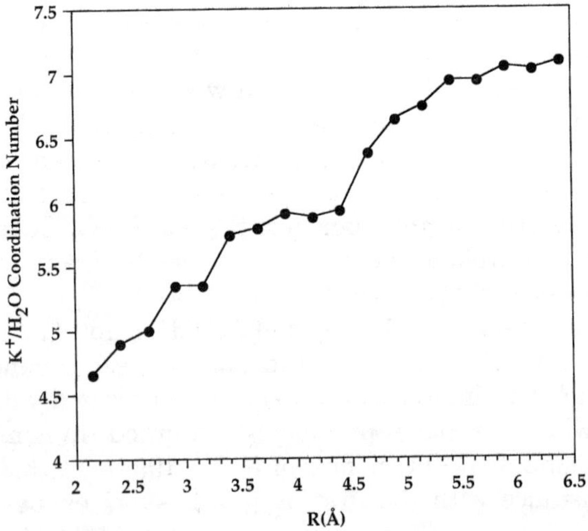

Figure 6. Simulation average coordination number for K^+ at distances sampled along potential of mean force profile. Coordination number is defined as number of water molecules in first solvation shell of K^+, obtained by integrating the K^+/O water radial distribution function to a distance of 3.5 Å.

coordination number is 5, with the two oxygens of 12DME acting as sixth and seventh ligating oxygens. These results are indicative of the sequential replacement of two waters from the first coordination shell of K^+ by the oxygens of 12DME.

Using equations (1) and (2), we can estimate the free energy of K^+/12DME binding (ΔA_{bind}). Integration of the pmf from 2 Å to the maximum at 3.5 Å gives a ΔA_{bind} = 0.5 ± 0.1 kcal/mol, indicating that K^+/12DME is not a strongly bound species in aqueous solution. As expected, this result is more stable than the ΔA_{bind} = 0.9 ± 0.1 kcal/mol that we previously reported for the K^+/dimethoxyethane pmf (Thompson, 1995b). Neither the dimethoxyethane nor 12DME results should be surprising, in spite of the fact that 18c6 avidly binds K^+. This is an example of the chelate effect, where an organized array of relatively weakly binding ligands (i.e., dimethoxyethane or 12DME) can constitute a larger ligand with a stable binding free energy, like 18c6. Cram (1988) used this observation in his synthesis of the spherands, which have very stable free energies of binding.

SUMMARY AND CONCLUSIONS

The pmf of K^+ with the bidentate ligand 1,2-dimethoxyethane (12DME) reveals a surprisingly rich amount of structural information on the interactions of the cation, neutral organic ligand, and solvent during the cation binding process. We find that a single H_2O stabilizes the 12DME OCCO torsion in a gauche conformation, in essence preorganizing it for cation binding. The initial interaction with the cation breaks this solvent stabilization, causing a distinctive conformational shift in the 12DME OCCO angle to a predominantly trans conformation. The initial K^+/12DME interaction is similar to the K^+ interaction with a dimethoxyethane (Thompson, 1995b). At shorter distances, K^+ does interact with both oxygens of the ligand, restoring it to a configuration similar to free 12DME, where K^+ has now replaced the water molecule previously bound by the ligand. Even though the free ligand is preorganized by solvent, there may be a significant energy effect during complexation due to disruption of the preorganization as the cation initially desolvates the ligand. This has implications for the free energy of binding for K^+/18c6, where the cation may cause a local disruption in the 18c6 preorganized macrocycle ring and, in effect, "roll into" the crown cavity sequentially, first by "dimethoxyethane-like" then by "12DME-like" interactions.

Indeed, a highly concerted process for cation binding, where multiple oxygens of the ligand simultaneously push out multiple waters from the first solvation shell of the cation, may have a higher energy barrier to complexation than a sequential binding process. The K^+/18c6 pmf result of Kowal and Geiger (1995) seems to indicate this; their simulation protocol tends to force a more concerted approach to cation binding than the simulation protocol used by Dang and Kollman (1995) and by us. These results will be useful for understanding cation complexation by larger multidentate ligands such as the crown-ethers.

ACKNOWLEDGMENTS

This work was performed under the auspices of the Division of Chemical Sciences, Office of Basic Energy Research, U.S. Department of Energy under Contract DE-AC06-76RLO 1830 with Battelle Memorial Institute, which operates the Pacific Northwest Laboratory, a multiprogram national laboratory. The author gratefully acknowledges Liem Dang for useful discussions and Eric Glendening for a critical reading of this manuscript. We would also like to acknowledge Thom Dunning and Bruce Garrett for encouragement.

REFERENCES

Allen, MP, and DJ Tildesley. **1987**. Computer Simulations of Liquids; Oxford University Press Inc.: New York.

Argus 3.0, and MA Thompson. **1994**. Pacific Northwest Laboratory, Richland, WA. d3f012@pnlg.pnl.gov.

Berendsen, HJC, JPM Postma, WF van Gunsteren, A DiNola, and JR Haak. **1984**. J. Chem. Phys. 81, 3684.

Berendsen, HJC, JR Grigera, and TP Straatsma. **1987**. J. Phys. Chem. 91, 6269.

Cram, DJ. **1988**. Angew. Chem. Int. Ed. Engl. 27, 1009.

Dang, LX, and PA Kollman. **1990**. J. Am. Chem. Soc. 112, 5716.

Dang, LX, and PA Kollman. **1995**. J. Phys. Chem. 99, 55.

Dang, LX, JE Rice, J Caldwell, and PA Kollman. **1991**. J. Am. Chem. Soc. 113, 2481.

De Jong, F, and DN Reinhoudt, DN. **1980**. Advances in Physical Organic Chemistry. 17, 279.

Field, MJ, PA Bash, and M Karplus. **1990**. J. Comp. Chem. 11, 700.

Glendening, ED, D Feller, MA Thompson. **1994**. J. Am. Chem. Soc. 116, 10657.

Ha, YL, and AK Chakraborty. **1993**. J. Phys. Chem. 97, 11291.

Kowall, T, and A Geiger. **1994**. J. Phys. Chem. 98, 6216.

Kowall, T, and A Geiger. **1995**. J. Phys. Chem. 99, 5240.

Liu, H, F Muller-Plathe, and WF van Gunsteren. **1995**. J. Chem. Phys. 102, 1722.

Marrone, TJ, DS Hartsough, and KM Merz Jr. **1994**. J. Phys. Chem. 98, 1341.

Mootz, D, A Albert, S Schaefgen, D Staben. **1994**. J. Am. Chem. Soc. 116, 12045.

Parker, D. **1983**. Adv. Inorg. Chem. Radiochem. 27, 1.

Pople, JA, DP Santry, and GA Segal. **1965**. J. Chem. Phys. 43, S129.

Prue, JE. **1969**. Journal of Chemical Education. 46, 12.

Ranghino, G, S Romano, JM Lehn, and G Wipff. **1985**. J. Am. Chem. Soc. 107, 7873.

Ryckaert, JP, G Ciccotti, and HJC Berendsen. **1977**. J. Comp. Phys. 23, 327.

Samec, Z, and P. Papoff. **1990**. Anal. Chem. 62, 1010.

Smith, DE, and LX Dang. **1994**. J. Chem. Phys. 100, 3757.

Smith, GD, RL Jaffe, and DY Yoon. **1995**. J. Am. Chem. Soc. 117, 530.

Thompson, MA. **1995a**. J. Phys. Chem. 99, 4794.

Thompson, MA. **1995b**. J. Am. Chem. Soc. 117, 11341.

Thompson, MA, ED Glendening, and DF Feller. **1994**. J. Phys. Chem. 98, 10465.

van Eerden, J, WJ Briels, S Harkema, and D Feil. **1989**. Chem. Phys. Letts. 164, 370.

van Gunsteren, WF, PK Weiner, and AJ Wilkinson. **1993**. Computer Simulation of Biomolecular Systems; Eds.; ESCOM: Leiden, The Netherlands Vol. 2.

Wipff, G, P Weiner, and P Kollman. **1982**. J. Am. Chem. Soc. 104, 3249.

Zwanzig, RW. **1954**. J. Chem. Phys. 22, 1420.

MACROSCOPIC ELECTROSTATICS WITH ATOMIC DETAIL (MEAD): APPLICATIONS TO BIOMACROMOLECULES

D. Bashford

Department of Molecular Biology, The Scripps Research Institute, La Jolla, California

Key Words: *Electrostatics, protein, peptide, solvation, hydrogen bond*

ABSTRACT

The model we call Macroscopic Electrostatics with Atomic Detail (MEAD) has been used to calculate the energetics of proteins, peptides, and small molecules in solution. In this model the solute and solvent regions are assigned different dielectric constants, and the boundary between them is defined by the details of atomic coordinates and radii. The electrostatic potential is found by numerical solution of the Poisson-Boltzmann equation. We report a number of tests, applications, and improvements to the method. We have compared MEAD calculations of solvent modifications of amide-amide hydrogen-bonding interactions and peptide solvation with analogous microscopic calculations, using explicit water molecules. Local conformational changes have been incorporated into calculations of pK_a values of protein side chains, resulting in improved accuracy and establishment of a framework for studying pH-dependent conformational changes. MEAD has also been used to incorporate solvent effects in density functional quantum mechanical calculations, and to calculate solvation energies and absolute pK_a values of small molecules.

INTRODUCTION

Models that treat the solvent as a dielectric continuum while retaining the detailed shape of the solvent-solute interface (Miertus et al., 1981; Warwicker and Watson, 1982) have come into increasing use for calculations of solvation energy (Lim et al., 1991; Sitkoff et al., 1994; Honig and Nicholls, 1995), pK_a values in proteins (Bashford and Karplus, 1990, Yang et al., 1993), and reaction fields for quantum mechanical calculations (Tomasi and Perisco, 1994; Cramer and Truhlar, 1994). We present here some examples of tests and applications of a model of this general type. The particular electrostatic model used is based on the approximation that the aqueous solvent's response to solute charges can be adequately represented as the dielectric response of a continuum region of dielectric constant 80. Furthermore, those aspects of solute polarizability that are not handled by more explicit representations can also be represented by a dielectric constant; typically, one chosen between 1 and 4. Nonspecific salt effects can be treated by a linearized Boltzmann term in the electrostatic equations. Both the boundary between regions of different dielectric constants and the boundary of the ion-accessible region are determined by the detailed atomic coordinates and radii. We call this type of methodology Macroscopic Electrostatics with Atomic Detail (MEAD).

In such a model the electrostatic potential is determined by solving the Poisson or Poisson-Boltzmann equation:

$$\vec{\nabla} \epsilon (\vec{r}) \vec{\nabla} \phi (\vec{r}) - \kappa^2 (\vec{r}) \epsilon (\vec{r}) \phi(\vec{r}) = -4\pi\rho (\vec{r}), \tag{1}$$

where ϕ is the Electrostatic potential, $\varepsilon(\vec{r})$ the dielectric constant, is the charge distribution, and $\kappa(\vec{r})$ the Debye-Hückel constant in ion-accessible regions, and zero elsewhere. Since the molecular surface defining the boundary of the dielectric is generally irregular, numerical methods are needed to solve this equation. Our computer program suite, MEAD, uses a finite-difference method.

We present an overview of our laboratory's recent development and application of this methodology from the perspective of its uses as a tool for quantitative calculations on functional groups in biomacromolecules.

INTERACTIONS

Before applying MEAD-type methods to biomacromolecules where there are many simultaneous interactions, one would like to test whether the solvation of and interactions between small molecules and peptides can be accurately modeled. Figure 1 shows the basic model of solvation used. The process is broken down into the solvation of a hypothetical, completely chargeless molecule and the work of charging or discharging the molecule in solvent or vacuum, respectively. It is assumed that all of the effects of the polarity of the solute are accounted for in the charging terms which are to be calculate according to the MEAD model. The application of this method to small-molecule solvation (Gilson et al., 1987, Sitkoff et al., 1994) and the closely related problem of small-molecule pK_a values (Lim et al., 1991) have been reported in the literature and will not be reviewed here.

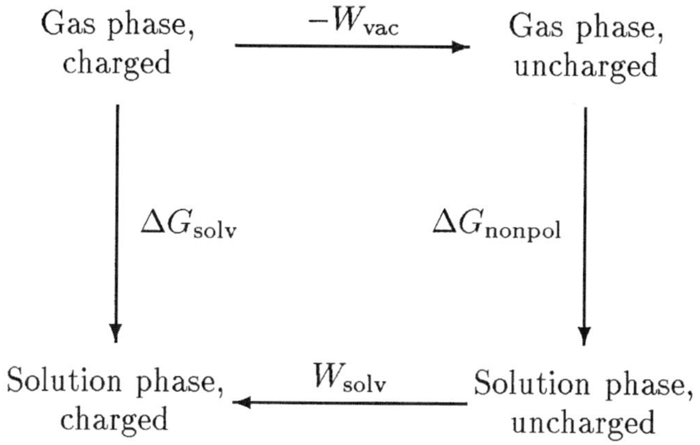

Figure 1. Thermodynamic cycle used for the continuum solvation calculations.

We have recently focused on solvent effects on interactions between small molecules and between different parts of a peptide by calculating potentials of mean force (PMF) along pathways that involve making and breaking hydrogen bonds (Ösapay et al., 1996). Experimental PMF are generally not available, but microscopic calculations of such PMF have been made by molecular dynamics simulations, using the CHARMM forcefield (Sneddon et al., 1989, Tobias et al., 1992, Young and Brooks, 1996) with explicit TIP3P water solvent. A directly analogous calculation can be made by evaluating the vacuum energies of a peptide conformer or H-bonding pair, using CHARMM and adding a solvation term calculated by MEAD. However, it is vital that the same charges be used in both the MM forcefield and in MEAD (Ösapay et al., 1996). A surface-area term of 5.42 cal/mole/Å^2 for nonpolar solvation (Sitkoff et al., 1994, Simonson and Brünger, 1994) has also been included in the calculations presented here but does not contribute significantly to the results.

The results for forming a single hydrogen bond in a formamide-formamide dimer and for forming two hydrogen bonds in an antiparallel-ß-sheet-like dimer of alanine dipeptides are shown in Figures 2 and 3. In both cases the vacuum potential has a deep minimum, but the MEAD-calculated solvent contribution increases as the close approach of the hydrogen-bonding pairs reduces solvent accessibility and, thus, the favorable interactions of the polar groups with the high dielectric of the solvent. The net effect of these opposing trends is to reduce the

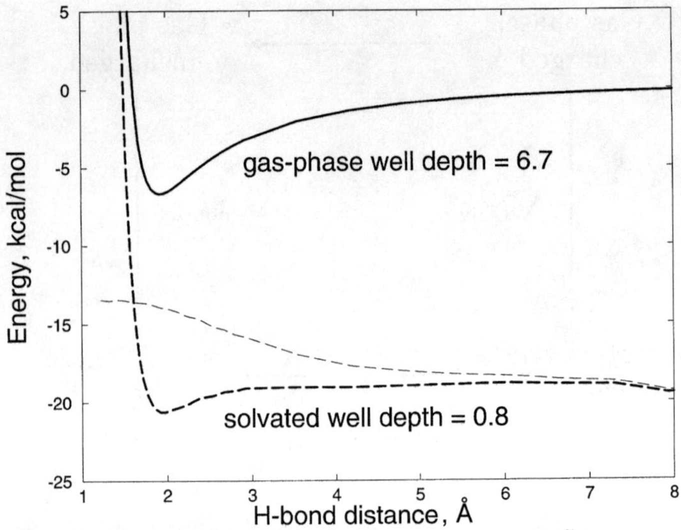

Figure 2. Energy profiles for formation of a linear dimer of formamides. Solid curve: gas-phase profile; light dashed curve: solvation contribution from continuum model; heavy dashed curve: net profile in water.

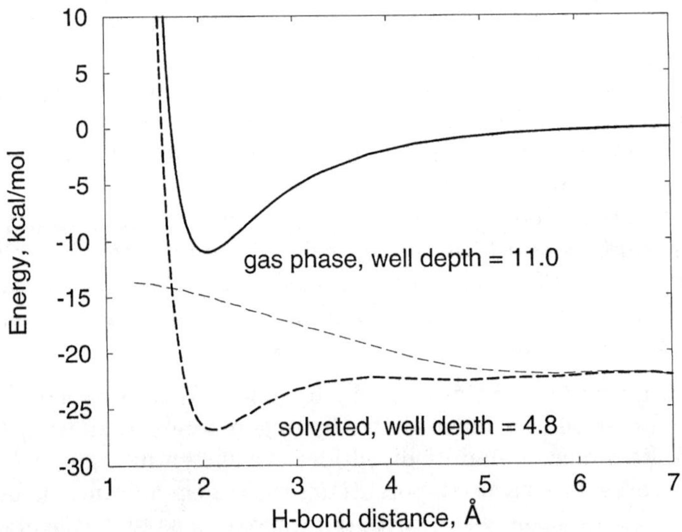

Figure 3. Energy profiles for formation of a linear dimer of alanine dipeptides. Solid curve: gas-phase profile; light dashed curve: solvation contribution from continuum model; heavy dashed curve: net profile in water.

well dept of the calculated PMF. These well depths are similar to the results of calculations using explicit TIP3P solvent and perturbation free-energy methods. Explicit solvent calculations for the formamide dimer give a well depth of 0.4 kcal/mol versus 0.8 kcal/mol in our results, and explicit solvent gives 5.6 kcal/mol for the alanine dipeptide dimer versus 4.8 kcal/mol here. A notable feature of these results is that the well depth per hydrogen bond is significantly larger for alanine dipeptide than for formamide, no matter which method of calculation is used. In the formamide case, the greater exposure to water leads to a weakening of the hydrogen bond, compared to the alanine dipeptide case.

The dimer calculations had no "floppy" degrees of freedom, but one would like to know how well a more flexible system is handled. Figures 4 and 5 compare microscopic and MEAD calculations for the unwinding of one residue from the C-terminal end of an Ala_{15} helix. In the microscopic calculations of Young and Brooks (1996), PMF were calculated, with the reaction path being the terminal backbone dihedral. Harmonic constraints on all H-bonds distances except the one being

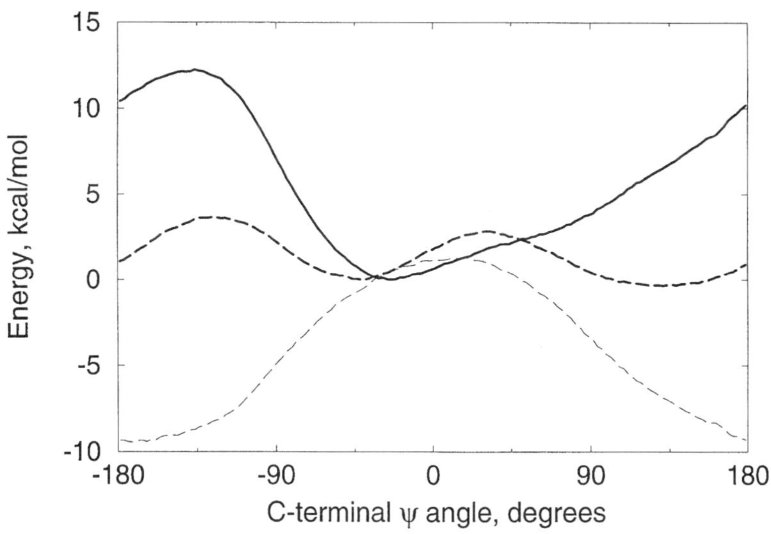

Figure 4. Potentials of Mean Force (PMF) contributions to the unwinding of one residue from the C-terminus of a helix as calculated with explicit TIP3P water solvent. Solid curve: vacuum PMF; heavy dashed curve: TIP3P-solvated PMF; light dashed curve: solvation contribution calculated by difference of solvated and vacuum PMF.

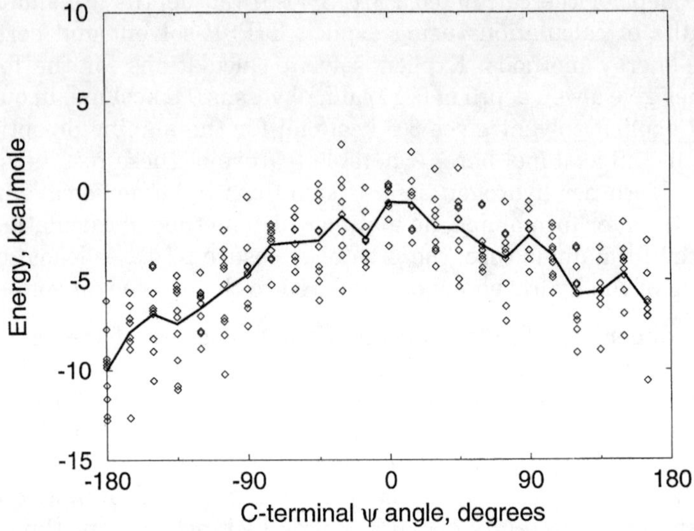

Figure 5. MEAD calculation of solvent contribution to PMF for unwinding of one residue from the C-terminus of a helix; analogous to the solvation contribution shown in Figure 4. Diamonds: solvation energies calculated from individual snapshots. Solid curve: mean of snapshots.

broken were used to keep the rest of the helix intact. Figure 4 shows the PMF calculated in a TIP3P-solvated system and in a vacuum. The difference is the solvent contribution as determined microscopically. For a MEAD calculation of the solvent contribution curve, 10 snapshots from each angle window of the vacuum PMF calculation were taken, and their solvation energies were calculated using MEAD. The results, shown in Figure 5, compare quite well to the microscopic calculations.

These comparisons and comparisons of several other similar interacting systems lead us to the conclusion that MEAD-type methods often given results quite similar to more computationally intensive microscopic models of solvation, although in some cases, the magnitude of the effects calculated by MEAD is somewhat less. It is not yet clear whether this reflects limitations of the MEAD model or the effects of long-range cutoffs in the microscopic calculations. We also conclude that limited flexibility can be dealt with by sampling, although the sampling methods could be more sophisticated than those shown here.

TITRATION WITH CONFORMATIONAL FLEXIBILITY

Bashford and Karplus (1990) showed that the MEAD method can also be extended to the titration of ionizable side chains in proteins (Figure 6). In this case, a model compound pK_a value is taken as a starting point, and one must calculate the difference in the work of altering the titrating charges from their deprotonated to their protonated states (Figure 6). Three kinds of energy terms are involved: (1) self-energy: the interaction of the titrating charges with the polarization that they have induced in their own surroundings; (2) background terms: the interaction of the titrating charges with nontitrating charges; (3) site-site terms: the interactions between titrating sites (protein only). The first two terms contribute to the "intrinsic pK_a" that a group in the protein would have if all other ionizable groups were held in some standard state; typically, their charge-neutral state.

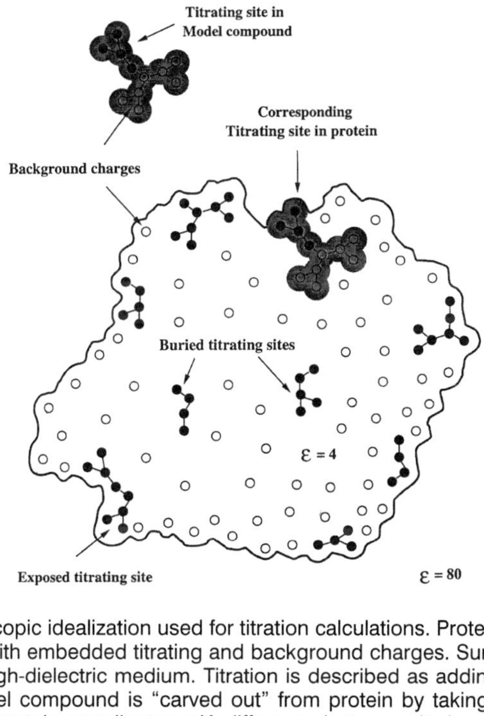

Figure 6. Macroscopic idealization used for titration calculations. Protein is treated as low-dielectric object with embedded titrating and background charges. Surrounding solvent is represented as high-dielectric medium. Titration is described as adding proton charge to titrating site. Model compound is "carved out" from protein by taking model compound coordinates from protein coordinates. pK_a difference between site in protein and in model compound is due to differences in electrostatic work required to alter titrating charges in model compound and protein.

Most calculations of this kind have assumed that the protein remains "frozen" in its crystallographically determined structure even as its side chains are undergoing titration (Bashford and Karplus, 1991 Yang et al., 1993). However, even the early studies showed that conformational change could be important. The pK_a values calculated for two different crystal forms of lysozyme differed significantly for several sites (Bashford and Karplus, 1990). In calculations on bacteriorhodopsin, the uncertainties of the structural model were dealt with by a rough ensemble approach. It was found that, for one arginine side chain, protonation states in agreement with experiment could be found only if the side chain was moved toward the active site. This observation led to a proposal that the model structure be modified (Bashford and Gerwert, 1992).

Recently, You and Bashford (1995) proposed a more systematic method of including conformational variation in protein titration calculations. Unlike past approaches, it takes account of the fact that different structural forms may dominate the ensemble, depending on the charge state of the side chain (Figure 7). This approach should not only improve the accuracy of pK_a calculations but should also allow for calculations of pH-driven conformational change.

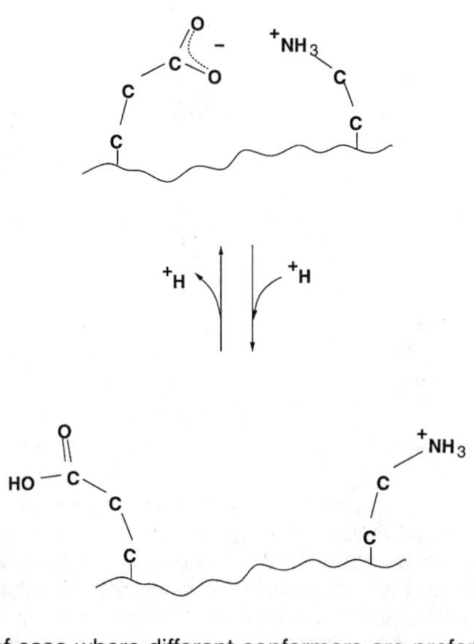

Figure 7. Illustration of case where different conformers are preferred in different protonation states of side chain.

If we consider only the problem of calculating the intrinsic pK_a for a single site, and we assume that there are N_c conformers available to the molecule, a canonical ensemble over the conformers and protonation states gives

$$pK_{intr} = pK_{model} - \frac{1}{2.303}$$

$$x \left[\ln\sum_{n=1}^{N_c} e^{-\Delta G_{env}(A_{p,n})/RT} - \ln\sum_{n=1}^{N_c} e^{-\Delta G_{env}(A_{p,n}H)/RT} - \ln\sum_{n=1}^{N_c} e^{-\Delta G_{env}(A_{m,n})/RT} + \ln\sum_{n=1}^{N_c} e^{-\Delta G_{env}(A_{m,n}H)/RT} \right], \quad (2)$$

where AH and A represent the protonated and deprotonated forms of the group, respectively; the subscripts p and m indicate protein and model compound, respectively; and the subscript n labels conformers. The environmental contribution to the relevant free-energy differences, ΔG_{env}, has the form

$$\Delta G_{env}(A_{p,n}) = E_{p,n} + \Delta G_{Born}(A_{p,n}) + \Delta G_{Back}(A_{p,n}), \quad (3)$$

where the three terms on the righthand side are nonelectrostatic energy, self-energy, and background energy, respectively.

The interaction W_μ between any two titrating sites, μ and v, is treated in an averaged way:

$$W_{\mu\nu} = \sum_{i=1}^{N_{Q_\mu}} \left[Q^{(h)}_{\mu,i} - Q^{(d)}_{\mu,i} \right] \left[\langle \phi_p(r_i;Q^{(h)}_\nu) \rangle - \langle \phi_p(r_i;Q^{(d)}_\nu) \rangle \right], \quad (4)$$

where the Q_i are the charges for the atoms, i, of site μ in the protonated or deprotonated (superscript h or d) state, and the ϕ are the potentials due to charges on another site, v. This implies that the conformational changes that influence pK_{intr} at one site are uncorrelated with those at another. In other words, the conformational changes are mostly localized—for example, sidechain dihedral changes. To include site-site conformational correlations without special approximations in an N-site molecule with N_c conformers per site (and two protonation states per site) would introduce a combinatorial complexity of order N^{2Nc}. We leave the problem of site-site conformational correlations to future studies.

Once all the pK_{intr} and $W_{\mu\nu}$ have been calculated, the degree of protonation at any pH can be found by a Boltzmann average over protonation states:

$$\theta_i = \frac{\sum_x^N x_i \exp\left[\sum_\mu^N x_\mu 2.303(pK_{intr,\mu} - pH) - \frac{1}{RT}\sum_{\mu,\nu}^N W_{\mu\nu}(x_\mu + q_\mu^\circ)(x_\nu + q_\nu^\circ)\right]}{\sum_x^N \exp\left[\sum_\mu^N x_\mu 2.303(pK_{intr,\mu} - pH) - \frac{1}{RT}\sum_{\mu,\nu}^N W_{\mu\nu}(x_\mu + q_\mu^\circ)(x_\nu + q_\nu^\circ)\right]} . \quad (5)$$

We carried out calculations for leu lysozyme in which the conformational variation at each site consisted of changes in the χ_1 and χ_2 side chain dihedral angles of Asp and Tyr residues and the χ_2 and χ_3 angles of Glu residues. Thirty-six conformers of each flexible chain were included. The nonelectrostatic energies were evaluated using the CHARMM forcefield (Brooks et al., 1983). This work has been described in detail in You and Bashford (1995). The results in Table 1 show that the agreement with experiment is significantly better when conformational freedom is accounted for.

REACTION FIELDS FOR QUANTUM MECHANICS

In dealing with enzyme active sites, one would like to be able to combine the capacity of quantum chemical methods for structural and energetic detail with some reasonable representation of the effects of the solvent and/or protein environment. Macroscopic models of the solvent reaction field in small-molecule quantum calculations have long been in use, including some that include the detailed shape of the solute (Tomasi and Perisco, 1994; Cramer and Truhlar, 1994). Recently, we have described a method of coupling density functional theory (DFT) with MEAD calculations (Chen et al., 1994). The energy functional used in DFT can be written

$$E = \sum_i \langle\psi_i|h(1)|\psi_i\rangle + \frac{1}{2}\int \rho_{tot}(r)\Phi(r,r')\rho_{el}(r')drdr' + E_{xc}[\rho_{tot}], \quad (6)$$

where h(1) is the one-electron Hamiltonian, E_{xc} is an exchange-correlation term, and $\Phi(r, r')$ is the green function—the potential at r due to a unit charge at r′. For vacuum calculations, Φ is simply the Coulomb

Table 1. pK_a calculations for leu lysozyme. Experimental (Exp) pK_a values compared to calculations based on the triclinic crystal structure (Xtal), or on energy-minimized structure, or the flexible protocol of Equations (2) to (4). Bottom lines give root-mean-square (RMS) deviation from experiment.

Site	Exp pK_a	Xtal	Minimized	Flexible
GLU-7	2.60 - 3.10	1.99	2.62	3.00
GLU-35	6.1 - 6.3	5.96	4.21	5.67
ASP-18	2.58 - 2.74	2.75	2.68	2.74
ASP-48	1.2 - 2.0	0.61	3.95	3.67
ASP-52	3.60 - 3.76	6.98	6.79	4.14
ASP-66	0.4 - 1.4	1.12	1.95	2.65
ASP-87	1.92 - 2.22	0.84	2.77	2.84
ASP-101	4.02 - 4.16	7.84	4.89	3.78
ASP-119	3.11 - 3.29	3.19	3.39	3.37
TYR-20	10.3	13.98	12.87	12.06
TYR-23	9.8	11.65	10.34	10.78
TYR-53[(a)]	12.1	20.66	18.60	17.07
ASP & GLU RMS		1.69	1.40	0.76
TYR RMS		5.49	4.05	3.10
All site RMS		2.41	2.36	1.68

[(a)]As noted by Bashford and Karplus (1990), the calculation cannot be expected to reproduce the experimental pK_a for TYR-53 since the deprotonation of this residue is coupled to the unfolding of the protein.

operator $1/r$. But in the context of the MEAD model, Φ for a solvated system is the green function of the Poisson equation. One finds

$$E = E^\circ + \frac{1}{2} \int \rho_{tot}(r) \left[\phi_{solv}(r) - \phi_{vac}(r)\right] dr, \qquad (7)$$

where E° is the vacuum energy functional, and the difference between ϕ_{solv} and ϕ_{vac} is the reaction field calculated by MEAD. An iteration to self-consistency is used to compute the solvation energy and the solute electronic structure in the solvent environment (Figure 8). When combined with DFT calculations of gas-phase proton affinities, solvation calculations of acids and their conjugate bases can be used to calculate absolute pK_a values. Absolute redox potentials can be calculated by an analogous procedure.

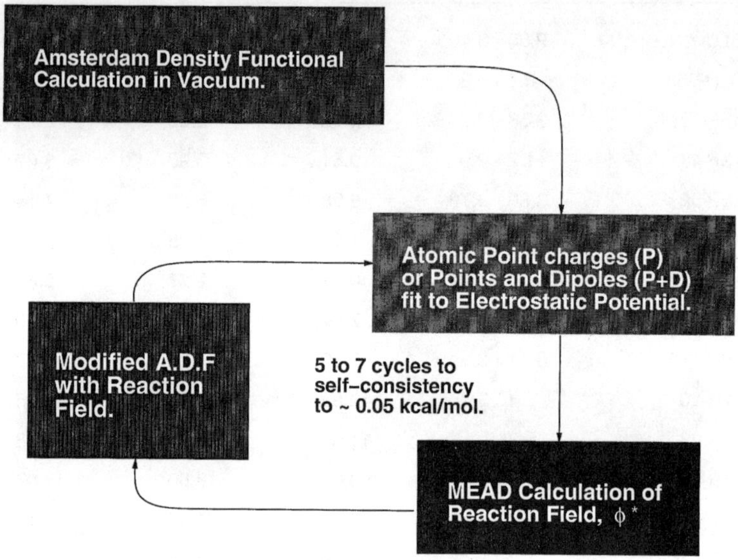

Figure 8. Cycle of calculations for including MEAD-calculated reaction fields in density function quantum mechanics calculations in self-consistent way.

Results for some small-molecule pK_a calculations are displayed in Table 2 and compared to experiment in Figure 9. (See Richardson et al., 1996 for further details.)

FUTURE DIRECTIONS

Several aspects of the multiconformation methods could be improved. The lack of site-site conformational correlation has been pointed out above; although a fully correlated treatment does not appear practical, limited inclusion of correlations for neighboring sites should be possible. The efficiency of the methods could be improved by building on an observation that we made during the calculations reported here. Specifically, that for most protein sidechains, only a few of the 36 conforma-

Table 2. pK_a values for various acids.

Acid	Calculated	Experimental
Water	16.4	15.7
Formic acid	1.5	3.8
Acetic acid	4.9	4.8
Methanol	22.5	16.0
Methyl-ammonium	12.2	10.6
Imidazolium	8.7	7.0
Mercaptoethanol	9.3	10.5
Phenol	10.7	10.0
Methyl-guanidinium	13.4	13.4
Acetamidium	–5.0	–0.4
Acetic acid	4.9	4.8
Chloroacetic acid	0.8	2.9
Dichloroacetic acid	–2.0	1.2
Trichloroacetic acid	–3.0	0.7

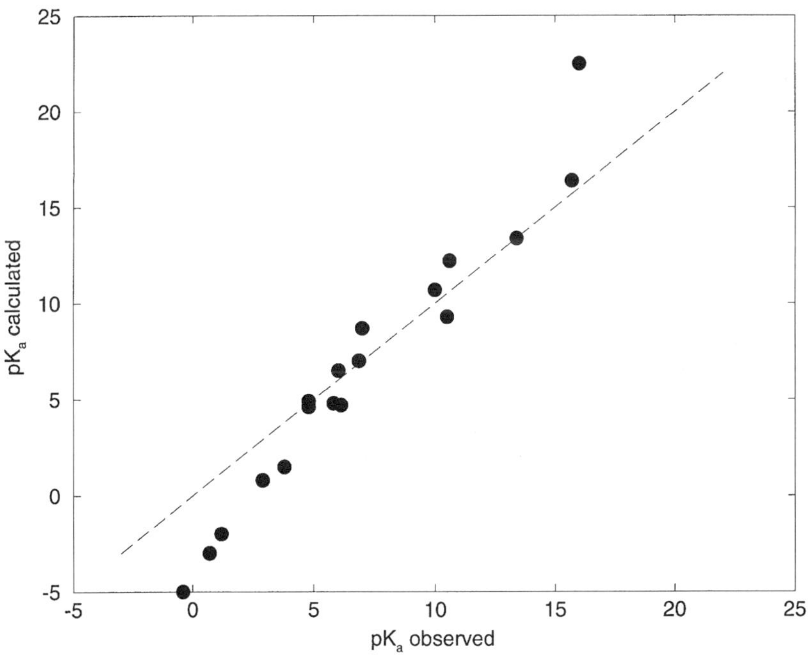

Figure 9. Comparison of calculated versus experimentally observed values of pK_a for various small molecules. Data points are from Table 2. Dashed line is pK_a calculated = pK_a observed.

tions sampled contribute significantly to the calculated results. Therefore, a reasonably accurate prescreening algorithm could greatly reduce the number of conformations that need to be examined. Both these ideas are currently being pursued in our research group.

A natural follow-up to the successful treatment of solvent effects in small-molecule quantum calculations is to use MEAD methods to model the effects of both the solvent and protein environment on protein active centers whose properties are to be calculated by quantum mechanics. Formally, this is very much like the methodology of the previous section, except that the "solvent" is now replaced by the protein-solvent system. A MEAD program that implements this idea has been written and is currently being used for calculations on iron-sulfur centers in proteins.

ACKNOWLEDGMENTS

It is a pleasure to acknowledge my collaborators, Charles L. Brooks, III, David A. Case, Jun L. Chen, Wendy Chen, Louis Noodleman, Klara Ösapay, Chun-Yang Peng, William H. Richardson, Tony You, and William S. Young. Support was provided by the National Institutes of Health (GM45607 and GM38794).

REFERENCES

Bashford, D and K Gerwert. **1992**. Electrostatic calculations of the pK_a values of ionizable groups in bacteriorhodopsin, J. Mol. Biol. 224:473-486.

Bashford, D and M Karplus. **1990**. pK_a's of ionizable groups in proteins: Atomic detail from a continuum electrostatic model. Biochemistry 29:10219-10225.

Brooks, BR, RE Bruccoleri, BD Olafson, DJ States, S Swaminathan and M Karplus. **1983**. CHARMM: A program for macromolecular energy, minimization, and dynamics calculations. J. Comp. Chem. 4:187-217.

Chen, JL, L Noodleman, DA Case, and D Bashford. **1994**. Incorporating solvation effects into density functional electronic structure calculations. J. Phys. Chem. 98:11059-11068.

Cramer, C and D Truhlar. **1994**. Development and biological applications of quantum mechanical continuum solvation models, pp. 9-54. In: P Politzer and J Murray, eds., Quantitative Treatments of Solute/Solvent Interactions. Elsevier, Amsterdam.

Gilson, MK, KA Sharp and BH Honig. **1987**. Calculating the electrostatic potential of molecules in solution: Method and error assessment. J. Computat. Chem. 9:327-335.

Honig, B, A Nicholls. **1995**. Classical electrostatics in biology and chemistry. Science 268:1144-1149.

Lim, C, D Bashford and M Karplus. **1991**. Absolute pK_a calculations with continuum dielectric methods. J. Phys. Chem. 95:5610-5620.

Miertus, S, E Scrocco and J Tomasi. **1981**. Electrostatic interaction of a solute with a continuum: A direct utilization of *ab initio* molecular potentials for the provision of solvent effects. Chem. Phys. 55:117-129.

Ösapay, K, WS Young, D Bashford, CL Brooks III, and DA Case. **1996**. Dielectric continuum models for hydration effects on peptide conformational transitions. J. Phys. Chem. 100:2698-2705.

Richardson, WH, C-Y Peng, D Bashford, L Noodleman and DA Case. **1996**. Incorporating solvation effects into density functional theory: Calculations of absolute acidities. Int. J. Quant. Chem (in press).

Simonson, T and AT Brünger. **1994**. Solvation free energies estimated from macroscopic continuum theory: An accuracy assessment. J. Phys. Chem. 98:4683-4694.

Sitkoff, D, KA Sharp and B Honig. **1994**. Accurate calculation of hydration free energies using macroscopic solvent models. J. Phys. Chem. 98:1978-1988.

Sneddon, SF, DJ Tobias, and CL Brooks III. **1989**. Thermodynamics of amide hydrogen bond formation in polar and apolar solvents. J. Mol. Biol. 209:817-820.

Tanford, C and JG Kirkwood. **1957**. Theory of protein titration curves. I. General equations for impenetrable spheres. J. Am. Chem. Soc. 79:5333-5339.

Tobias, DJ, SF Sneddon, and CL Brooks, I. **1992**. Stability of a model beta-sheet in water. J. Mol. Biol. 227:375-380.

Tomasi, J and M Persico. **1994**. Molecular interactions in solution: An overview of methods based on continuous distributions of solvent. Chem. Rev. 94:2027-2094.

Warwicker, J and HC Watson. **1982**. Calculation of the electric potential in the active site cleft due to α-helix dipoles. J. Mol. Biol. 157:671-679.

Yang, A-S, MR Gunner, R Sampogna, K Sharp, and B Honig. **1993**. On the calculations of pK_as in proteins. Proteins 15:252-265.

You, T and D Bashford. **1995**. Conformation and hydrogen ion titration of proteins: A continuum electrostatic model with conformational flexibility. Biophys. J. 59:1721-1733.

Young, WS and CL Brooks III. **1996**. A microscopic view of helix propagation: N- and C-terminal growth in alanine helices. J. Mol. Biol. 259:560-572.

HOMOLOGY GUIDED PROTEIN FOLDING: APPLICATION TO PROTEIN THERAPEUTIC DESIGN

S. Srinivasan and S. Sudarsanam

Department of Protein Chemistry, Immunex Corporation, Seattle, Washington

Key Words: *homology modeling; distance geometry; CD40 ligand; TNF family; loop modeling*

ABSTRACT

Experimentally determined 3-D structures of proteins are often used to derive protein folding rules that are, in turn, used to predict 3-D structures of structurally uncharacterized proteins. These methods, known as *de novo* folding of proteins, have met with limited success. However, since not all known proteins fold in unique tertiary topology, it is possible to model one protein using the 3-D structure of another as a template. This approach, known as homology modeling, is often used to build 3-D models of proteins bearing strong sequence identities to a template for use in structure-function studies and in structure-based drug design. A systematic approach to modeling closely related members of a protein family, first put forward by Greer, has now been implemented in many commercial software packages. Recently, according to Chothia's analysis, all the functional proteins from human genes will fold into one of less than a thousand novel folds. If this is true, a method to fold distantly related proteins on one of the known folds is just as challenging as *de novo* folding of proteins because such a method should account for the minor structural differences imposed by the amino acid substitutions in the core region. We present an automated method to fold proteins on a given template by using homologous distances derived from a template, together with a database of dimeric building blocks derived from the loop regions of high-resolution crystal structures.

INTRODUCTION: WHY HOMOLOGY FOLDING?

Almost a decade before the first proteins were structurally characterized, the geometry of common α-helical and β-sheet conformations was predicted by Pauling and Corey (1951) and Pauling et al. (1951). The three-dimensional structures of most of the proteins in the current Protein Data Bank (PDB; Bernstein et al., 1977) consist predominantly of α-helical and β-sheet conformations. The successful prediction of common secondary structural modules of proteins, such as a-helix and b-sheet, from simple potential energy computation led to the optimism that one day it would be possible to predict protein folds *de novo* from amino acid sequences. However, in spite of the growth in protein structure databases and an exponential increase in computing power, protein folding continues to challenge biologists.

While efforts to solve the protein-folding problem continue, the number of structurally characterized proteins using x-ray and nuclear magnetic resonance (NMR) techniques has grown exponentially. The growth trend in structural and sequence databases indicates that the repertoire of tertiary structures is more limited than would appear from the diversity in the sequences of proteins, that is, 1,000 (Chothia, 1992). At present there are about 250 unique folds characterized by x-ray and NMR methods (Orengo et al., 1994; Hobohm and Sander, 1994). Because there is a relatively small number of protein families, tertiary structural templates for about one-fourth of all proteins are already known, and the protein-folding problem is reduced to matching a sequence to its native template. However, in the absence of any significant sequence homology, it is not trivial to establish the one-to-many relationship between structure and sequence databases. In the case of divergent proteins, one may resort to other predictive methods to identify folding topology, including *de novo* prediction of folding topology, inverse protein-folding approaches and experimentally determined functional relationships.

More recently, experimentally determined functional relationships have been used to predict tertiary structural similarities in cases where they cannot be established with confidence using only homology at the sequence level. Members of helical cytokines and the TNF ligand family are examples of successful function-based structure predictions (Bazan, 1992; Farrah and Smith, 1992). However, for such predictions to be of practical value, it is important that three-dimensional (3-D) models be built using structural homology.

One of the prerequisites to any homology modeling method is a structurally meaningful sequence alignment between a model protein and one or more structurally characterized proteins. For divergent proteins, it is not possible to obtain structurally meaningful sequence alignments because of sparse sequence identities. Thus, even when the tertiary topology of a novel protein can be predicted with confidence it is not trivial to generate useful 3-D models. In the absence of any sequence identity between a novel protein and a selected template, one is forced to use the topological constraints of the template to obtain structurally meaningful sequence alignment. One way to accomplish this is to build models using a first-pass sequence alignment and use the structural inconsistencies of the model to improve the sequence alignment. Some of the anticipated structural inconsistencies while homology modeling divergent proteins include volume restrictions for substituted amino acids in the interior and incompatible distances between the two cysteines involved in disulfide cross-link. Homology modeling methods in the past are inherently inefficient in detecting such structural inconsistencies. In the next section, we give a background of homology-modeling methods and discuss why these methods are ill-suited and impractical for modeling divergent proteins.

BACKGROUND TO HOMOLOGY-MODELING METHODS

Until recently, not much progress had been made in the methodology development since the work of Browne et al. (1969), who generated a model of a-lactalbumin based on the 3-D structure of egg-white lysozyme. However, methods developed by Kabsch (1976) and McLachlan (1982) were used to quantitate the otherwise visually obvious structural similarities between two proteins. Using these methods and superimposing 3-D structures of members of a protein it has been shown that by carefully selecting backbone atoms of only certain discontinuous regions along the sequence, these structures could be superimposed to a very low root mean square (rms) deviation. These regions are known as structurally conserved regions (SCR), and the regions connecting two SCR are known as structurally variable regions (SVR). The topologically equivalent atoms of all the structurally characterized serine proteases can be perfectly overlaid by superimposing only the backbone atoms of the SCR. Greer (1981, 1991) took advantage of this observation to put forward a systematic approach to build models in a computer, utilizing the rapid growth in computer technology and protein structural database. Despite the success of these methods in modeling highly homologous proteins, the use of these methods to model divergent proteins is limited.

Using crystallographic techniques, it has been shown that divergent proteins with no detectable sequence homology retain as much as 60% structural homology (Chothia and Lesk, 1986). For homology modelers this is a mixed blessing because while accuracy of homology-built models for proteins with 50% or more sequence identity is still under intense debate, the possibility of modeling proteins with as little as 15 to 20%

sequence identity to a template is challenging. An even tougher test of homology-modeling methods comes from the important observation by Chothia and Lesk (1986); the rms deviation of divergent proteins can be as large as 3 Å, unlike what is found by superimposing members of serine proteases. It is not practical to achieve correlated adjustments of the backbone atoms in the SCR using homology-modeling methods that operate in Cartesian space (Greer, 1991). Hence, for homology-modeling methods to be useful for modeling divergent proteins, it is important that we learn how to model SCR of divergent proteins.

STATE-OF-THE-ART METHODS IN HOMOLOGY MODELING

Distance-geometry-based homology-modeling methods are important in the evolution of homology modeling methods. This is because use of distance space instead of Cartesian space allows for a holistic approach to homology modeling. In spite of the wide use of distance geometry methods to build models consistent with NMR-derived distances, their power in homology modeling was not widely recognized until recently. In a method developed by Srinivasan et al. (1986), all the inter-C-α and inter-C-β distances of conserved amino acids were used to embed models of bungarotoxin. The main contribution of this work was to demonstrate the power of distance geometry to mix and match structural constraints from a number of not so similar templates. Havel and Snow (1991) used a distance geometry approach to simulate atomic coordinates of all atoms of Kazal-type trypsin inhibitor by using only inter-C-α and inter-C-γ distances of only identical amino acids from the template structure. Because of the partial use of homologous distances in both methods, several additional protocols were used to retain the integrity of the constraints derived from template proteins. In both methods the economic use of distance constraints resulted in undesirably large rms deviation between model and template even for highly homologous proteins.

A more holistic and elegant approach was subsequently put forward by Havel (1993) and Srinivasan et al. (1993): holistic, because these methods allow for integration of steric, topological, and homology constraints obtainable from various sources; elegant, because when a complete set of homologous distances is used to embed models, no additional protocols are required to retain the structural integrity of models (Srinivasan et al., 1993). In the distance-geometry-based approaches the atomic positions for all atoms, including the loop regions, are assigned by embedding the metric matrix. Even for atoms in the loop regions the embedded coordinates are consistent with standard chemical constraints albeit far from native conformations because of lack of long-range conformational constraints in loop regions. The method proposed by Srinivasan et al. (1993) has been implemented in a software package, FOLDER.

Although the development of FOLDER was initiated for modeling divergent proteins, at the very least it can be used to model highly homologous proteins. FOLDER has been extended to build models of divergent proteins by using a divergence factor to relax the bounds of the distances derived from the template (Sudarsanam et al., 1994). Using this modification, relative orientation of secondary structural regions can be relaxed while holding each SCR rigid. This is an important aspect of homology-modeling methods for modeling divergent proteins; otherwise, the constraints of the template will be incorrectly overimposed on the model structure. However, we observed that these subtle rearrangements in the SCR in our test cases did not move the model closer to its crystal structure any more than the unrelaxed model. This is because, while the divergence factor allows for the increased flexibility between SCR necessary to repack the protein interior, no additional constraints unique to model protein were added to compensate for this increased flexibility in the SCR. Additional constraints unique to model protein could be obtained by predicting native conformations of loops of the model protein, using the conformational preference for amino acids in the loop region. To this end, we have developed a method to sample various conformations of the backbone atoms of the loop regions biased by the amino acid sequence of the loop region, while being consistent with the geometry of the anchoring SCR (Sudarsanam et al., 1995).

Here, we have utilized this approach to predict a model for CD40 ligand. Moreover, we have applied the biological knowledge of the TNF ligand/receptor system to predict a quaternary model required for biological activity of CD40 ligand. Finally, we demonstrate the power of this homology modeling approach in engineering a chimeric CD40 ligand trimer structure which is expressed and biologically active.

HOMOLOGY MODELING OF CD40 LIGAND

More often than not, sequence identities among receptors of a ligand family are detectable suggesting divergence of these receptor genes from a common ancestor (Cosman, 1993, 1994). However, the structural similarities among the ligands of these receptor superfamilies are not readily detectable at the sequence level in spite of their structural similarities (Rozwarski et al., 1994). Hence, the predictions of tertiary structural similarities among divergent members of a protein family can be made based on sequence homology between their cognate receptors or binding proteins (Bazan, 1992). The discovery of sequence homology between TNF receptor and other known cell surface proteins suggested that signalling by these receptor molecules was triggered by yet unknown ligands structurally similar to TNF (Smith et al., 1990). Based on the sequence homology between CD40 and TNF receptor a search for CD40 ligand was initiated in anticipation that a ligand to CD40 might have important immunological functions just like TNF (Spriggs et al., 1992). A comparison of amino acid sequences of CD40 ligand and TNF resulted in the identification of sequence motifs that are suggestive of similarities in their tertiary structures (Farrah and Smith, 1992).

Using FOLDER we built models of CD40 ligand consistent with the sequence alignment shown in Figure 1A. The crystal structure of TNF (Eck and Sprang, 1989) was used as a template. The distance constraints for the only nonhomologous disulfide were provided as

```
Humtnf:  -----KPVAHVVANPQAEGQ---LQWLNRRAN---ALLANGVELRDN----QLVVPSE
Humlta:  -----KPAAHLIGDPSKQNS---LLWRANTDR---AFLQDGFSLSNN----SLLVPTS
Gldfsl:  -----RKVAHLTGNPHSRSIP--LEWEDTYGT---ALIS-GVKYKKG----GLVINET
Hufasl:  -----RSVAHLTGKSNSRSMP--LEWEDTYGI---VLLS-GVKYKKG----GLVINET
Hcd401:  -----QIAAHVISEASSKTTSV-LQWAEKGYYTM-SNN--LVTLENG---KQLTVKRQ
STRAND:  -----BBBBBBBB---------------------bbbbbccccc-----CCCCC---

Humtnf:  GLYLIYSQVLFKGQGCP----STHVLLTHTISRIAVSY-QTKVNLLSAIKSPCQRETP
Humlta:  GIYFVYSQVVFSGKAYSPKATSSPLYLAHEVQLFSSQY-PFHVPLLSSQKMVYP----
Gldfsl:  GLYFVYSKVYFRGQSCN--N----QPLNHKVYMRNSKY-PEDLVLMEEKRLNY-----
Hufasl:  GLYFVYSKVYFRGQSC------NNLPLSHKVYMRNSKY-PQDLVMMEGKMMSY-----
Hcd401:  GLYYIYAQVTFCSNRE-----ASSQAPFIASLCLKS-PGRFERILLRAANTHSSAKP-
STRAND:  DDDDDDDDDD↑DD-----------EEEEEEEEEE------FFFFFFFFFF-------

Humtnf:  EGAEAKPWYEPIYLGGVFQLEK-GDRLSAEIN-RPDYLL--FAESGQVYFGIIAL---
Humlta:  -GLQEP-WLHSMYHGAAFQLTQ-GDQLSTHTD-GIPHLV---LSPSTVFFGAFAL---
Gldfsl:  CTTGQI-WAHSSYLGAVFNLTS-ADHLYVNIS-QLSLIN---FEESKTLFGLYKL---
Hufasl:  STTGQM-WARSSYLGAVFNLTS-ADHLYVNVS-ELSLVN---FEESQTFFGLYKL---
Hcd401:  -----C-GQQSIHLGGVFELQP-GASVFVNVT-DPSQVS---HGTGFTSFGLLKL---
STRAND:  -----↑-GGGGG↑GGGGGGG----HHHHHHHH---------------II↑IIIIII---

                  H224                                 S255
```

Figure 1A. Sequence alignment of selected members of TNF ligand family (Humtnf: TNF-α; Humita: TNF-β; Gldfsl: *Fas* ligand in gld mice; Hufasl: human *Fas* ligand; Hcd401: Human CD40-ligand; STRAND: β-strands in the 3-D structure.) Cysteines involved in disulfide cross-links in TNF-α and CD40 ligand are shown connected. H224 and S255 of CD40 ligand are marked.

described in Srinivasan et al. (1993). If one were to constrain the backbone atoms of the two cysteines as in the template, the disulfide crosslink in the model would be between two adjacent strands within a sheet. Besides, the two cysteines involved in the disulfide are not adjacent amino acids of the two strands but are unusually shifted by two amino acids away from the adjacent positions. Because of this, in one of our models, where the backbone of both cysteines in CD40 ligand was constrained as in TNF, dihedral angle $C\beta_i$-$S\gamma_i$-$S\gamma_j$-$C\beta_j$ was found to be –141°, which is very strained, based on the statistical preference for this angle in PDB (Sowdhamini et al., 1989). If one were to retain the allowed geometry for the disulfide cross-link, the backbone atoms of residues near cysteines would deviate considerably away from the template as shown in Figure 1B. Since the strands containing both the cysteines are involved in the oligomerization site in TNF crystal structure, we further explored the conservation in the quartenary structure of CD40 ligand and TNF.

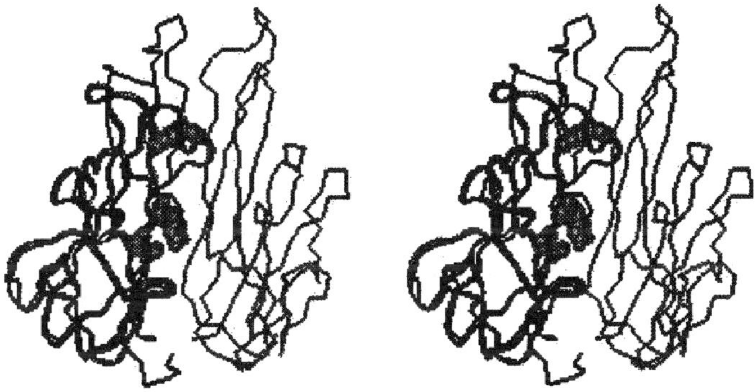

FIGURE 1B. Stereo diagram of C-α trace of two out of three molecules of TNF as found in the trimeric form in crystal structure of TNF (Eck and Sprang 1989). Shown in ribbon is model of CD40 ligand built using FOLDER and superimposed on one of the two molecules of TNF-α. Shown in CPK representation from top to bottom are the nonhomologous disulfide, H224, and S255.

It has been shown, using crystallographic studies, that TNF molecules form compact trimers (Eck and Sprang, 1989). Although conservation in quaternary structures among divergent protein is not well documented, the physiological relevance of the trimeric form for members of TNF family has been revealed recently by crystal structure analysis of ligand-receptor complex (Banner et al., 1993). In the crystal structure, three LT-β molecules (a member of the TNF family) form compact trimer like the TNF trimer (Eck and Sprang, 1989), and the three receptor chains interact with the three symmetric subunit interfaces of LT-β molecules. Also, the three receptor chains make no contacts with each other, suggesting that ligand trimerization is a prerequisite for receptor cross-linking and ligand-mediated signaling. A similar mode of ligand-receptor interaction between CD40 and CD40 ligand can be hypothesized based on sequence homology between CD40 and TNF receptors. Based on this hypothesis, Peitsch and Jongeneel (1992) proposed a model of CD40 ligand and predicted that CD40 ligand would form compact trimers just as seen for TNF and LT-β. If this were true, one would expect soluble forms of CD40 ligand to initiate receptor cross-linking in a manner analogous to TNF and LT-β. However, the soluble form of CD40 ligand was not biologically active in spite its observed binding to CD40-Fc (Fanslow et al., 1994). Since CD40 ligand, unlike TNF and LT-b, is only found on cell surfaces, it makes sense that cell-bound CD40 ligands are more likely to find other CD40 ligand molecules in the right orientation for oligomerization than their soluble counterparts. Thus, the trimeric quaternary structure of soluble CD40 ligand may not be as stable as TNF, contradicting the prediction that CD40 ligand forms compact trimers as TNF does (Peitsch and Jongeneel, 1992).

In order to understand the oligomerization state of soluble CD40 ligand we compared structural and amino acid changes along the oligomerization sites. The strand containing two cysteines involved in the disulfide crosslink disrupts the strands, allowing the backbone to swing out to allow for proper geometry for the disulfide crosslink. The structural changes in these strands, because of the disulfide, reflect in structural changes in the subunit interface between CD40 ligand and TNF. Also, a comparison of the oligomerization sites on CD40 ligand model and other members of the TNF family reveal less hydrophobic substitution in CD40 ligand at key sites involved in the stabilization of intersubunit interactions. For example, at position 255 in CD40 ligand a *Ser* is substituted for a highly conserved *Phe* and at position 224 *His* is substituted for a conserved *Phe* (Figure 1A). The absence of *Fas*-mediated apoptosis in gld mice is also attributed to a point mutation of *Leu* at position 255 in *Fas* ligand, another member of TNF family (Takahashi et al., 1994). The amino acid numbering is consistent with that of Spriggs et al. (1992). Thus, the structural and nonconservative amino acid substitutions in the subunit interface of CD40 ligand, when compared to other bioactive soluble ligands of this family, suggest potential loss of stability of the quaternary structure of CD40 ligand in its soluble form.

PROTEIN THERAPEUTIC DESIGN

If, indeed, soluble CD40 ligand loses the native quartenary structure of cell-bound forms, it should be possible to induce bioactivity in soluble forms of CD40 ligand if a cell-bound configuration can be mimicked in the soluble form. To this end, we engineered a soluble agonist for CD40 ligand by tethering the extracellular region of the CD40 ligand molecule to a mutated trimeric *Ile* zipper (Harbury et al., 1994) in the N-terminal, referred to here as CD40LT (CD40 ligand trimer). In Figure 1C we show that the geometry of the C-terminal end of the *Ile* zipper is compatible with the N-terminal of the trimeric TNF crystal structure. We have shown that the zipper-tethered molecule restores biological activity (Fanslow et al., 1994) irrespective of the number of amino acids in the space between the zipper and the folding domain. This construct suggests for the first time that more than two molecules of CD40 ligand are required for signaling.

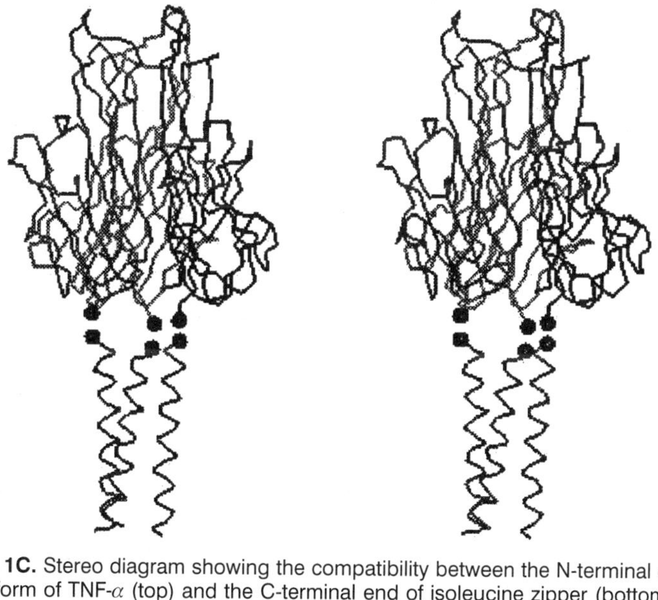

FIGURE 1C. Stereo diagram showing the compatibility between the N-terminal end of the trimeric form of TNF-α (top) and the C-terminal end of isoleucine zipper (bottom) that are tethered to engineer soluble and bioactive CD40 ligand.

CONCLUSION

FOLDER, the distance-geometry-based homology-modeling method developed by our group, can generate models consistent with a template protein, and it allows for incorporation of structural constraints unique to the model protein. The models generated by FOLDER have no short contacts even when amino acids forming the interior core of the model protein are not conserved with the template. No rigorous energy minimization steps are necessary, as this step will not add new information to the model. FOLDER is the first homology-modeling package that offers a way to impose geometrical constraints for nonhomologous disulfides in the model protein. The advantage of FOLDER is that it produces models that are faithful to the input distances and hence preserves the subtle but important structural differences that are key to the diversity in biology. While structural homology is the basis for generating models, it is the structural differences that lead to identification of important characteristics that are unique to the model protein.

Very recently, the crystal structure of CD40 ligand was published (Karpusas et al., 1995) and shown to exist in trimeric quaternary form at high (crystallographic) concentrations, as would be expected for cell-bound CD40 ligands that are biologically active. The trimeric structure of *Escherichia coli*-expressed CD40 ligand at high concentrations is not surprising because it has been shown that *E. coli*-expressed CD40 ligand is biologically active at 20 μg ml^{-1}, suggesting trimeric forms at higher concentrations (Mazzei et al., 1995). However, in the crystal structure, the β-strands containing the cysteines are disrupted in the crystal structure of CD40 ligand just as predicted by our modeling efforts. Hence, it can be concluded that the trimeric quaternary structure of CD40 ligand is not stable in soluble form at concentrations useful for therapeutic use. However, the CD40LT engineered by us is active at 200 ng ml^{1} and is suitable for therapeutic use (Fanslow et al., 1994).

Thus, while we continue to develop our method, we realize that we are far from offering a robust solution to loop and SCR modeling for divergent proteins. In the meantime, using the current version of FOLDER, we have demonstrated that useful and accurate models can be built for use in engineering proteins. In this context, it is worth mentioning that the accuracy of a model or structure depends on its usability in real-life problems and should not always be measured in terms of resolution.

ACKNOWLEDGMENT

The authors wish to acknowledge Dr. Carl J. March for his valuable contributions to the success of this project. The authors also acknowledge Anne C. Bannister for her editorial help.

REFERENCES

Banner, DW, A D'Arcy, W Janes, R Gentz, HJ Schoenfeld, C Broger, H Loetscher, and W Lasslauer. **1993**. Crystal structure of the soluble human 55 kd TNF receptor-human LT-b complex: implications for TNF receptor activation. Cell 73:431-445

Bazan, JF. **1992**. Unraveling the structure of IL-2. Science 257:410-412.

Bernstein, FC, TF Koetzle, GJB Williams, EF Meyer, MD Brice, JR Rojers, O Kennard, T Shimanouchi, and M Tasumi. **1977**. The Protein Data Bank: A computer-based archival file for macromolecular structures. J. Mol. Biol. 112:535-542.

Browne, WJ, ACT North, DC Phillips, K Brew, TC Vanaman, and RL Hill. **1969**. A possible three-dimensional structure of bovine a-lactalbumin based on that of hen's egg-white lysozyme. J. Mol. Biol. 42:65-86.

Chothia, C. **1992**. One thousand families for the molecular biologist. Nature 357:543-544.

Chothia, C and AM Lesk. **1986**. The relation between the divergence of sequence and structure in proteins. EMBO J. 5:823-826.

Cosman, DA. **1993**. The hematopoietin receptor superfamily. Cytokine 5:95-106

Cosman, DA. **1994**. A family of ligands for the TNF receptor superfamily. Stem Cells 12:440-455

Eck, MJ and SR Sprang. **1989**. The structure of tumor necrosis factor-a at 2.6 Å resolution. J. Biol. Chem. 264:17595-17605.

Fanslow, WC, S Srinivasan, R Paxton, MG Gibson, MK Spriggs, and RJ Armitage. **1994**. Structural characteristics of CD40 ligand that determine biological function. Semin. Immunol. 6:267-278

Farrah, T and CA Smith. **1992**. Emerging cytokine family. Nature 358:26.

Greer, J. **1981**. Comparative model-building of the mammalian serine proteases. J. Mol. Biol. 153:1027-1042.

Greer, J. **1991**. Comparative modeling of homologous proteins. Methods Enzymol. 202:239-252.

Harbury, PB, PS Kim, and T Alber. **1994**. Crystal structure of an isoleucin-zipper trimer. Nature 371:80-83.

Havel, TF. **1993**. Predicting the structure of the flavodoxin from *Escherichia coli* by homology modeling, distance geometry and molecular dynamics. Mol. Simulations 10:175-210.

Havel, TF and ME Snow. **1991**. A new method for building protein conformations from sequence alignments with homologues of known structure. J. Mol. Biol. 217:1-7.

Hobohm, U and C Sander. **1994**. Enlarged representative set of protein structures. Protein Sci. 3:522-524.

Kabsch, W. **1976**. A solution for the best rotation to relate two sets of vectors. Acta Crystallogr. A32:922-923.

Karpusas, M, YM Hsu, JH Wang, J Thompson, S Lederman, L Chess, and D Thomas. **1995**. 2Å crystal structure of an extracellular fragment of CD40 ligand. Structure 3:1031-1039.

Mazzei, GJ, MD Edgerton, C Losberger, S Lecoanet-Henchoz, P Graber, A Durandy, JF Gauchat, A Bernard, B Allet, and JY Bonnefoy. **1995**. Recombinant soluble trimeric CD40 ligand is biologically active. J. Biol. Chem. 270:7025-7028.

McLachlan, AD. **1982**. Rapid comparison of protein structures. Acta Crystallogr. A38:871-873.

Orengo, CA, DT Jones, and JM Thornton. **1994**. Protein superfamilies and domain superfolds. Nature 372:631-634.

Pauling, L and RB Corey. **1951**. Conformation of polypeptide chains with favored orientations around single bonds: Two new pleated sheets. Proc. Natl. Acad. Sci. USA 37:729-740.

Pauling, L, RB Corey, and HR Branson. **1951**. The structure of proteins: Two hydrogen-bonded helical configurations of the polypeptide chain. Proc. Natl. Acad. Sci. USA 37:205-211.

Peitsch, MC and CV Jongeneel. **1992**. A 3-D model for the CD40 ligand predicts that it is a compact trimer similar to the tumor necrosis factors. Int. Immunol. 5:233-238.

Rozwarski, DA, AM Gronenborn, GM Clore, JF Bazan, A Bohm, A Wlodawer, M Hatada, and PA Karplus. **1994**. Structural comparisons among the short-chain helical cytokines. Structure 2:159-173

Smith, CA, C Davis, D Anderson, L Solam, MP Beckmann, R Jerzy, SK Dower, D Cosman, and RG Goodwin. **1990**. A receptor for tumor necrosis factor defines an unusual family of cellular and viral proteins. Science 248:1019-1023

Sowdhamini, R, N Srinivasan, B Shoichet, DV Santi, C Ramakrishnan, and P Balaram. **1989**. Stereochemical modeling of disulfide bridges. Criteria for Introduction into proteins by site-directed mutagenesis. Protein Eng. 3:95-103.

Spriggs, MK, RJ Armitage, L Strockbine, KN Clifford, BM Macduff, TA Sato, CR Maliszewski, and WC Fanslow. **1992**. Recombinant human CD40 ligand stimulates B cell proliferation and immunoglobulin E secretion. J. Exp. Med. 176:1543-1550.

Srinivasan, S, M Shibata, and R Rein. **1986**. Multistep modeling of protein structure: Application to bungarotoxin. Int. J. Quantum Chem. Symp. 13:167-174.

Srinivasan, S, CJ March, and S Sudarsanam. **1993**. An automated method for modeling proteins on known templates using distance geometry. Protein Sci. 2:277-289.

Sudarsanam, S, CJ March, and S Srinivasan, S. **1994**. Homology modeling of divergent proteins. J. Mol. Biol. 241:143-149

Sudarsanam, S, RF DuBose, CJ March, and S Srinivasan. **1995**. Modeling protein loops using a Φi+1, Ψi dimer database. Protein Sci. 4:1412-1420.

Takahashi, T, M Tanaka, CI Brannan, NA Jenkins, NG Copeland, T Suda, and S Nagata. **1994**. Generalized lymphoproliferative disease in mice, caused by a point mutation in the *Fas* ligand. Cell 76:969-976.

NUCLEIC ACIDS: STRUCTURE, FUNCTION, AND HEALTH RISK IMPLICATIONS

750-MHZ NMR STUDIES OF CHEMICALLY MODIFIED DNA STRUCTURE/DYNAMICS: SEQUENCE CONTEXT EFFECTS AND PROTEIN RECOGNITION

M. A. Kennedy, K. McAteer, J. M. Lingbeck, B. D. Thrall,
D. L. Springer, M. G. Kubinec, and P. D. Ellis.

Pacific Northwest National Laboratory, Richland, Washington

Key Words: *DNA repair, sequence context effects, NMR spectroscopy water of hydration*

ABSTRACT

Our interest is in the structural basis for protein recognition of DNA damage. We are using solution-state NMR to study protein structure as well as sequence context effects on DNA secondary structure and dynamics. We are using this information to understand how DNA sequence context modulates the topological presentation of DNA lesions and how this is in turn affects damage recognition by DNA repair proteins. In one example, we have observed that the minor groove widths in poly:G DNA are remarkably similar to those typically associated with B´DNA in poly:A regions. Based on a screening of several sequences containing G inserts into poly:A segments, we find that guanine can be accommodated into a narrow minor groove structure. We will report on the structural details that allow this to occur. Narrow minor grooves in poly:G DNA explain binding patterns of the carcinogen benzo[a]pyrene diol epoxide to DNA.

Another intriguing observation of how sequence context bears on DNA secondary structure and dynamics is the occurrence of slow, large-amplitude base motion at many TpA steps in DNA. We are systematically studying DNA sequence context to determine if "rules" can be established for when large-amplitude, slow motion can be expected at TpA steps in DNA. We will also report on our preliminary studies on the structure of zinc-finger DNA binding domains that are involved in human DNA repair proteins.

INTRODUCTION

An understanding of protein recognition of DNA damage and the ensuing repair processes is fundamental to cancer research. We are using high-resolution NMR spectroscopy to gain further insight into sequence context effects on DNA secondary structure and dynamics and relating these results to DNA damage and repair. Our studies focus on the potent carcinogen benzo[a]pyrene diol epoxide (BPDE), which has been shown to bind (1) nonrandomly to DNA with preferential binding to guanines; (2) with distinguishable binding patterns, depending on sequence context; and (3) with enhanced binding in poly-guanine regions (Thrall et al., 1994). In poly-G regions, a ladder pattern of quantitative formation of BPDE-adduct is observed, with up to a fourfold greater yield at the 5′ end, as shown in Figure 1.

Structural studies of BPDE covalently bound to DNA oligomers have shown that the predominant adduct binds in the minor groove and, upon binding, the minor groove widens by ~3.7 Å to accommodate the BPDE molecule (Cosman et al., 1992). The structural basis for patterns of BPDE binding in poly-G DNA segments was investigated using NMR spectroscopy.

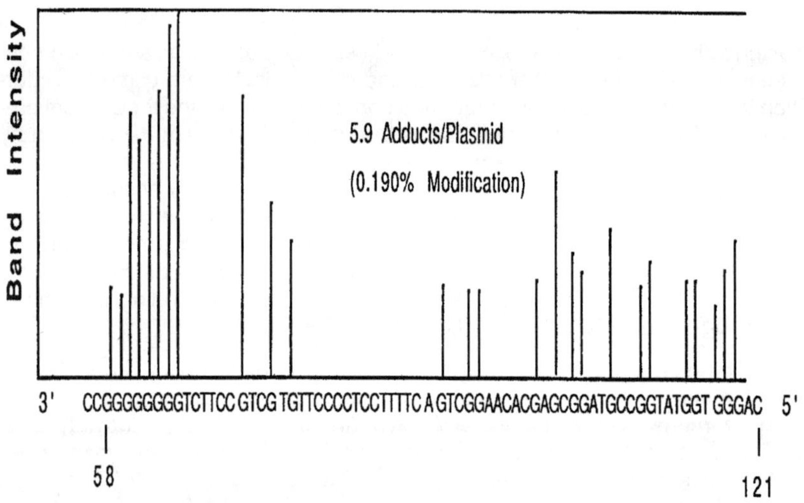

Figure 1. Sequence-dependent adduction of plasmid DNA by benzo[a]pyrene diol epoxide (BPDE). Plasmid pGEM-5S modified with an average of 5.9 adducts/plasmid was replicated using Sequenase, separated by denaturing gel electrophoresis, and exposed to autoradiographic film for 2.5 hr. Average adduction level at each guanine was determined by densitometry. Results reveal a characteristic pattern of adduction in the poly-dG region indicated by bases 58-65, where the adduction level is progressively less at the 3′ end than at the 5′ end of the guanine stretch.

The minor groove width in DNA is known to vary significantly with sequence. The most striking example of minor groove width variations with DNA sequence is the compression and distinctively narrow minor groove of poly-A runs in DNA (Chuprina et al., 1991; Lipanov et al., 1987. In order to test the hypothesis that variations in the minor groove width influence BPDE adduct yields, we decided to study the minor groove structure of a series of eight self-complementary DNA oligonucloetides containing guanine inserted into poly-A regions. Before embarking on these structural studies, we measured BPDE adduction levels in the sequences listed below and found a fourfold difference in the adduction yield, depending on whether the guanine occurred at the 5´ end (2.52% in 5´-TATAGAAATTTCTATA-3´) or 3´ end (0.65% in 5´-TATAAAAGCTTTTATA-3´) of the poly-A tract. The width of the minor groove in poly-A tracts can be indirectly determined by NMR spectroscopy by measuring the distance between the adenine-H2 proton (n) and the H1 proton of the sugar (m+1) on the complementary strand. The distance is measured by quantitating the nuclear Overhauser effect crosspeak between these two protons in a two-dimensional NOESY spectrum (Wüthrich, 1986). By taking this design approach, we were able to examine the effect of guanine on minor groove structure in poly-A DNA. The following sequences were studied:

(1) 5´-CAAAAATTTTTG-3´

(2) 5´-CGAAAATTTTCG-3´

(3) 5´-CAGAAATTTCTG-3´

(4) 5´-CAAGAATTCTTG-3´

(5) 5´-CAAAGATCTTTG-3´

(6) 5´-CAAAAGCTTTTG-3´

(7) 5´-CAGGAATTCCTG-3´

(8) 5´-CGGGAATTCCCG-3´

Sequence (1) is a control which contains five consecutive adenine nucleotides and is known to have a narrow minor groove structure. In the other sequences, we were able to infer the effect of replacing an adenine with a guanine on the minor groove structure of the DNA molecule by measuring the minor groove distance at the adenine bases surrounding the substituted guanine(s) and comparing these distances to those in the control sequence. As shown in Figure 2, the measured distances indicate that the minor groove narrows significantly in the control sequence at the 3 end of the poly-A tract. Further examination of

Figure 2 shows that in all the sequences studied, the substitution of guanine for adenine did not alter the pattern of narrowing the minor groove from 5´ to 3´ direction that was observed in the control sequence. Even in sequence (8), where three guanines are substituted for the first three adenine bases in the control sequence, the minor groove width at the two adenines, while ~1 Å wider than the control, is still significantly narrower than in B-type DNA. Therefore, the pattern of compression of the minor groove width observed in poly-A DNA appears to be preserved in mixed poly-A:poly-G sequences. This helps to understand why BPDE adduction is greater at the 5 end of poly-G runs compared to the 3´ end. At the 5´ end of a poly-G tract, where the minor groove width is normal compared to that of B-type DNA, less perturbation to the secondary structure of the DNA would be required to accommodate the ~3.7 Å widening of the minor groove upon formation of the BPDE adduct compared to the larger structural rearrangement that must occur at the 3´ end where the minor groove structure is exceptionally narrow.

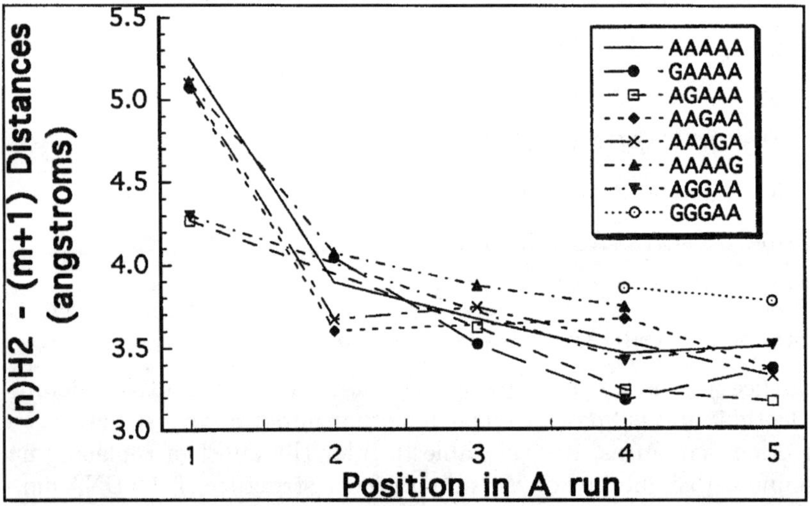

Figure 2. The cross-strand distances from the (n)A-H2 to (m+1)H1´ distances are plotted as a function of the position of the adenine in sequences 1-8. The distances were determined from the initial buildup rate of NOESY crosspeaks taken at several mixing times and collected at 750 MHz. The determination of unknown distances was based on calibration compared to the cross-relaxation rates between the H5-H6 cytosine protons, whose distance is fixed at 2.5 Å.

INDIRECT DETERMINATION OF MINOR GROOVE WIDTHS IN DNA USING NMR

Of more general interest, the results show that guanine can be accommodated in a relatively narrow minor groove topology in DNA. In order to determine structurally how guanine can be accommodated into narrow minor groove topologies, we determined the structure of sequences (4) and (8) by NMR spectroscopy. The structural models shown in Figure 3 clearly exhibit structural differences that are evident from observation of the kink in the ribbon representation of the DNA backbone that occurs at the GpA step. Structural details of these two sequences will be presented elsewhere. Visually, it is apparent that the minor groove is greater in the GGGAA sequence than in the AAGAA

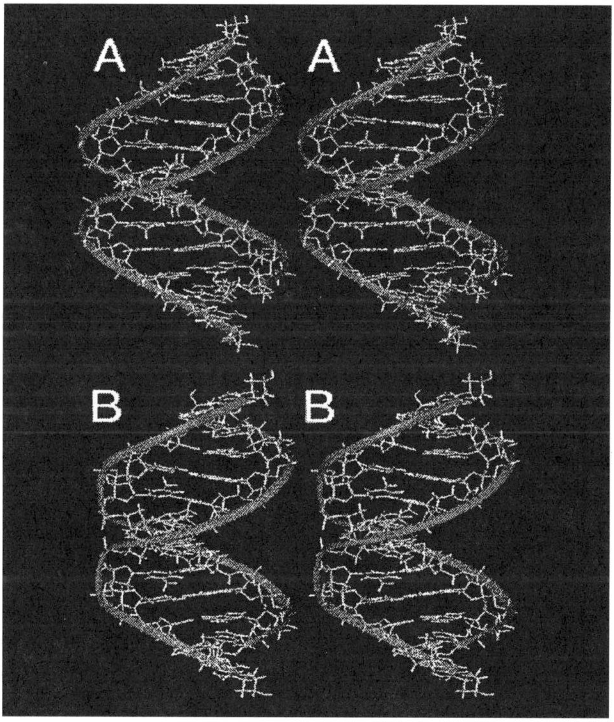

Figure 3. Structures derived from NMR data collected at 750 MHz for **A**, 5′-CAAGAATTCTTG-3′ and **B**, 5′-CGGGAATTCCCG-3′, shown in cross-eyed stereo representation. The DNA backbone is represented by a solid oval ribbon to illustrate how the minor groove varies with sequence and also to illustrate the kink observed at the GpA step.

sequence. Direct measurement of the minor groove width as (n+2) phosphorus to (n-2) phosphorus in the two model structures (Figure 4) shows that the minor groove width still reaches a minimum value at the ApT step in both sequences. It is interesting to note that in the AAGAA sequence the minor groove width is compressed by ~3 Å, from ~12 Å to ~9 Å, whereas in the GGGAA sequence the minor groove width is compressed by only 2 Å, from ~12Å to ~10Å. This indicates that while guanines can be accommodated into narrow minor groove topologies similar to those observed in poly-A DNA, the magnitude of the compression is not as great.

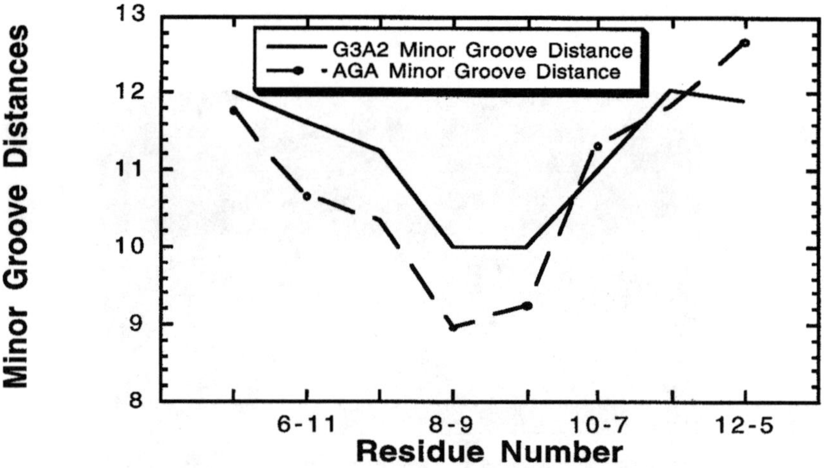

Figure 4. (n+2) Phosphorus to (n-2) phosphorus distance, plotted as a function of residue number. Distances were measured from structures derived from NMR data collected at 750 MHz and refined using restrained molecular dynamics and back-calculation.

The observed compression of the minor groove in sequences containing multiple guanines helps to understand the ladder pattern of BPDE adduction but does not help explain the enhanced binding at the 5′ end of such sequences. The chemical reaction that occurs when BPDE forms a covalent DNA adduct requires that the epoxide opens, and a bond is formed between the C10 atom of the BPDE and the exocyclic 2-amino nitrogen atom of guanine. A potential competing reaction is the hydrolysis of the epoxide to an unreactive tetraol. Therefore, the presence or absence of water in the local microenvironment of the reaction could influence whether the fate of the epoxide is tetraol or adduct. Guanines in DNA have been shown by others to disrupt the spine of hydration in the minor groove, i.e., regions high in guanine content can be considered dry." Therefore, in these dry" regions of DNA, the formation of the covalent adduct would be favored compared to the hydrolysis that could occur in a more wet" microenvironment such as occurs in poly:A DNA. In the latter, a spine of hydration in DNA is thought to stabilize the secondary structure associated with exceptionally narrow minor groove topologies. This could explain the enhanced adduct yields in poly-G regions of DNA which led us to investigate the extent of water of hydration in the sequences listed above and to examine the role of water in BPDE adduction to DNA.

NMR STUDIES OF MINOR GROOVE HYDRATION IN DNA

In order to directly explore the hydration of the minor groove in these sequences, NMR approaches have been used. Hydration water near DNA protons can be detected by 2-D NOESY and 2-D ROESY NMR experiments since NOE crosspeaks can be detected between the protons of water molecules that are close in space to DNA protons (Otting and Wüthrich, 1989, 1991; Kubinec and Wemmer, 1992). Crosspeaks can also arise from chemical exchange processes, and the NOESY experiment cannot differentiate between the two types of crosspeaks since both are observed as positive peaks in spectra. Fortunately, the 2-D ROESY NMR experiment can discriminate between crosspeaks arising from chemical exchange and crosspeaks that are a result of cross-relaxation. A negative sign of a ROESY crosspeak relative to the diagonal peaks shows that the magnetization transfer is by direct cross-relaxation, since both chemical exchange and spin-diffusion would give rise to positive crosspeaks. By comparison of NOESY and ROESY data, water molecules bound to DNA can be determined, and the NMR experiment allows differentiation between water molecules that are exchanging rapidly with the bulk solvent and water molecules that undergo much slower exchange. One-dimensional cross-sections through the 2-D NOESY and ROESY spectra of [d(CGGGAATTC-CCG)$_2$] taken along F2 at the F1 chemical shift of the water resonance are shown in Figure 5. The only protons that exhibit a strong interaction with water (bound with a lifetime of >2ns) are A5 H2 and A6 H2.

The other positive peaks in the NOESY spectrum either remain positive in the corresponding ROESY spectrum or are slightly negative, indicating short residence times for water bound to these protons, which reside in the major groove as expected. These results indicate that the minor groove structure is not significantly perturbed when A-tracts follow a G-tract. However, since guanines contain an exocyclic amino group at the C2 position in the minor groove, compared to a nonexchangeable proton in the case of adenine, we have not been able to directly detect long-lived water molecules in the minor groove involving G:C base pairs. We are currently investigating these interactions further.

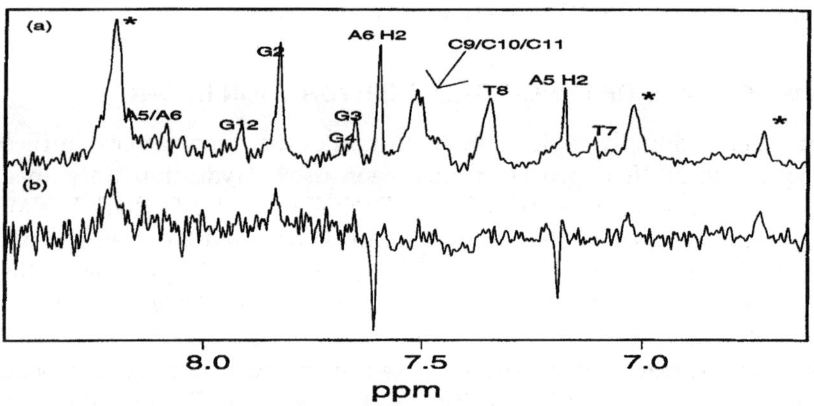

Figure 5. One-dimensional cross-sections through **A**, the two-dimensional NOESY and **B**, the two-dimensional ROESY spectra of [d(CGGGAATTCCCG)$_2$] taken along F2 at the F1 chemical shift of the water resonance. The peaks marked with an asterisk are a result of chemical exchange.

REFERENCES

Chuprina, VP, OY Fedoroff, and BR Reid. **1991**. New insights into the structure of A_n tracts and B′-B′ bends in DNA. Biochemistry 30:562-568.

Cosman, M, C De Los Santos, R Fiala, BE Hingerty, SB Singh, V Ibanez, LA Margulis, D Live, NE Geacintov, and S Broyde. **1992**. Solution conformation of the major adduct between the carcinogen (+)-antibenzo[a]pyrene diolexpoxide and DNA. Proc. Natl. Acad. Sci. USA, 89:1914-1918.

Kubinec, MG and DE Wemmer. **1992**. Solution NMR of water on protein surfaces. Curr. Opin. Struct. Biol. 2:828-831.

Lipanov, AA and VP Chuprina. **1987**. The structure of poly(dA):poly (dT) in a condensed state and in solution. Nucl. Acids Res. 15:5833-5844.

Otting, G and KJ Wüthrich. **1989**. Studies of protein hydration in aqueous solution by direct NMR observation of individual protein-bound water molecules. Am. Chem. Soc. 111:1871-1875.

Otting, G, E Liepinsh, and K Wüthrich. **1991**. Protein hydration in aqueous solution. Science 254:974-980.

Thrall, BD, DB Mann, MJ Smerdon, and DL Springer. **1994**. Nucleosome structure modulates benzo[a]pyrene diolepoxide. Biochemistry 33:2210-2216.

Wüthrich, K. **1986**. NOE-observable ´H-´H distances in nucleic acides, pp. 203-223, Chapter 11. In: *NMR of Proteins and Nucleic Acids*. Wiley, New York.

MODELING HYDRATION AND TAUTOMERIZATION OF NUCLEIC ACID BASES IN RELATION TO DNA PROPERTIES

V. I. Poltev,[1] E. J. Gonzalez,[1,2] A. V. Teplukhin,[3] G. G. Malenkov,[4] R. Rein,[5] M. Shibata,[5] and J. H. Miller[6]

[1]Institute of Theoretical and Experimental Biophysics, Russian Academy of Sciences, Pushchino, Moscow Region, Russia

[2]Autonomous University of Puebla, Puebla, Mexico

[3]Institute of Mathematical Problems of Biology, Russian Academy of Sciences, Pushchino, Moscow Region, Russia

[4]Institute of Physical Chemistry, Russian Academy of Sciences, Moscow, Russia

[5]Roswell Park Cancer Institute, Buffalo, New York

[6]Pacific Northwest National Laboratory, Richland, Washington

Key Words: *DNA hydration, Monte Carlo simulation, base tautomerization, potential functions, biosynthesis fidelity*

ABSTRACT

Hydration properties of individual nucleic acid bases were calculated and compared with the available experimental data. Poltev and Malenkov potential functions reproduce well experimental enthalpies of single-base, single-water complex formation and base hydration energies. The most energetically favorable single-water positions near nucleic acid bases involve the formation of two distorted H bonds.

Hydration of rare tautomers of nucleic acid bases can be energetically more favorable, less favorable, or nearly the same as the normal ones. Hydration patterns of rare tautomers differ greatly from those of normal tautomers. Monte-Carlo calculations explain the predominance of guanine's keto form in water solutions despite nearly equal energies of keto and enol tautomers in vacuum. Interactions with only one or two water molecules are sufficient to produce this shift in guanine's tautomeric equilibrium.

Hydration simulation of DNA duplexes shows that all the water bridges in the minor groove of the duplexes that have a substantial probability of forming are arranged across the groove. In the major groove, water bridges occur, with nearly equal probability, both across and along the groove.

INTRODUCTION

The importance of DNA hydration has been well established since the earliest fiber diffraction studies (Franklin and Gosling, 1953). Both B-A conformational transitions and transitions between B-family conformations depend on DNA-water interactions. X-ray crystallographic studies reveal ordered water arrangements near some DNA atoms (Drew and Dickerson, 1981). In some cases, single water molecules play a role at the interface between DNA and protein (Feng et al., 1994). Thus, we have substantial reason for investigating the hydration of DNA subunits.

Tautomeric transitions were suggested as a pathway of nucleic acid biosynthesis errors by Watson and Crick (1953). Transitions of guanine or thymine to enol form and adenine or cytosine transition to imin form can result in the formation of a guanine-thymine or adenine-cytosine base pair of similar shape to that of Watson-Crick A:T and G:C pairs. However, it is now generally believed that mispairing of bases in their normal tautomeric forms is the main source of base substitution errors. Such mispairs were not only predicted by computations (Chuprina and Poltev, 1983; Rein et al.,1983) but also revealed in both crystal and solution experiments (Patel et al., 1984; Kennard, 1985; Prive et al., 1987).

Although no distinct experimental evidence exists for involvement of tautomerization in mispairing errors, some data suggest reexamination of the role of rare tautomers in the processes of nucleic acid functioning. The case of guanine keto-enol (laktam-laktim) tautomerization is of special interest because keto and enol tautomers exist in nearly equal concentrations when isolated in an inert gas matrix (Sheina et al., 1987; Szczepaniak et al., 1988). In aqueous solutions, monomer crystals, and nucleic acid duplexes, the keto/enol concentration ratio is on the order of 10^4 to 10^5 (Saenger, 1984). Thus, the influence of hydration on guanine tautomeric equilibrium is an established experimental fact. The reasonable suggestion has been made (Petruska et al.,1986) that water is eliminated from the polymerization site to enhance base-pairing selectivity during DNA biosynthesis. Since the keto and enol tautomers of guanine will have nearly equal stability in a non-polar environment and the enol tautomer should pair with thymine, the question arises, "Why does the guanine pair nearly exclusively with cytosine during DNA biosynthesis." Thus, an explanation is needed both of the shift of tautomeric equilibrium of guanine on the transition from nonpolar media to water solutions and of the high selectivity of guanine pairing.

This paper describes results of hydration simulations of nucleic acid bases in various tautomeric forms in relation to the problems mentioned above as well as general features of hydration shell structure in the grooves of DNA duplexes. Some preliminary results of this work have been published elsewhere (Poltev et al., 1995a,b,c).

COMPUTATIONAL METHODS

Unless otherwise indicated, intermolecular interaction energies were calculated with Poltev and Malenkov potential functions (PM PF) as described earlier (Poltev et al., 1984). For comparison with other potentials and with experimental data, two other PF sets were used: that of Weiner and Kollman, WK (Weiner et al., 1986), which together with Jorgensen's TIP3P water model (Jorgensen et al., 1983) are used in the AMBER program, and optimized potentials for liquid simulation (OPLS) developed by Jorgensen (Pranata et al., 1991). The same algorithms and computer programs were used in all cases after relevant changes were made in the routines for energy calculations. Geometries of the base molecules were the same in all the calculations. Positions of the force centers and the partial charges in the water molecules were those suggested by the authors of the models. A new oxygen type (hydroxyl O) for calculation of enol tautomers was added (Poltev et al., 1995a) to PM PF.

Monte-Carlo simulation of hydration (Metropolis algorithm, NVT ensemble) was performed using the same procedure as in earlier works (Poltev et al., 1984, 1992). Geometric criteria for H-bond formation were used: H...O (H...N) and O...O (O...N); distances should be simultaneously less than 2.4 and 3.2Å, respectively, without any angular or energy requirements (Poltev et al., 1984). A "one-water bridge" results when a water molecule forms H bonds with two or more hydrophilic centers. The specific case of one water hydrogen forming H bonds with two acceptors of H bonds (bifurcated H bond) is considered separately. Two hydrophilic atoms H-bonded with two water molecules forming an H bond with one another, are called a two-water bridge.

When calculating single-base hydration, periodic boundary conditions were superimposed on water molecules only because interactions of the base with water and base molecules in neighbor unit cells are small enough to be neglected. The size of the unit cell for the system of 400 water molecules corresponds to a water density of 0.995 g/cm^3 and takes into account the inherent volume of the base molecules as observed in crystals. No boundary conditions were imposed when calculating the hydration of a DNA duplex, but water molecules were not allowed to be further than 10 Å from the center of structural units of the DNA (base, sugar, or phosphate).

SINGLE-BASE, SINGLE-WATER INTERACTIONS

To find the sites of preferential hydration of bases we investigated the dependence of single-base, single-water interaction energy on their mutual arrangement. The computations were extended to rare tautomeric forms of bases, i.e., enol forms of 9-methylguanine, 1-methylthymine and 1-methyluracil, and imin forms of 9-methyladenine, 1-methylcytosine and 1,5-dimethylcytosine. For normal tautomeric forms, the interaction energy in the most favorable configuration can be compared with experimental data on enthalpies of water-base complex formation. Thus, we performed these computations using the three sets of potential functions described in Methods. These results were presented earlier in more detail (Poltev et al., 1995b,c); we only summarize the main conclusions here.

All the local minima correspond to in-plane arrangement of a base and water oxygen. The deepest minima for all the bases correspond to the formation of two distorted H-bonds. These minima for pyrimidines refer to water H-bonding with the atoms involved in Watson-Crick pairing. For purines, the deepest minima correspond to H-bonds of water with the atoms exposed in the DNA major groove. Both water arrangement and interaction energies in minima vary with tautomeric form not only for sites involved in proton transfer but for other hydrophilic sites as well.

Table 1 represents interaction energy minima for normal forms of 4 DNA bases calculated via three sets of potential functions. The rather good consistency obtained with PM and WK potential functions can be seen for many minima, while results obtained with J potentials differ more substantially. Calculations with PM potentials show that substitution of a base hydrogen atom by the methyl group produces a slight (about 0.1 or 0.2 kcal/mol) change in the value of local minima of interaction energies with a water molecule not H-bonded to this hydrogen. This allows the results of Table 1 to be compared with mass-spectrometric values of the enthalpy of water-base complex formation for a set of methylated bases (Table 2) and demonstrates generally good agreement between the values calculated with PM PF and the experimental data. The agreement of the results calculated with WK PF is slightly less satisfactory (e.g., for 2,2,9-trimethylcytosine), and J PF underestimate absolute values for many cases, so no further computations were not performed with them.

Table 1. H-bonds and water-base interaction energies in local minima, calculated with three sets of potential functions.

Minimum	H-bonds	PM[a]	WK[a]	J[a]
		9-Methyladenine		
A	N7,H6[b]	-10.81	-10.80	-7.76
B	N1,H6[c]	-9.08	-9.78	-7.58
C	N3	-6.99	-6.01	-5.16
		1-Methylthymine		
A	O4	-6.34	-5.46	-6.40
B	O4,H3	-7.95	-8.11	-7.48
C	O2,H3	-6.98	-8.01	-6.46
D	O2	-5.93	-5.58	-6.28
		9-Methylguanine		
A	N7,O6	-12.53	-9.98	-9.37
B	O6,H1	-10.94	-11.04	-11.11
C[c]	H1,H2[c]	-10.52	-10.16	-9.62
D	N3,H2[b]	-10.40	-9.77	-7.81
		BF1-Methylcytosine		
A	H2[b]	-6.70	-7.30	-6.24
B	N3,H2[b]	-11.81	-11.58	-9.92
C[b]	N3,O2	-11.80	-9.87	-8.75
D	O2	-7.57	-6.92	-7.33

[a]PM, WK and J are energy values (kcal/mol) in local minima for Poltev-Malenkov, Weiner-Kollman, and Jorgensen potentials, respectively.
[b]This minimum for WK and J potentials corresponds to a single H-bond with N3.
[c]This minimum for J potentials corresponds to a single H-bond with H1.

Table 2. Experimental enthalpy values of single-water, single-base complex formation in vacuum for some methylated bases in comparison with data calculated with three sets of potential functions.

Base	Experimental	PM[(a)]	WK[(a)]	J[(a)]
m^1 Thy	-10.4 ± 0.9	-7.9 (B)	-8.1 (B)	-7.5 (B)
m^1 Cyt	-11.4 ± 0.8	-11.8 (B)	-11.6 (B)	-9.9 (B)
$m^{1,4,4}$ Cyt	-11.8 ± 0.9	-11.8 (C)	-9.9 (C)	-8.8 (C)
m^9 Ade	-10.6 ± 1.0	-10.8 (A)	-10.8 (A)	-7.8 (A)
$m^{6,6,9}$ Ade	-8.3 ± 0.8	-7.0 (C)	-6.0 (C)	-5.1 (C)
$m^{2,2,9}$ Gua	-14.1 ± 1.0	-12.5 (A)	-11.0 (B)	-11.1 (B)

[(a)]PM, WK and J are energy values (kcal/mol) in local minima for Poltev-Malenkov, Weiner-Kollman, and Jorgensen potentials, respectively.
[(b)]Letters in parentheses are minima from Table 1.

MONTE-CARLO HYDRATION SIMULATIONS OF NORMAL TAUTOMERS OF NUCLEIC ACID BASES

Experimental hydration enthalpies are available for three natural bases, 9-methyladenine, -23.0 kcal/mol (Zielenkiewicz et al., 1984), 1-methylthymine, -23.4 kcal/mol, and 1-methyluracil, -21.3 kcal/mol (Teplitsky et al., 1980) . These values were obtained as differences in the heats of sublimation and dissolution. PM PF result in hydration energies of -26.2, -22.5 and -20.4 kcal/mol, respectively, for the three bases listed above. Computations with WK PF give values of -22.0, -14.4 and -15.7 kcal/mol, respectively. Thus, although the PM value for 9-methyladenine is slightly larger than the experimental result, which is reproduced well by WK PF, overall agreement with the experiment is significantly better when PM PF are used since WK PF underestimate the absolute values for 1-methylthymine and 1-methyluracil. The main contribution to the absolute value of the difference in hydration energies calculated for a particular base via PM versus WK potentials comes from water-base interactions. On the other hand, the relative differences among hydration energies calculated for the three bases arise mainly from differences in water-water interactions. The relative values of hydration enthalpies of thymine and uracil are reproduced via PM PF, not by WK PF.

Thus, the better agreement of calculated results with experimental data when PM PF are used, both for single-base, single-water interactions and for fully hydrated bases, suggests that these PF are those most appropriate for calculations of DNA hydration.

TAUTOMERIZATION OF GUANINE IN RELATION TO MISPAIRING

As was mentioned in the introduction, guanine tautomerization is an especially interesting problem in relation to errors in nucleic acid biosynthesis. Therefore, we consider guanine in more detail and designate isomers of the enol tautomer by H6->N7 and H6->N1, which means guanine enol isomers with the hydrogen at O6 facing N7 and N1, respectively. We use similar designations for other enol and imin isomers. Mutual arrangement of 9-methylguanine and water molecules in some characteristic positions are displayed in Figure 1 while energy values are shown in Table 3. These data demonstrate the following features of single-base, single-water interactions: (1) water positions in minima with two distorted H-bonds, (2) the deepest minimum for water H-bonded to N7 and O6 atoms, and (3) different arrangements and different energy values in minima for different tautomers.

Table 4 shows the results of Monte-Carlo simulation of hydration of the three guanine isomers. The differences in hydration energy of –5.6 and –6.8 kcal/mol between the keto and enol tautomers are consistent with an equilibrium constant of 10^4 under the reasonable assumption that the entropy difference between the two tautomers is not large enough to significantly affect the tautomerization constant. The relatively less intrinsically stable enol isomer H6->N7 is more favorably hydrated. These results provide an explanation for the shift of guanine tautomeric equilibrium in water solution to keto form, in spite of nearly equal intrinsic stability of keto and enol tautomers.

The results of single-water, single-base energy computations allow us to answer another question related to guanine tautomerization: "Why does guanine pair nearly exclusively with cytosine during DNA biosynthesis in spite of the nearly equal stability of the two tautomers in a nonpolar environment?" Since the interaction energy of guanine's keto tautomer with a single water molecule is more favorable than that of enol by 2.5 kcal/mol, the presence of only one or two water molecules will cause the keto form to predominate at the replication site.

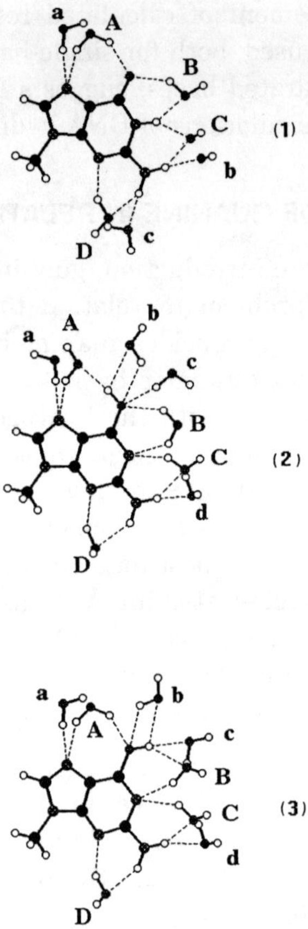

Figure 1. Mutual arrangement of 9-methylguanine and water molecules in local energy minima (upper-case letters) and some other characteristic positions (lowercase letters), selected so that water-base H-bonds would be practically linear or that water and base O-H bonds would be parallel: (1) keto tautomer, (2) enol H6->N7 isomer, (3) enol H6->N1 isomer.

Table 3. Interaction energies (kcal/mol) of water molecules with guanine isomers[(a)]

Isomers	A	B	C	D	a	b	c	d
N1h	-12.5	-10.9	-10.5	-10.4	-9.1	-9.0	-7.3	
H6->N7	-10.4	-8.0	-9.5	-8.6	-7.3	-2.8	-5.8	-6.3
H6->N1	-10.0	-8.9	-8.2	-8.7	-7.9	-3.1	-6.4	-5.6

[(a)]Letters denote mutual arrangements between water and base molecules shown in Figure 1.

Table 4. Hydration characteristics of keto tautomer (N1H) and two isomers of enol tautomer of 9-methylguanine, as calculated with Poltev-Malenkov potential functions.

	E[(a)]	Et[(a)]	Eb[(a)]	Ew[(a)]	Nb[(a)]
N1H	-36.8[(b)]	-3609.1	-73.1	-3536.0	9.38
H6->N1	-30.0	-3602.3	-58.2	-3544.1	8.53
H6->N7	-31.2	-3603.5	-57.8	-3546.1	8.71

[(a)]E, hydration energy; Et, total energy;Eb and Ew, water-base and water-water contributions; Nb, number of water-base H bonds.
[(b)]Energy values are in kcal/mol.

TAUTOMERIZATION OF OTHER NUCLEIC ACID BASES

It is important to determine if keto tautomers are always more favorably hydrated than enol forms in order to understand the role of rare tautomers in both spontaneous mutagenesis and in predicting tautomeric equilibrium, as well as possible mispairing of modified bases.

Thymine (and uracil) tautomerization can occur by proton transfer from N3 to O4 or O2, giving rise to two isomers. Enol forms were not detected either in water solutions or in inert gas matrix isolation, which indicates that the tautomeric equilibrium is shifted nearly completely to the diketo form. Computations reveal more favorable hydration of 1-methylthymine enol tautomers as compared to that of the diketo form. The hydration energy difference between diketo and the most favorable (as estimated from quantum mechanical computations) enol form (H4->N3) is 8.7 kcal/mol. This difference is more pronounced for other enol forms whose intrinsic stability is lower. Thus, the reason for preferential existence of thymine in the diketo form is its higher intrinsic stability, hydration makes the difference in concentrations of keto and enol forms less pronounced, but the stability of the diketo form is still so high that no enol tautomers were observed experimentally.

The hydration energies of amin and imin tautomers of 1-methylcytosine are nearly equal; for the H4->N3 isomer, the difference is only 0.5 kcal/mol, in favor of the imin form. Although water-base interactions are more favorable for the amin tautomer (the difference is 3.3 kcal/mol), water-water interactions have a compensating effect. The same is true for 1,5-dimethylcytosine; however, 9-methyladenine hydration favors the imin tautomer, and the reason for amin preference in water solution is its higher intrinsic stability. Thus, hydration of rare tautomers of bases can be less favorable, more favorable, or equal to normal tautomers. All of the molecular structure, not just a type of tautomeric transition, is important for hydration energetics. No conclusions can be made *a priori* for base derivatives.

GENERAL FEATURES OF HYDRATION SHELL STRUCTURES OF THE MINOR AND MAJOR GROOVES OF B-DNA DUPLEXES

The question addressed in this section is, "Are there any general features of the structure of hydration shells in the duplex grooves?" DNA hydration, nucleotide sequence, and conformation are all interrelated. The occurrence of B conformation with narrowed minor groove of An:Tn tracts is a good example of this phenomenon. But the question in which we are interested now is, "What features are common for all the B-like conformations of any sequence?"

We performed Monte-Carlo simulations of several duplexes with modified thymines. Bulky substituents at the C5 position of thymine excluded water molecules from part of major groove but did not significantly change the hydration of the rest of the groove. We illustrate our conclusions by the results for the duplex d(GGTTATTAATGG):d(CCAAT$_p$AAT$_a$TACC), where subscripts p and a denote pyrene and aniline residues instead of the methyl group. The conformation of this duplex is displayed in Figure 2, and schemes of water bridges in the grooves are presented in Figure 3. The main general feature of all the duplexes considered is the "unidirectionality" of one- and two-water bridges in the minor groove. All the bridges formed with a probability higher than 5% are arranged across the groove for any base step. This is not a consequence of more favorable distances between hydrophilic atoms across the groove compared to those along the groove. Intrastrand distances between purine N3 and adjacent pyrimidine O2 are no less favorable for water-bridging than interstrand N3/O2 distances of the same two base pairs. The only relationship between distance and bridge formation was an increase in one-water bridge formation probability and decrease in a two-water bridge probability when the interatomic distance is shortened.

As for hydration shell structure in the major groove, we observed bridges arranged in both directions, across the groove and along the groove. Again, we have observed no direct relationship between the distance between hydrophilic atoms and the probability of a one- or two-water bridge formation. This probability in the major groove, and in the minor one as well, depends on overall groove structure in the vicinity of an atom pair.

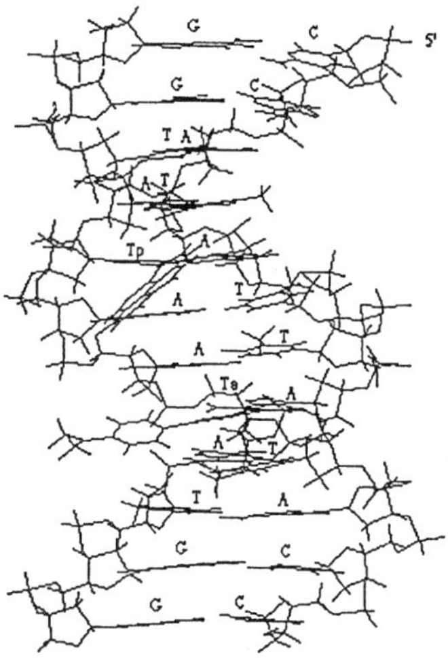

Figure 2. Conformation of DNA duplex. Pyrene and aniline residues bulge out of major groove but do not disturb overall groove hydration.

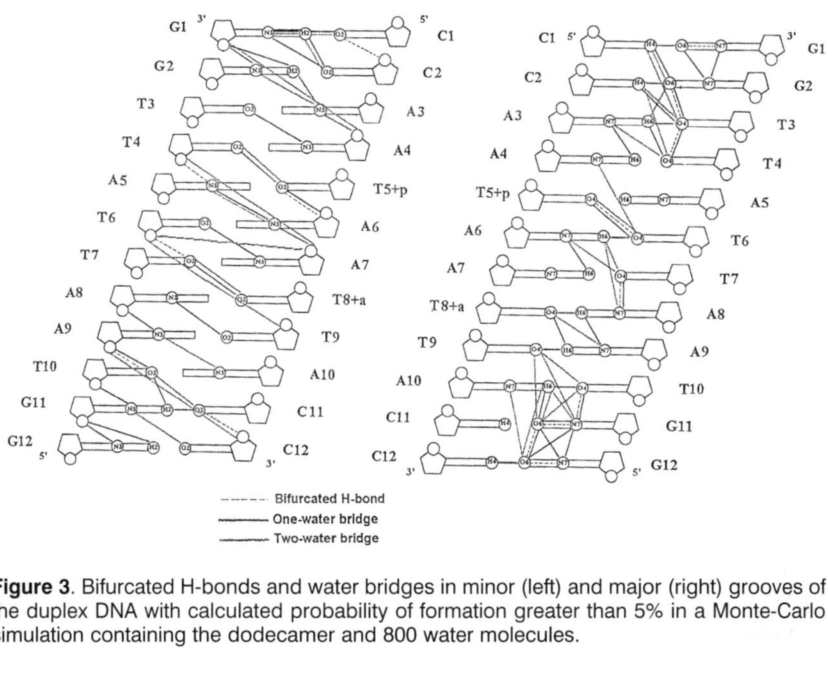

Figure 3. Bifurcated H-bonds and water bridges in minor (left) and major (right) grooves of the duplex DNA with calculated probability of formation greater than 5% in a Monte-Carlo simulation containing the dodecamer and 800 water molecules.

ACKNOWLEDGMENTS

The research described was made possible in part by Grant No. NLE300 from the International Science Foundation and the Russian Government and by support from the Office of Health and Environmental Research (OHER) of the U.S. Department of Energy under contract DE-AC06-76RLO 1830.

REFERENCES

Chuprina, VP and VI Poltev. **1983**. Possible conformations of double helical polynucleotides containing incorrect base pairs. Nucl. Acids Res. 13:5205-5222.

Drew, HR and RE Dickerson. **1981**. Structure of a B-DNA dodecamer. III. Geometry of hydration. J. Mol. Biol. 151:535-556.

Feng, J-A, RC Johnson, and RE Dickerson. **1994**. Hin recombinase bound to DNA: The origin of specificity in major and minor groove interactions. Science 263:348-355.

Franklin, R and RG Gosling. **1953**. The molecular structure of thymonucleate fibers. I. The influence of water content. Acta Crystallogr. 6: 673-677.

Jorgensen, WL, J Chandrasekhar, JD Madura, RW Impey, and ML Klein. **1983**. Comparison of simple potential functions for simulating liquid water. J. Chem. Phys. 79:926-935.

Kennard, O. **1985**. Structural studies of DNA fragments: G:T wobble base pair in A, B, and Z-DNA. G:A base pair in B-DNA. J. Biomol. Struct. Dyn. 3:205-226.

Patel, DJ, SA Kozlowski, S Ikuta, and K Itakura. **1984**. Dynamics of DNA duplexes containing internal G·T, G·A, A·C and T·C pairs: hydrogen exchange at and adjacent to mismatch sites. Fed. Proc. 43:2663-2672.

Petruska, J, LC Sowers, and MF Goodman. **1986**. Comparison of nucleotide interactions in water, proteins, and vacuum: model for DNA polymerase fidelity. Proc. Natl. Acad. Sci. USA 83:1559-1562

Poltev, VI, TI Grokhlina, and GG Malenkov. **1984**. Hydration of nucleic acid bases studied using novel atom-atom potential functions. J. Biomol. Struct. & Dyn. 2:413-429.

Poltev, VI, AV Teplukhin, and GG Malenkov. **1992**. Computational investigation of the role of hydration in nucleic acid structure and function. Int. J. Quant. Chem. 42:1499-1514.

Poltev, VI, EJ Gonzalez, and AV Teplukhin. **1995a**. Possible role of rare tautomers of nucleic bases in mutagenesis:Effect of hydration on tautomer equilibrium. Mol. Biol. (Mosc.) 29:365-375; English Translation in Mol. Biol. 29:213-219.

Poltev, VI, EJ Gonzalez, AV Teplukhin, GG Malenkov, and JH Miller. **1995b**. Comparative study of three systems of potential functions for simulation of nucleic acid hydration. Russian Chemical Bulletin 44:1333-1338.

Poltev, VI, GG Malenkov, EJ Gonzalez, AV Teplukhin, R Rein, M Shibata, and JH Miller. **1995c**. Modeling DNA hydration: Comparison of calculated and experimental hydration properties of nucleic acid bases. J. Biomol. Struct. & Dyn. 13:

Pranata, J, SJ Wierschke, and JL Jorgensen. **1991**. OPLS potential functions for nucleotide bases. Relative association constants of hydrogen-bonded base pairs in chloroform. J. Am. Chem. Soc. 113:2810-2819.

Prive, GG, U Heinemann, S Chandrasegaran, L-S Kan, ML Kopka, and RE Dickerson. **1987**. Helix geometry, hydration, and G.A mismatch in a B-DNA decamer. Science 238:498-504.

Rein, R, M Shibata, R Garduno-Juares, T Kieber-Emmons. **1983**. Structure of mispairs leading to substitution mutations, pp. 269-288. In: *Structure and Dynamics Nucleic Acids and Proteins*, E. Clementi and R.H. Sarma (eds.). Adenine Press, New York.

Saenger, W. **1984**. *Principles of Nucleic Acid Structure*. Springer-Verlag, New York.

Sheina, GG, SG Stepanian, ED Radchenko, and Yu P Blagoi. **1987**. IR spectra of guanine and hypoxanthine isolated molecules. J. Mol. Struct. 158:275-292.

Sukhodub, LF. **1987**. Interaction and hydration of nucleic acid bases in vacuum. Experimental study. Chem. Rev. 87:589-606.

Szczepaniak, K, M Szczepaniak, and WB Person. **1988**. Infrared studies and the effect of ultraviolet irradiation on the tautomers of 9-methylguanine isolated in an argon matrix. Chem. Phys. Lett. 153:39-44.

Teplitsky, AB, IK Yanson, OT Glukhova, A Zielenkiewcz, W Zielenkiewicz, KL Wierzchowski. **1980**. Thermochemistry of aqueous solutions of alkylated nucleic acid bases. III. Enthalpies of hydration of uracil, thymine and their derivatives. Biophys. Chem. 11:17-21.

Watson, JD, and FHC Crick. **1953**. Genetical implications of the structure of deoxyribose nucleic acid. Nature 171:964-967.

Weiner, SJ, PA Kollman, DT Nguyen, and DA Case. **1986**. An all atom force field for simulation of proteins and nucleic acids. J. Comp. Chem. 7:230-252.

Zielenkiewcz, A, W Zielenkiewicz, LF Sukhodub, OT Glukhova, AB Teplitsky, and KL Wierzchowski. **1984**. Thermochemistry of aqueous solutions of alkylated nucleic acid bases V. Enthalpies of hydration of N-methylated adenines. J. Solution Chem. 13:757-765.

PROTEIN-DNA COMPLEXES: STRUCTURE, FUNCTION AND RECOGNITION

AN ATOMIC VIEW OF ADDITIVE MUTATIONAL EFFECTS IN A PROTEIN STRUCTURE

M. M. Skinner and T. C. Terwilliger

Structural Biology Group, Los Alamos National Laboratory, Los Alamos, New Mexico

Key Words: *Protein crystallography/protein structure/protein engineering/ protein-nucleic acid interactions*

ABSTRACT

Substitution of a single amino acid in a protein will often lead to substantial changes in properties. If these properties could be altered in a rational way, then proteins could be readily generated with functions tailored to specific uses. When amino acid substitutions are made at well-separated locations in a single protein, their effects are generally additive. Additivity of effects of amino acid substitutions is very useful, because the properties of proteins with any combination of substitutions can be inferred directly from those of the proteins with single changes. It would therefore be of considerable interest to have a means of knowing whether substitutions at a particular pair of sites in a protein are likely to lead to additive effects.

The structural basis for additivity of effects of mutations on protein function was examined by determining crystal structures of single and double mutants in the hydrophobic core of gene V protein. Structural effects of mutations were found to be cumulative when two mutations were made in a single protein. Additivity occurs in this case because the regions structurally affected by mutations at the two sites do not overlap, even though the sites are separated by only 9 Å. Structural distortions induced by mutations in gene V protein decrease rapidly, but not isotropically, with distance from the site of mutation. It is anticipated that cases where structural and functional effects of mutations will be additive could be identified simply by examining whether the regions structurally affected by each component mutation overlap.

ADDITIVITY OF MUTATIONAL EFFECTS ON PROTEIN FUNCTION

It is now widely appreciated that effects of mutations on a property of a protein molecule are sometimes additive (Wells, 1990). This means, for example, that if a particular mutation increases the stability of the protein by 1 kcal/mol, and if a second mutation increases the stability by 2 kcal/mol, then a protein containing both mutations might be stabilized by about 3 kcal/mol. In cases studied so far, additivity of this type is generally observed if the amino acid side chains that are altered in the two mutants are not in direct contact with each other and the amino acid side chains involved are uncharged (Hecht et al., 1986; Matsumura et al., 1989; Sandberg and Terwilliger, 1991; Stearman et al., 1988; Wells, 1990; Zhang et al., 1991). Our laboratory provided a demonstration of this with a set of 16 double mutants at two sites in the gene V protein in which the two sites were not in direct contact (Sandberg and Terwilliger, 1991). Other examples where the effects of mutations are nearly additive include effects on DNA binding specificity of l repressor (Nelson and Sauer, 1985), substrate specificity of subtilisin (Carter and Wells, 1987; Wells et al., 1987a,b,c), and cofactor binding affinity of glutathione reductase (Scrutton et al., 1990).

Recently, the deviations of effects of multiple mutations at well-separated sites from exact additivity in a number of proteins were examined in detail (LiCata and Ackers, 1995). In several cases a pattern of either super- or subadditivity was found, suggesting some systematic kind of interaction among the mutated residues, although the nature of this interaction is unknown. It appears, in any event, that small nonadditive effects might result from mutations at distant sites. A special case of such nonadditive effects were the mutants of staphylococcal nuclease, in which nonadditive effects on protein stability were evidently due to interactions occurring in a partially folded or unfolded state of the protein (Green and Shortle, 1993). In contrast to well-separated mutations, mutations at sites that are in contact with one another often show strongly nonadditive behavior (Hurley, 1994; Hurley et al., 1992; Mildvan et al., 1992). For example, altering the side chains facing each other at positions 33 and 47 in gene V protein, increasing the bulk of one and decreasing the bulk of the other, results in strongly subadditive effects on stability (Sandberg and Terwilliger, 1993).

USING ADDITIVITY TO SIMPLIFY THE ENGINEERING PROBLEM

Single mutations in a protein generally affect many different properties of that protein. This often complicates engineering of a protein, where changing just one property may be desirable, or where secondary effects of a mutation (such as decreased stability, for example) render the desirable effects (such as increased affinity for a substrate) less useful. If mutations affect several different properties of a protein in an additive fashion, however, then combinations of mutations could be constructed by rational design that would affect one property of a protein without substantially affecting another (Sandberg and Terwilliger, 1993). This could be used to engineer the specificity of gene V protein-DNA interactions, for example, while simultaneously maintaining the stability of the protein. In effect, combinations could be found in which the individual mutations cancel each other's undesirable effects while adding to each other's desired effects. If effects on several different properties of a protein are all additive, then the properties of a mutant with several mutations can be thought of as a "vector sum" of the properties of the wild type, where the vectors represent the changes in these properties from each mutation. This observation is quite important because it means that if we had a large "basis set" of mutants that were characterized with respect to several properties, we could design a new mutant with a desired combination of these properties, just by choosing a set of mutations for which these vectors add to the desired values.

The purpose of this work was to determine to what extent this approach is feasible when applied to structural changes in a protein in addition to the functional effects of mutations. If the additivity of effects of mutations on both structure and function is often very good and we are able to define conditions where it is and is not applicable, then our technique for specifically engineering individual properties of a protein may have very general applicability. It could be used to engineer proteins with altered substrate specificity but without decreased stability, for example, or to engineer proteins with increased catalytic activity but without altered substrate specificity.

GENE V PROTEIN AS A MODEL SYSTEM FOR PROTEIN ENGINEERING

The model system for this study is the gene V protein of bacteriophage f1 (Figure 1). Gene V protein is a small, 87-residue protein that binds tightly and cooperatively to single-stranded DNA and RNA (Alberts et al., 1972; Baas, 1985; Cavalieri et al., 1976; Pratt and Ehrdahl, 1968). Its natural function is to bind to a single-stranded DNA produced during infection of *Escherichia coli* by bacteriophage f1, preventing its conversion into double-stranded DNA, and, instead allowing its packaging into new phage particles (Baas, 1985). The gene V protein is a dimer, and we have recently determined its crystal structure at high resolution (Skinner et al., 1994). This structure shows which residues in the dimeric protein are close together and which are not, and, thus, which is useful in our interpretation of reasons for deviations from additivity in effects of mutations on protein stability. In addition to the structure of the protein we have recently determined, we previously isolated 350 single-amino-acid substitution mutants of the gene and characterized their function *in vivo* (Terwilliger et al., 1994; Zabin et al., 1991).

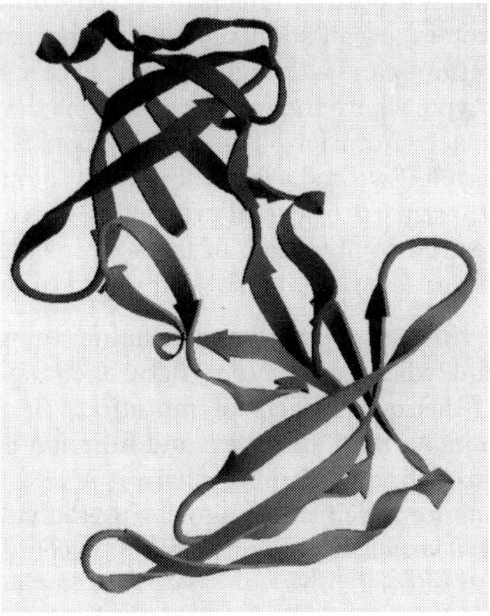

Figure 1. Gene V protein dimer (Adapted from Skinner et al., 1994). The gene V protein is a symmetric dimer. The 2-fold axis of symmetry is approximately perpendicular to the page at the center of the figure.

STRUCTURAL BASIS FOR ADDITIVITY OF MUTATIONAL EFFECTS ON THE PROPERTIES OF A PROTEIN

We have purified and characterized the stabilities and DNA-binding affinities of a collection of single and double mutants of gene V protein in which many of the mutations were at two buried sites, Val35 and Ile47 (Sandberg et al., 1995; Sandberg and Terwilliger, 1991, 1993; Terwilliger, 1993). Mutations at these sites result in closely additive effects on the properties of the resulting proteins, and we intended to determine whether there was a clear structural basis for this additivity. We were able to obtain high-quality crystals of 9 mutants of gene V protein with single and double amino acid substitutions at positions Val35 and Ile47 (Skinner and Terwilliger, 1996; Zhang et al., 1996). At position 35 the single mutant V35I was obtained, and at position 47, we obtained structures of the single mutants I47V, I47L, and I47M. The double mutants obtained were V35I/I47V, V35I/I47L, V35I/I47M, V35I/I47F, and V35A/I47L. Diffraction data of excellent quality (R_{sym} of approximately 4% on F for each dataset) were collected to a resolution of 1.8 Å from each of these crystals, all of which were closely isomorphous to crystals of wild-type gene V protein (space group *C2*).

From analysis of $(F_{mut} - F_{wt})\mathbf{e}^{i\phi c}$ and $(2F_{mut} - F_{wt})\mathbf{e}^{i\phi c}$ difference Fourier maps (obtained using phases calculated from the wild-type gene V protein model) it was clear that the structural changes caused by these mutations could be readily identified, even though these structural shifts were typically on the order of 0.1 to 0.2 Å. Figure 2 illustrates a small part of the $(F_{mut} - F_{wt})\mathbf{e}^{i\phi c}$ difference Fourier for the V35A/I47L double mutant in the neighborhood of position 35. The refined models for the wild-type and mutant structures are shown. The shifts in position of atoms near the V35A mutation are clearly indicated by the pattern of positive and negative difference density on either side of the atoms.

Additivity of Structural Effects of Mutations

Upon examining the difference Fouriers discussed above and the refined structures of single and double mutants of gene V protein, it was clear that structural changes in gene V protein caused by mutations at positions 35 and 47 are remarkably additive (Skinner and Terwilliger, 1996; Zhang et al., 1995). There are two ways in which this additivity is exhibited; the first is illustrated in Figure 3. This figure shows the coordinate shifts in the side chain of Leu 81, a residue that is midway between residues 35 and 47 and that is affected by both the V35I and I47L mutations. The coordinate shifts in this side chain caused by the two mutations are very nearly additive in the double mutant.

A second reason for the additivity of structural effects is that the regions affected by the two mutations are at least partially distinct (Skinner and Terwilliger, 1996; Zhang et al., 1996). In this case, one mutation affects one region, the other affects another, and therefore it is not surprising that the effects are additive. This is also the case for gene V protein mutants at positions 35 and 47. Figure 4 shows positive and negative peaks of difference electron density which indicate the shifts in atomic positions that occur in the V35I single mutant. The peaks of difference density are clustered near the corresponding sites of mutation, with few structural changes extending far from the site. The regions affected by each mutation are not spheres about the mutated

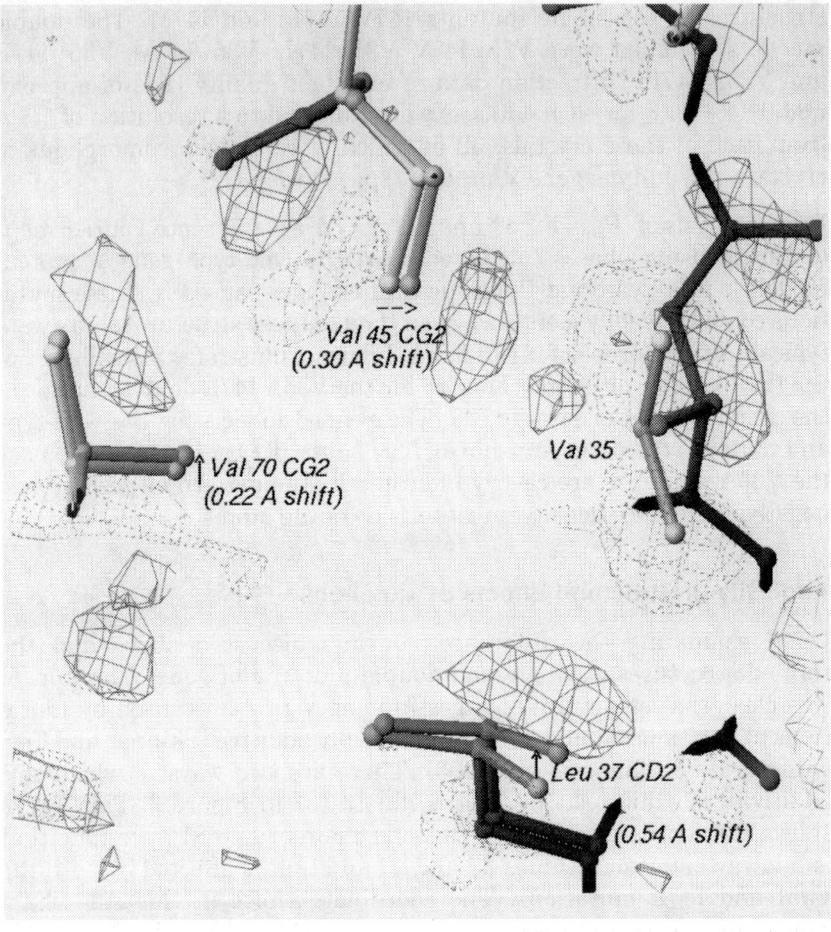

Figure 2. V35A/I47L double mutant difference Fourier. Relevant part of the wild-type gene V protein structure is shown in white, the mutant in gray. Positive difference density is shown in solid contours, negative in dashed contours; contours are drawn at $\pm 2.5\sigma$. Coordinate shifts for selected atoms are shown.

sites. Instead, the effects are highly nonisotropic, evidently depending on which regions of the protein molecule have to adjust in order to accommodate the newly introduced mutated side chain. This is true in general for the gene V protein mutant structures we have examined (Skinner and Terwilliger, 1996; Zhang et al., 1996). Structural changes are relatively local, but are nonisotropically distributed about the site of mutation.

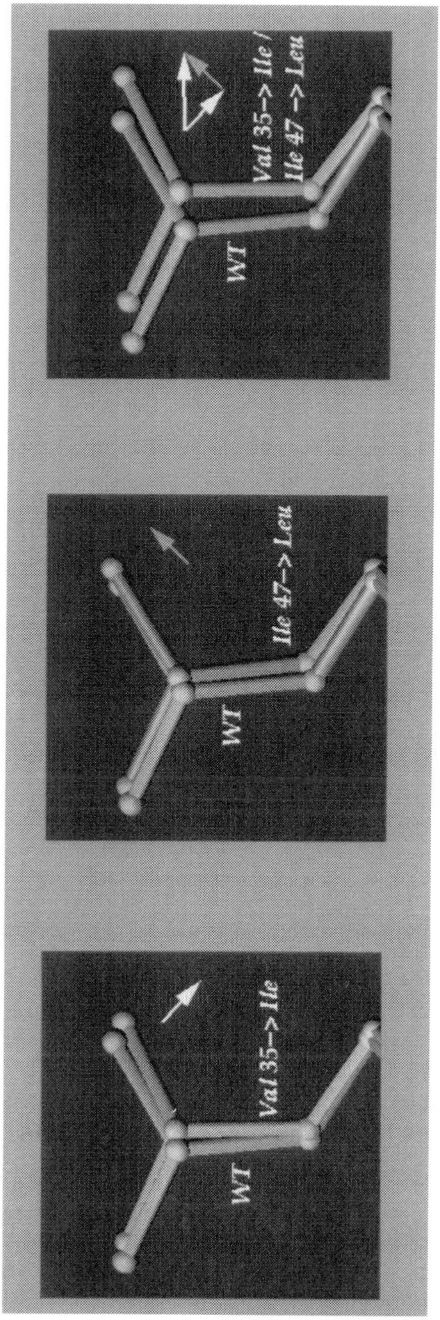

Figure 3. Additivity of structural changes in single and double mutants of gene V protein. Position of side chain of Leu 81 in WT, V35I, I47L, and V35I/I47L mutants is shown. Directions of coordinate shifts are indicated by arrows.

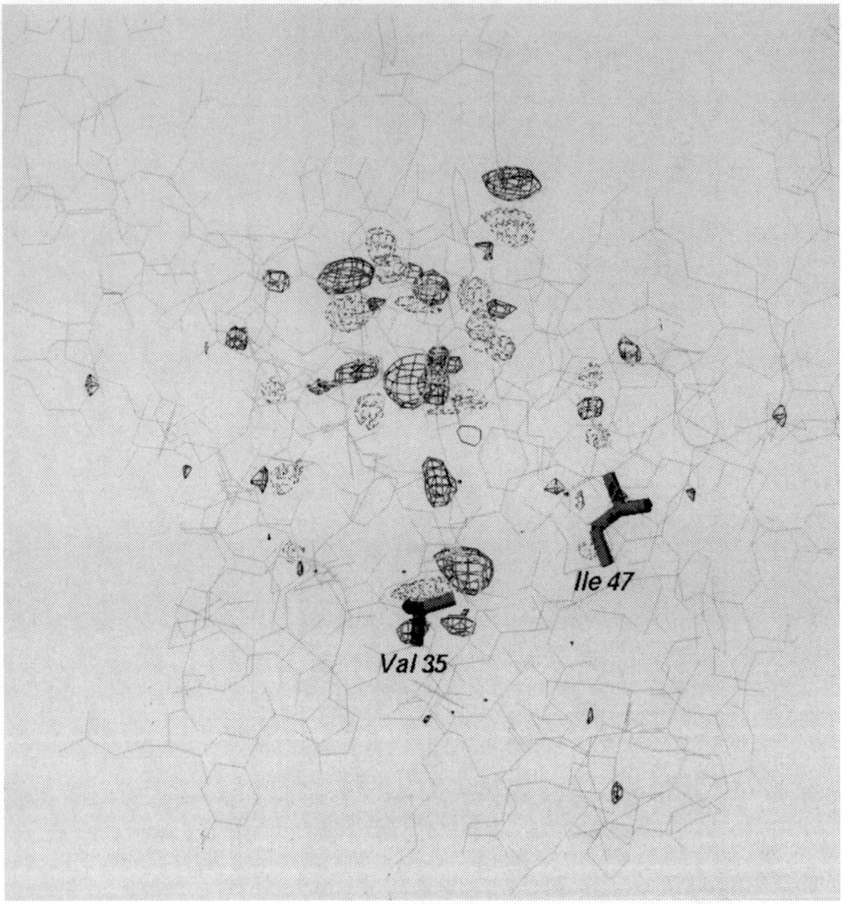

Figure 4. Difference Fourier map for V35I mutant. Relevant part of the wild-type gene V protein structure is shown as a line drawing. Positive difference density is shown in solid contours, negative in dashed contours; contours are drawn at $\pm 4\sigma$.

Using the Region Structurally Affected by Single Mutations to Identify Pairs of Mutations that will Show Additive Effects

The local and highly nonisotropic nature of structural effects caused by mutations in the core of gene V protein suggested a simple way to identify cases where structural and functional effects of multiple mutations are probably additive. Comparing the structure of each mutant with a single amino acid substitution with that of the wild-type protein, the region influenced structurally by that mutation can be identified, either qualitatively as in Figure 4, or more quantitatively. Pairs of mutants in which the regions of influence do not substantially overlap can be expected to result in generally additive effects (as double mutants at positions 35 and 47 do in gene V protein). In locations in the protein where the regions of influence do overlap, the structural changes might (as in Figure 3) or might not show additive effects, and the extent of nonadditivity might depend on the extent of structural changes in the individual mutants in the region of overlap. Addressing these issues will be an important area of future research in this field.

ACKNOWLEDGMENTS

The authors thank the National Institutes of Health and the Laboratory Directed Research and Development program of Los Alamos National Laboratory for generous support.

REFERENCES

Alberts, B, L Frey, and H Delius. **1972**. Isolation and characterization of gene 5 protein of filamentous bacterial viruses. J. Mol. Biol. 68:139-152.

Baas, PD. **1985**. DNA replication of single-stranded *Escherichia coli* DNA phages. Biochim. Biophys. Acta 825:111-139.

Carter, P and JA Wells. **1987**. Engineering enzyme specificity by "substrate-assisted catalysis." Science 237:394-399.

Cavalieri, SJ, KE Neet, and DA Goldthwait. **1976**. Gene 5 protein of bacteriophage fd: a dimer which interacts cooperatively with DNA. J. Mol. Biol. 102: 697-711.

Green, SM and D Shortle. **1993**. Patterns of nonadditivity between pairs of stability mutations in staphylococcal nuclease. Biochemistry 32:10131-10139.

Hecht, MH, JM Sturtevant, and RT Sauer. **1986**. Stabilization of l repressor against thermal denaturation by site-directed Gly->Ala changes in a-helix 3. Proteins 1: 43-46.

Hurley, JH. **1994**. The role of interior side-chain packing in protein folding and stability, pp. 549-578. In: The Protein Folding Problem and Tertiary Structure Prediction. Birkhauser, Boston, Massachusetts.

Hurley, JH, WA Baase, and BW Matthews. **1992**. Design and structural analysis of alternative hydrophobic core packing arrangements in bacteriophage T4 lysozyme. J. Mol. Biol. 224:1143-1159.

LiCata, VJ and GK Ackers. **1995**. Long-range, small magnitude nonadditivity of mutational effects in proteins. Biochemistry 34:3133-3139.

Matsumura, M, WJ Becktel, M Levitt, and BW Matthews. **1989**. Stabilization of phage T4 lysozyme by engineered disulfide bonds. Proc. Natl. Acad. Sci. USA 86:6562-6566.

Mildvan, AS, DJ Weber, and A Kuliopulos. **1992**. Quantitative interpretations of double mutations of enzymes. Arch. Biochem. Biophys. 294:327-340.

Nelson, HCM and RT Sauer. **1985**. Lambda repressor mutations that increase the affinity and specificity of operator binding. Cell 42:549-558.

Pratt, D and WS Ehrdahl. **1968**. Genetic control of bacteriophage M13 DNA synthesis. J. Mol. Biol. 37:181-200.

Sandberg, WS and TC Terwilliger. **1991**. Energetics of repacking a protein interior. Proc. Natl. Acad. Sci. USA 88:1706-1710.

Sandberg, WS and TC Terwilliger. **1993**. Engineering multiple properties of a protein by combinatorial mutagenesis. Proc. Natl. Acad. Sci. USA 90:8367-8371.

Sandberg, WS, PM Schlunk, HB Zabin, and TC Terwilliger. **1995**. Relationship between *in vivo* activity and *in vitro* measures of function and stability of a protein. Biochemistry 34:11970-11978.

Scrutton, NS, A Berry, and RN Perham. **1990**. Redesign of the coenzyme specificity of a dehydrogenase by protein engineering. Nature 343:38-43.

Skinner, MM and TC Terwilliger. **1996**. Using additivity of mutational effects to simplify protein engineering. Proc. Natl. Acad. Sci. USA (in press).

Skinner, MM, H Zhang, DH Leschnitzer, H Bellamy, RM Sweet, CW Gray, RNH Konings, AH-J Wang, and TC Terwilliger. **1994**. Structure of the gene V protein of bacteriophage f1 determined by multiwavelength X-ray diffraction on the selenomethionyl protein. Proc. Natl. Acad. Sci. USA 91: 2071-2075.

Stearman, RS, AD Frankel, E Freire, B Liu, and CO Pabo. **1988**. Combining thermostable mutations increases the stability of l repressor. Biochemistry 27:7571-75-74.

Terwilliger, TC. **1993**. MAD phasing: Bayesian estimates of Fa. Acta Crystallogr. D50: 11-16.

Terwilliger, TC, HB Zabin, MP Horvath, WS Sandberg, and PM Schlunk. **1994**. In vivo characterization of mutants of the bacteriophage f1 gene V protein isolated by saturation mutagenesis. J. Mol. Biol. 236:556-571.

Wells, JA. **1990**. Additivity of mutational effects in proteins. Biochemistry 29:8509-8517.

Wells, JA, BC Cunningham, TP Graycar, and DA Estell. **1987a**. Recruitment of substrate-specificity properties from one enzyme into a related one by protein engineering. Proc. Natl. Acad. Sci. USA 84: 5167-5171.

Wells, JA, BC Cunningham, TP Graycar, DA Estell, and P Carter. **1987b**. On the evolution of specificity and catalysis in subtilisin. CSH Symp. Quant. Biol. 52:647-652.

Wells, JA, DB Powers, RR Bott, TP Graycar, and DA Estell. **1987c**. Designing substrate specificity by protein engineering of electrostatic interactions. Proc. Natl. Acad. Sci. USA 84:1219-1223.

Zabin, HB, MP Horvath, and TC Terwilliger. **1991**. Approaches to predicting effects of single amino acid substitutions on the function of a protein. Biochemistry 30:6230-6240.

Zhang, X-J, WA Baase, and BW Matthews. **1991**. Toward a simplification of the protein folding problem: A stabilizing polyalanine a-helix engineered in T4 lysozyme. Biochemistry 30:2012-2017.

Zhang, H, MM Skinner, WS Sandberg, AH-J Wang, and TC Terwilliger. **1996**. Context-dependence of mutational effects in a protein: the crystal structures of the V35I, I47V and V35I/I47V gene V protein core mutants. J. Mol. Biol. 259:148-159.

THE ROLE OF MOLECULAR RECOGNITION IN ACTIVATION AND REGULATION IN THE GROWTH HORMONE-PROLACTIN FAMILY OF HORMONES AND RECEPTORS

A.A. Kossiakoff, C. Schiffer, and A.M. de Vos

Department of Protein Engineering, Genentech, Inc.,
South San Francisco, California

Key Words: *Molecular recognition, cytokine receptor activation, growth hormone, prolactin*

ABSTRACT

The mechanisms that support communication between cells are just beginning to be understood at the molecular level. We have focused on the role played by molecular recognition in these processes. Toward this end, the structures of several hormone-receptor complexes of the endocrine family have been analyzed. These systems show that both the hormones and receptors make significant conformational adjustments to complement quite different binding surfaces between them. Although the binding surfaces are large and consist of many residues, it is generally the case that the binding energetics are developed by a small subset of these residues.

INTRODUCTION

The mechanism by which cells communicate is one of the most intensely studied problems in biology. This communication is generally thought to be mediated through cell-surface receptors, initiated by a ligand binding event at the extracellular portion of the receptor. The details at the molecular level of how this extracellular message is transduced through the membrane to the interior of the cell are only now beginning to emerge. In this paper we discuss some of the work we are doing, aimed at delineating the structural basis for the signal transduction process.

We have focused on the endocrine family of human hormones and receptors. These include growth hormone, placental lactogen and prolactin, and the growth hormone and prolactin receptors. The sequence homology of the extracellular domains (ECD) of these receptors classifies them as belonging to the hematopoietic receptor superfamily

(Bazan, 1990). They have a three-domain organization: the extracellular portion that binds the activating hormone, a transmembrane segment of about 25 amino acids, and a cytoplasmic domain. Biological response is triggered by hormone binding that drives receptor aggregation (Cunningham et al., 1991; De Vos et al., 1992). The aggregated receptor complex binds one or several tyrosine kinases belonging to the JAK family, which then transphosphorylate elements on themselves, the receptors, and associated transcription factors (Ihle et al., 1994).

The structure of the human growth hormone (hGH) bound to its receptor (hGH_R) first showed how the mechanism of receptor aggregation worked (De Vos et al., 1992). One hormone binds two receptors. This finding initially presented a puzzle: how do two identical receptor ECD bind to one hormone, considering that hGH has no semblance of symmetry that could support binding of each of the ECD in a similar fashion? (hGH is an asymmetric, four-helix bundle protein.) Surprisingly, the high-resolution structure revealed that the same regions of each individual receptor bind to two topographically very different regions on hGH. This is illustrated in Figure 1, which shows a molecular surface rendering of hGH. At binding Site 1, hGH has a large invagination, into which the contacting side chains of one of the receptor ECD insert. The surface area that becomes buried on the hormone is about 1300 $Å^2$. Binding Site 2 is much flatter, and the surfaces of both the hGH and the receptor (ECD2), appear not to be as complementary. In comparison, Site 2 buries only 850 $Å^2$ of hGH's surface. The difference in surface interactions presumably plays an active physiological role. It has been shown that the 2 ECD:1 hGH complex is formed in a stepwise manner. Binding Site 1 is a high-affinity site and is always occupied before binding Site 2 (Fuh et al., 1992).

The binding of the endocrine hormones to their respective receptors is a highly regulated process. hGH has the ability to activate the prolactin receptor ($hPRL_R$), whereas prolactin (hPRL) does not bind to the growth hormone receptor. Human placental lactogen (hPL) activates $hPRL_R$, although the hPRL and hPL have only limited homology (25%) (Nicoll et al., 1986). In contrast, hPL binds 2300-fold more weakly than does hGH to hGH_R, even though they have 85% sequence identity (Lowman et al., 1991a). hPRL and hPL can be recruited to bind to hGH_R by altering only about half a dozen critical residues (Lowman et al., 1991b; Cunningham et al., 1990). Growth hormones are highly conserved, and hGH can activate GH_Rs of other species. Surprisingly, the converse is not true: the GH of other species cannot activate the human receptor (Boutin et al., 1989).

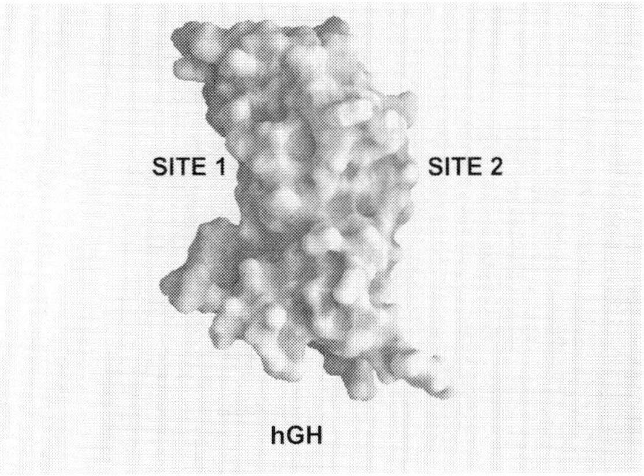

Figure 1. Solvent-accessible surface of human growth hormone (hGH), showing the topographically different Site 1 and Site 2 binding sites.

Taken together, these observations paint a puzzling picture. On the one hand, formation of the active 2:1 complex requires the GH_R and PRL_R to be versatile enough to bind two topographically highly diverse surfaces on the hormone. On the other hand, specificity is apparently dictated by relatively few key residues, because molecules very similar to hGH do not bind to hGH_R.

THE STRUCTURE OF HUMAN PLACENTAL LACTOGEN

We have recently solved the structure of hPL to 2.2 Å resolution (Kossiakoff, Somers, De Vos, unpublished). In most respects the structure of this hormone closely mirrors that of hGH. With this structure in hand, comparing the structures of hGH bound to hGH_R and the $hPRL_R$ indicates that several changes occur in the hormone between the two complexes. The most pronounced of these changes are in the minihelix 1 region and the loop connecting helices 1 and 2 (Figure 2). Noting the conformation of the minihelix, it is apparent that there is striking structural similarity between free hPL and hGH when it is bound to the $hPRL_R$. In fact, in these two structures the minihelix is melted out into a new type of loop structure. It is tempting to think, based on the structures, that changes in the minihelix and associated loop conformations are the principal determining factor in hPL's specificity toward the prolactin receptor.

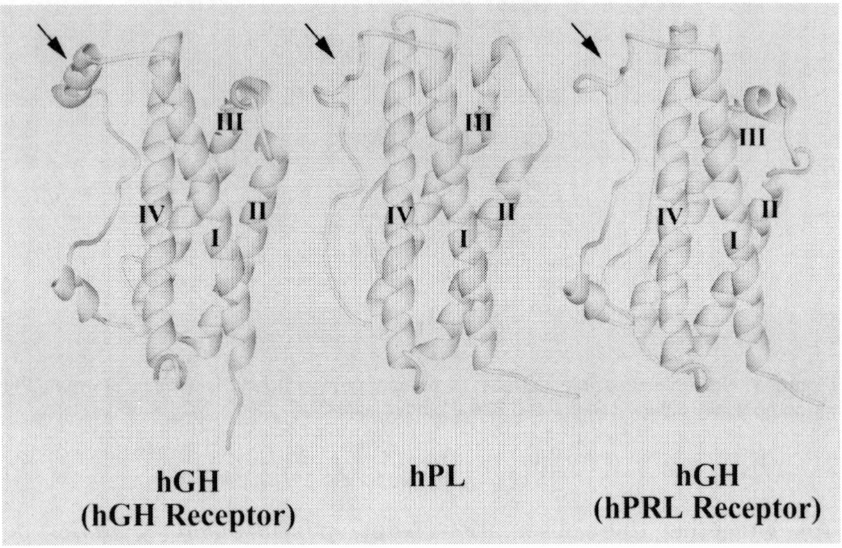

Figure 2. Ribbon diagrams of: human growth hormone (hGH) when bound to the hGH receptor (left panel), human placental lactogen (hPL) (middle panel), and hGH when bound to the prolactin receptor (right panel). Position of minihelix is shown by arrow in each panel.

THE STRUCTURE OF THE HGH "SUPERMUTANT"–RECEPTOR COMPLEX

A variant of hGH that binds 400 times tighter to its receptor was produced using a technique called "phage display." This method has been described in detail in some recent reviews (Wells and Lowman, 1992) and will not be discussed here. In outline, a protein of choice (in this case hGH) can be expressed on the surface of bacteriophage M13. M13 is the phage that is commonly used in *Escherichia coli* mutagenesis, and the gene for hGH has been inserted into the phage genome, where it can be easily mutated by standard techniques. Specific residues can be targeted for random mutagenesis; for hGH, 15 sites from 5 different libraries were mutated (Lowman et al., 1991a; Lowman and Wells, 1993). Each library had about 3 × 10^6 random mutants and, although not all combinations of the 20 amino acids at each site were sampled, the results certainly included a very extensive subset. The choice of which residues to mutate was made based on a previous alanine-scan mutagenesis (Cunningham and Wells, 1989, 1993). That analysis identified those residues that most influenced hormone binding, which were generally found on helix 4 and the loop connecting helix 1 and helix 2.

Figure 3 shows a ribbon diagram of hGH on which are marked the sites that were mutated during the phage display selections (Ultsch et al., 1994). Several characteristics are noteworthy. Three mutations were located in the Zn^{++} binding site: H18D, H21N, and E174S. The side chains of H18 and E174 act as ligands in the hGH-prolactin complex, and H21 plays a supporting role in the interaction by forming an H-bond to the carboxylate of E174 (Somers et al., 1994; Kossiakoff et al., 1994). The similar interaction in the hGH-hGHR does not require Zn^{++} binding and, in fact, mutagenesis data show that the coordinating side chains of H18, H21, and E174 have a negative influence on binding affinity in this latter system (Cunningham and Wells, 1993). Thus, it is not surprising that these residues were significantly altered in their character in the phage selections. Another point of interest is the finding that a few important changes that increased hormone affinity were residues (F10A and M14W) whose side chains were actually buried and did not make direct contact with the receptor itself. These are what are called "framework" residues, residues that alter the local scaffolding of the chain in such a way that presentation of neighboring side chains is affected slightly differently compared to the wild-type arrangements. The effect of framework residues has been appreciated in antibody engineering, where it is known that non-CDR residues can produce appreciable effects through indirect packing effects (Eigenbrot et al., 1993).

An accepted tenet of protein-protein interactions is that ionic interactions have a positive effect on binding energies. The *ala*-scan mutagenesis generally supported this notion (Cunningham and Wells, 1993). It was, therefore, very surprising that so many H-bonding interactions were eliminated in the phage selection mutagenesis: D171- receptor R43 (D replaced with S), R167- receptor E127 (R to N), Q46- receptor E120 (Q to W), and K41-receptor E127 (K to I). K41 and Q46 are contained in a segment of chain referred to as minihelix 1. The residues that interfaced the receptor (residues K41,Y42,L45 and Q46) were mutated together in one library. The clone that displayed the strongest binding (4.5 times wild-type) had a sequence of I, H, W, W at these positions. Residues 41 and 46 change from hydrophilic to hydrophobic, and 42, changed from hydrophobic to hydrophilic. Taken in the context of the sequence alone this is a puzzling result. How can one surface of an interface change stereochemical character to such a degree and remain complementary to the other surface?

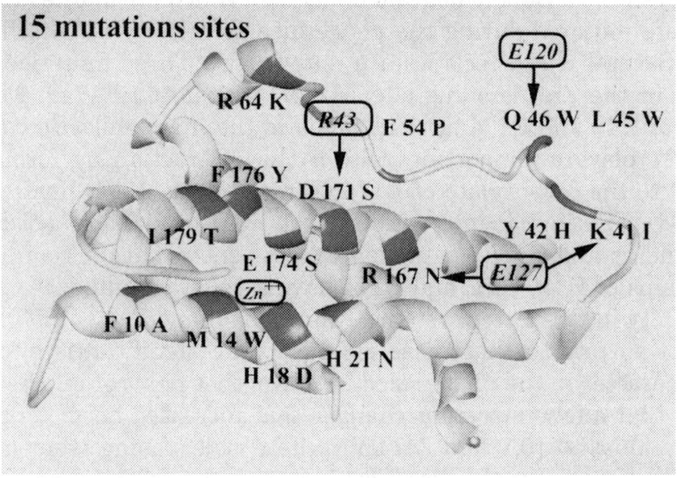

Figure 3. Ribbon representation of human growth hormone (hGH) with positions of sites of mutation for "supermutant" shown as strips and labeled. Circled residues are hGH receptor amino acids that make H-bonds in the hGH- hGHR complex to the wild-type hGH residues.

The situation described above is indeed a conundrum if the phage display mutagenesis is viewed simply as a procedure by which different amino acid side chains are displayed on a conformationally invariant main-chain backbone. Such is not the case, as is shown in Figure 4. In this figure three conformational states of minihelix 1 of hGH are shown. The first state is the helix in the context of the 2:1 complex (panel A). Panel (C) shows the helix in the free supermutant structure, and panel (B) is the minihelix conformation as it exists in the supermutant complex (Schiffer, Kossiakoff, unpublished). It is apparent that for both the hGH bound to its receptor and for the uncomplexed supermutant (panels A and C) this segment of chain remains formed as a helix. However, for the supermutant when it is complexed to the receptor, the minihelix has melted out, giving a more extended structure (panel B). The side chains of the mutated groups are disposed in quite different regions of space compared to where they would be if the segment of chain were to remain a helix. There is a very important lesson in this observation. Phage display mutagenesis not only probes sequence diversity but, in some types of tertiary structures, conformational diversity as well. Thus, conformational space was probed in the case of the minihelix and framework residues as a convolution of these two parameters, making the actual diversity available much larger than the computed 3×10^6 that was based on amino acid composition alone.

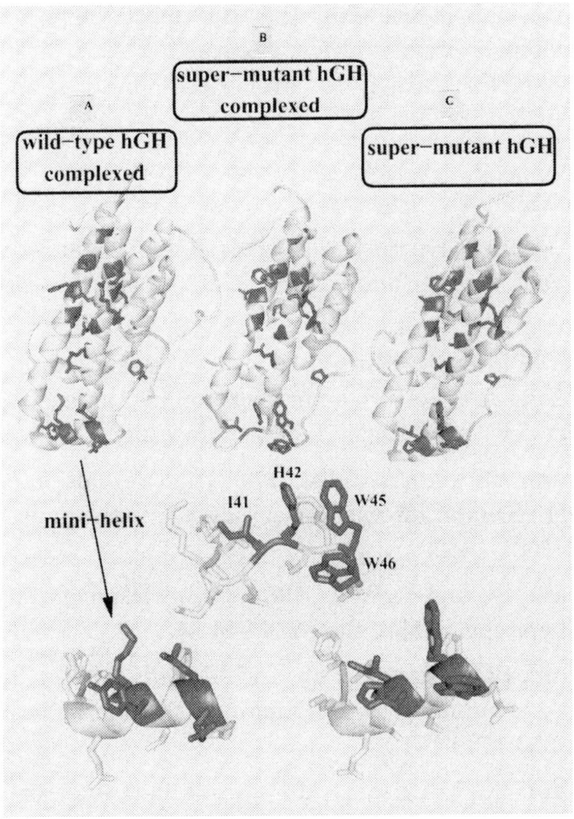

Figure 4. Changes in the minihelix conformation of: human growth hormone (hGH) complexed to hGHR (panel A); supermutant complexed to hGHR (panel B); and free supermutant (panel C, right). Note that for complexed supermutant, helix is melted out. Mutated residues on minihelix are shown.

CONCLUSIONS

Within this framework of hormone-receptor interactions there is an extraordinary adaptability among these molecules to synthesize binding epitopes to a wide variety of binding interfaces. The nature of the adjustments required to form the optimum binding surfaces suggests that recognition and binding are directed by an induced-fit mechanism. Distinct from the process of molecular recognition associated with the antibody-antigen paradigm, where binding is developed through sequence diversity, the endocrine receptors use essentially a constant set of residues to bind surfaces that are diverse both in sequence and in conformation. This is accomplished by employing conformational diversity, both local and global. This theme is most likely not limited to cytokine-receptor systems and we believe that other systems will be found that use this strategy to develop their binding and specificity determinants.

REFERENCES

Bazan, JF. **1990**. Structural design and molecular evolution of a cytokine receptor superfamily. Proc. Natl. Acad. Sci. USA 87:6934-6938.

Boutin, JM, M Edrey, M Shirota, et al. **1989**. Identification of a cDNA encoding a long form of prolactin receptor in human hepatoma and breast cancer cells. Mol. Endocrinol. 3:1455-1466.

Cunningham, BC and JA Wells. **1989**. High-resolution epitope mapping of hGH-receptor interactions by alanine-scanning mutagenesis. Science 244:1081-1085.

Cunningham, BC and JA Wells. **1993**. Comparison of a structural and a functional epitope. J. Mol. Biol. 234:554-563.

Cunningham, BC, S Bass, G Fuh, and JA Wells. **1990**. Zinc mediation of the binding of human growth hormone to the human prolactin receptor. Science 250:1709-1712.

Cunningham, BC, M Ultsch, AM De Vos, MG Mulkerrin, KR Clauser, and JA Wells. **1991**. Dimerization of the extracellular domain of the human growth hormone receptor by a single hormone molecule. Science 254:821-825.

De Vos, AM, M Ultsch, and AA Kossiakoff. **1992**. Human Growth Hormone and Extracellular Domain of its Receptor: Crystal Structure of the Complex. Science 255:306-312.

Eigenbrot, C, M Randal, L Presta, P Carter, and AA Kossiakoff. **1993**. X-ray structures of the antigen-binding domains from three variants of humanized anti-p185^{HER2} Antibody 4D5 and comparison with molecular modelling. J. Mol. Biol. 229:969-995.

Fuh, G, BC Cunningham, R Fukunaga, S Nagata, DV Goeddel, and JA Wells. **1992**. Rational design of potent antagonists to the human growth hormone receptor. Science 256:1677-1680.

Ihle, JN, BA Witthuhn, FW Quelle, K Yamamoto, WE Thierfelder, B Kreider, and O Silvennoinen. **1994**. Signaling by the cytokine receptor superfamily: JAKs and STATs. Trends Biochem. Sci. 19:222-227.

Kossiakoff, AA, W Somers, M Ultsch, K Andow, Y Muller, and AM De Vos. **1994**. Comparison of the intermediate complexes of human growth hormone bound to the human growth hormone and prolactin receptors. Protein Sci. 3:1697-1705.

Lowman, HB and JA Wells. **1993**. Affinity maturation of human growth hormone: Monovalent phage display. J. Mol. Biol. 234:564-578.

Lowman, HB, SH Bass, N Simpson, and JA Wells. **1991a**. Selecting high-affinity binding proteins by monovalent phage display. Biochemistry 30:10832-10838.

Lowman, HB, BC Cunningham, and JA Wells. **1991b**. Mutational analysis and protein engineering of receptor-binding determinants in human placental lactogen. J. Biol. Chem. 266:10982-10988.

Nicoll, CS, GL Mayer, and SM Russel. **1986**. Structural features of prolactins and growth hormones that can be related to their biological properties. Endocr. Rev. 7:169-203.

Somers, W, M Ultsch, AM De Vos, and AA Kossiakoff. **1994**. The X-ray structure of the growth hormone-prolactin receptor complex: Receptor binding specificity developed through conformational variability. Nature 372:1697-1705.

Ultsch, M, W Somers, AA Kossiakoff, and AM De Vos. **1994**. The crystal structure of affinity-matured human growth hormone at 2 Å resolution. J. Mol. Biol. 236:286-299.

Wells, JA and HB Lowman. **1992**. Rapid evolution of peptide and protein binding properties *in vitro*. Curr. Opin. Struct. Biol. 2:597-604.

HINGE-BENDING CLOSURE OF THE HYDROPHOBIC-BINDING SURFACE OF A TATA-BINDING PROTEIN MONOMER

K. Miaskiewicz and R. L. Ornstein

Environmental Molecular Sciences Laboratory, Pacific Northwest National Laboratory, Richland, Washington

Key Words: *TBP, TATA-box, protein dynamics hinge-bending, domain motions, protein-DNA interactions*

ABSTRACT

The TATA-box binding protein (TBP) is a universal transcription factor that initiates binding to TATA-box DNA sequence promoters. TBP homo- or hetero-dimer (TBP-DNA) complex crystal structures have been reported. The structure of the protein in these complexes is very similar and resembles a molecular saddle. Although TBP may prefer to form homodimers, the dimer must dissociate into monomers prior to binding DNA. We present the results of AMBER molecular dynamics simulations on a TBP monomer. Three similar trajectories were computed; each simulation shows large-scale hinge-bending that brings the two domains of TBP close to each other and results in closure of the DNA binding site.

INTRODUCTION

The TATA-box binding protein (TBP) is a universal transcription factor required by each of the three known eukaryotic RNA polymerases (Cormack and Struhl, 1992; Schultz et al., 1992; Meyers and Sharp, 1993). TBP has a highly conserved 180 amino acid carboxyl-terminal domain that binds DNA and interacts with a variety of transcription factors (reviewed in Nikolov and Burley, 1994) and with polymerase. Crystal structures of TBP-2 from *Arabidopsis thaliana* (Nikolov and Burley, 1994; Nikolov et al., 1992) and the C-terminal region of yeast TBP (Chasman et al., 1993) have been shown to form homodimers. The structure of TBP in the complex with DNA is very similar to its structure in the protein homodimer: the two halves of *apo*-protein in the dimer can be aligned on the corresponding parts in the complex with a root mean square positional deviation (RMSD) of under 0.6 Å (Kim and Burley, 1994). The main difference between the conformation of TBP in the homodimer and in the DNA complex is due to a small rotation in the linker region connecting the two intramolecular domains (Kim and Burley, 1994).

Solution studies are consistent with TBP dimer formation (Nikolov et al., 1992; Coleman et al., 1995; Kato et al., 1994). The TBP concentration within a rat liver cell nucleus has been reported to be in the range of 2–6 μM (Kato et al., 1994); thus TBP may rarely dissociate to monomers. Nevertheless, TBP binds to DNA as a monomer (Kim and Burley, 1994; Coleman et al., 1995; Kato et al., 1994; Perez-Howard et al., 1995; Kim et al., 1993a,b). How similar is the monomer structure to that of the individual TBP molecules in the dimer? Does the TBP monomer have a tendency to bind bent or straight DNA? Unfortunately, the TBP monomer structure has not been determined. In order to gain a fundamental understanding of the monomer structure and its role in DNA binding, we have performed molecular dynamics simulations on a (solvated) TBP monomer whose starting structure was taken from the crystal TBP-DNA complex.

MODEL AND METHODS

The starting structure of a monomer of *A. thaliana* TBP-2 was taken from the crystallographic structure of its complex with a DNA 14-mer (Kim and Burley, 1994; Kim et al., 1993). Positions of polar hydrogen atoms were optimized with NETWORK (Bass et al., 1992). Counterions (9 Na^+ and 20 Cl^-) were placed around accessible charged groups (but not around salt bridges) at a distance of 5 Å from the central heavy atom of any charged group (i.e., nitrogen of NH_3^+ or carbon of COO-), in order to neutralize the net system charge and to screen the individual charges. The system of protein and counterions was centered inside a 41-Å radius sphere of water. A semiharmonic potential with a 1-kcal/mol force constant was used to restrain the water molecules within the sphere boundary. The total number of atoms in the system was 27,482.

All molecular dynamics calculations were performed with the AMBER 4.0 program (Pearlman et al., 1991), using the all-atom force field of Kollman and coworkers (Weiner et al., 1986). Positions of waters and counterions were minimized, then subjected to 2 ps of molecular dynamics, and then additional minimization. Then the protein hydrogens and solvent were collectively allowed to minimize, followed by 2 ps of molecular dynamics, and then additional minimization. Then the whole system was allowed to minimize, followed by 4 ps of molecular dynamics with velocity reassignment. Then a 450-ps molecular dynamics production run was performed. The constant temperature algorithm was used with a coupling constant to the temperature bath of 0.15 ps for both solute and solvent. The SHAKE algorithm was used to restrain all the bonds, and a time step of 1.5 fs was applied. Rotational and translational motions were removed after every 30 steps.

To check the reproducibility of the results, the simulation was repeated two additional times: once with a slightly modified equilibration setup, and a second time with a 1-Å smaller radius for the water sphere (40 instead of 41 Å). In the former run, the equilibration process was extended for an additional 6 ps with velocity reassignment and with removal of translation and rotational motions every 10 steps (instead of every 15 steps). Since all three simulations resulted in very similar trajectory results, only the first simulation is discussed in detail.

RESULTS

The crystallographic structure of *A. thaliana* TBP, as it is observed when complexed to DNA, is shown in a ribbon representation in Figure 1. The protein is saddle-shaped with two very similar domains containing 88-89 amino acids each and related by approximately two-fold symmetry. Each domain is formed by two α-helices and a five-stranded antiparallel β-sheet. We refer to them as domain-1 (N-terminal half) and domain-2 (C-terminal half). The individual structural elements of domain-2 are "primed" while those of domain-1 are not. The loops connecting S2 and S3 and S2′ and S3′ are referred to as stirrups (Nikolov et al., 1992). The underside of the saddle, together with the two stirrups, is the site of DNA binding (Kim and Burley, 1994; Kim et al., 1993a,b).

The solvated monomer structure of TBP undergoes dramatic structural reorganization during the molecular dynamics simulations. The RMSD from the starting crystallographic structure for the whole protein is 7 Å, but it is much lower for the individual domains: ~ 3 Å (Figure 2A). The RMSD values appear to stabilize during the last 100 to 150 ps, indicating that the system has equilibrated. RMSD for the individual α-helices are less than 1.5 Å (Figure 2B). Helices H1 and H1 are extremely "quiet," with an RMSD from the starting structure around 0.3 Å. Some movement is observed for H2 around 150 ps and for H2 around 400 ps; in both cases the increased RMSD is due to reversible helix-bending (kinking).

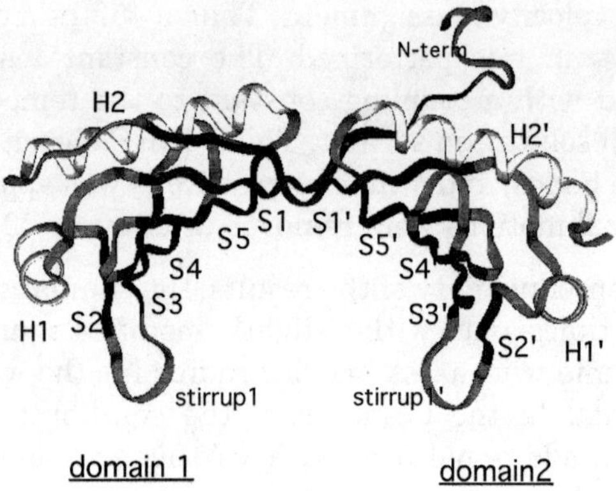

Figure 1. Secondary structure elements in TBP-2 from *Arabidopsis thaliana* . Strands are denoted with **S**, helices with an **H**. Elements belonging to domain-2 are primed.

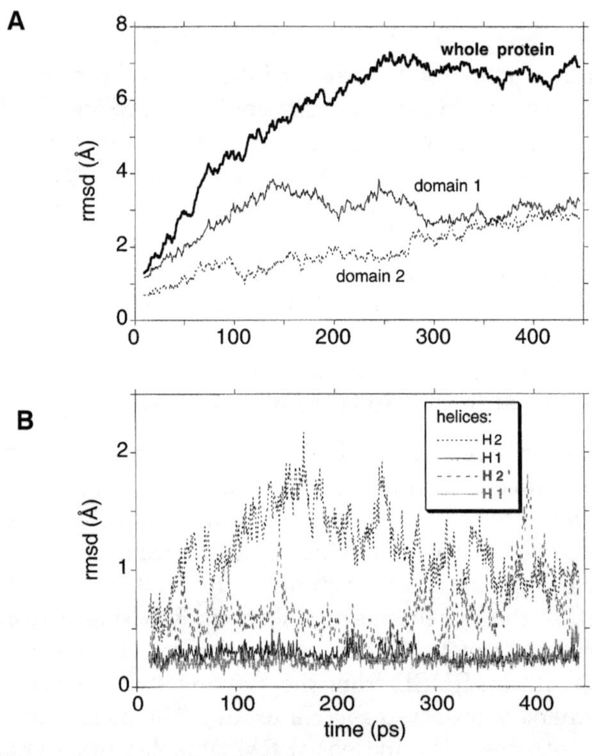

Figure 2. Root mean square positional deviation away from starting structure calculated with the protein backbone atoms (N, C, O, Cα).

The nature of the deviation from the crystal structure can be readily understood from the snapshots in Figure 3. A hinge-bending motion brings the two domains close to each other after about 200 ps, resulting in a closure of the DNA binding surface. The distance between the centers of mass of the two domains shortens by 7 Å (Figure 4), and the distance between some residues shortens by as much as 22 Å (Figure 5). TBP is bent by 55 during the simulation, compared to the crystal structure, as determined by the angle between centers of mass of the two domains and the center of mass of the hinge (Figure 6). The hinge-bending described has been reproduced in all three simulations on TBP. Figures 4-6 show that, while the time required for closing differs slightly from simulation to simulation, the final conformation of TBP is almost identical.

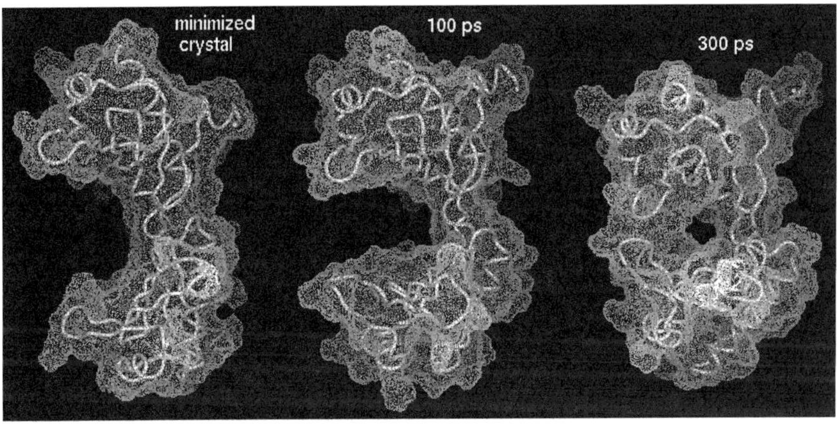

Figure 3. Three snapshots taken from simulation. Proteins are represented by ribbons with solvent accessible surfaces displayed.

Closure of the binding region can also be described quantitatively by considering solvent accessible surface (SAS) area (Figure 7). The SAS area of the whole protein increases in comparison to the crystal structure, as is often observed in simulations of proteins in a liquid water environment. However, the SAS area of the DNA binding region (underside of the molecular saddle) decreases substantially during the simulation, after some initial increase. As illustrated by the lower plot in Figure 6, the reduction of SAS is observed mainly for the stirrup regions. On the other hand, the most gains in SAS are seen around the linker region connecting the two domains of TBP.

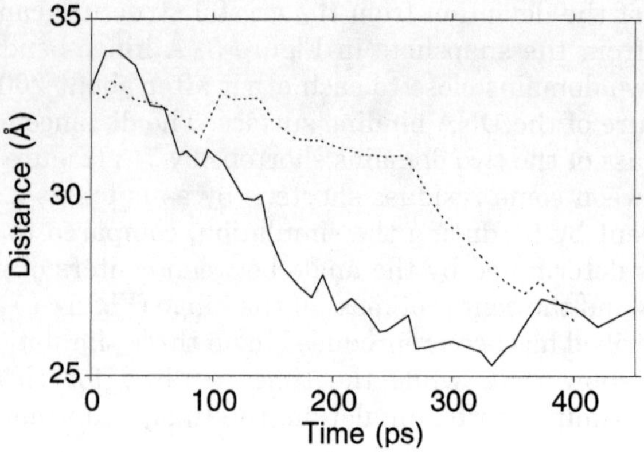

Figure 4. Separation distance between center of mass of the two domains taken as the center of mass of residues 34-103 and 124-194, respectively. Data shown for two simulations (solid and dashed lines).

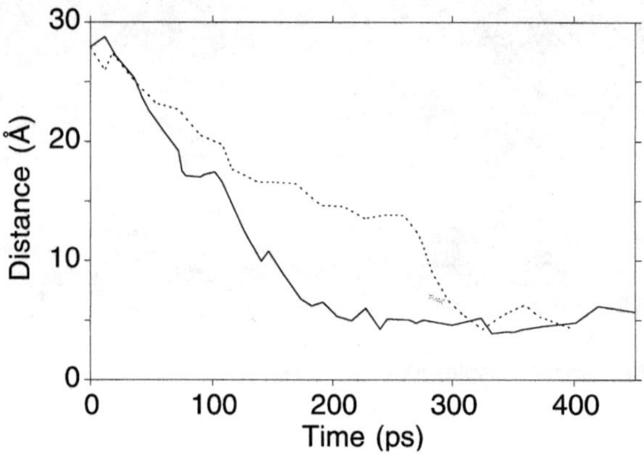

Figure 5. Separation distance between alpha carbons of residues ARG56 and LEU147, which are located in the middle of the two opposite stirrups.

Figure 8 depicts the extent of cross-correlation of the atomic motions during the first 200 ps of the simulation. Highly correlated motions are observed within each of the two separate domains, while significant interdomain correlation is not observed. The two domains are essentially independent.

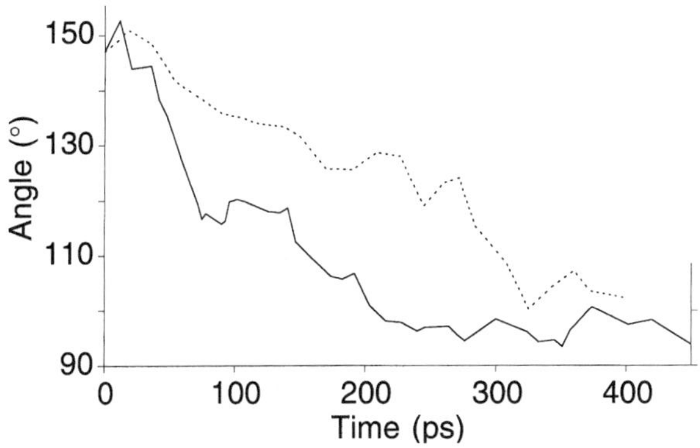

Figure 6. Time course variation of hinge-bending angle measured as the angle formed between the center of mass of domain 1 (residues 34-103), center of mass of hinge (residues 25-30 and 114-119), and center of mass of domain 2 (residues 124-194).

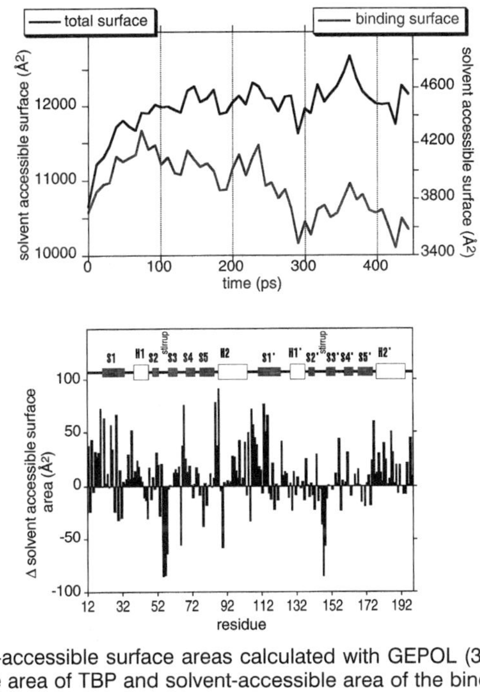

Figure 7. Solvent-accessible surface areas calculated with GEPOL (32). Upper plot: total solvent-accessible area of TBP and solvent-accessible area of the binding region, defined as including S5, S5′, S2, S2′, S3, S3′, S4, S4′ strands, stirrups, parts of S1 and S1′ strands an some of the connecting loops. Lower plot: difference between average solvent-accessible area in 250- to 450-ps time period and accessible area in starting structure, plotted for individual residues of TBP.

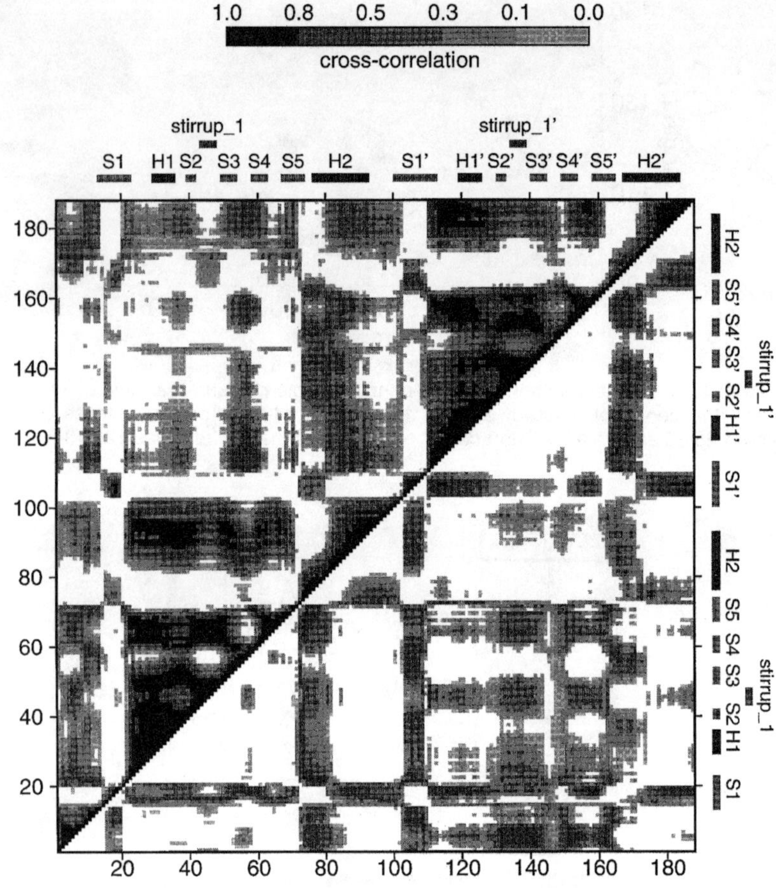

Figure 8. Cross-correlation map during first 200 ps. Cross-correlation takes values in -1 to 1 range (Ichiye and Karplus, 1991). Left upper corner of plot shows positive values (i.e., correlated motion); right lower corner displays negative values (i.e., anticorrelation).

DISCUSSION

The molecular saddle shape of TBP is characteristic of its bound form with DNA or in homodimers. A recent molecular dynamics simulation on the TBP-DNA complex indicates that TBP does not undergo large-scale dynamic motions in the complex (Miaskiewicz and Ornstein, 1996). The present solvated-TBP monomer simulations, however, indicate the involvement of large-scale hinge-bending motions that bring the two domains of TBP close to each other and results in closure of the underside of the molecular saddle, i.e., the DNA binding region. Hinge-bending of TBP, although previously hypothesized (Gerstein et al., 1994), has not previously been reported.

Large-scale interdomain motions are quite often seen in proteins built from multiple domains (reviewed in Gerstein et al., 1994). TBP hinge-bending, as seen in the present study, is similar to the motion described previously for lactoferrin (Gerstein et al., 1993), a transport protein that uses domain closure to recognize and sequester iron. TBP and lactoferrin are structurally similar. The linker (hinge) region is similar in both proteins, i.e., a two stranded β-sheet. For both proteins, the axis of rotation passes through the two β-strands linking the domains. In addition, the magnitude of bending is similar for these two proteins. Upon binding iron, the two domains in lactoferrin rotate by 53°, closing the binding site (Gerstein et al., 1993), while the rotation of the two domains observed in the present simulations on TBP is 55°. Similar large-scale domain closure has also been seen in maltodextrin-binding protein (Sharff et al., 1992) and in histidine-binding protein (Yao et al., 1994). Large-scale hinge-bending motions have been seen in recent simulations of several proteins and suggest that flexible regions between intramolecular domains may undergo considerable "relaxation" in the absence of crystal packing forces. For instance, we have observed large hinge-bending closure for an "open" structure of the M6I mutant of T4 lysozyme (Arnold and Ornstein, in press), while cytochrome P450BM3 (Paulsen and Ornstein, 1995) and haloalkane dehalogenase (Arnold and Ornstein, unpublished data) undergo large-scale opening of the active-site region.

Analysis of energy contributions (results not shown) indicates that the observed interdomain closing in TBP is achieved without a substantial change to the internal molecular mechanics energy of the protein. This observation is in line with the results of calculations performed for L-arabinose binding protein, showing that the hinge-bending for this protein involves only very modest change in the internal protein energy (Mao et al., 1982).

The flexible hinge and interdomain motions could play an important functional role in TBP. It has been suggested that the interdomain twist observed in the DNA complex facilitates TBP binding to the DNA minor groove (Kim and Burley, 1994). Recent measurements have shown that various DNA sequences, when bound to DNA, experience very different bending, which correlates with the overall binding affinity (Starr et al., 1995). DNA bending as in the DNA complex appears to be induced by the inherent tendency of TBP to undergo hinge-bending that brings the domains in close contact, as observed in the present study. Thus, the resulting conformation of the complex may be a compromise between two forces: hinge bending in TBP and the rigidity of the bound DNA.

Is the observed "closed" conformation in an equilibrium with the "open" form, or is it rather the stable form of monomeric TBP? On the picosecond time scale, only relatively small hinge-bending and twisting equilibrium motions can be expected (Gibrat and Go, 1990; Tirion and Ben-Avraham, 1993). However, large-scale hinge-bending or twisting motions that can be considered as an equilibrium of conformational states, e.g., open <-> closed, have been observed in proteins on a microsecond to a millisecond time scale (Thomas et al., 1979; Yoshimura et al., 1984; Li and Montelione, 1995; Careaga et al., 1995). This time scale is not yet accessible by theoretical simulations, thus the question of eventual equilibrium between open and closed forms of TBP cannot be addressed by our calculations. However, the rigidness of the closed conformation expressed by relatively small atomic motions and reproduced by all three performed simulations may suggest that the closed structure is the stable conformation of monomeric TBP.

It is possible that our calculations describe an inactive form of TBP, which is formed irreversibly either by very high dilution (S.K. Burley, personal communication) or by heat (S.K. Burley, personal communication; Nakajima et al., 1988). It has also been reported that even rough-handling" of the TBP (e.g., quick thawing, repeated freeze/thaw cycles etc.) can significantly alter its properties (Perez-Howard et al., 1995). Transformation to the globular closed form described in this work could explain the sensitivity of TBP to different physical factors.

ACKNOWLEDGMENTS

We thank Dr. Stephen K. Burley for providing the coordinates for crystallographic structures of TBP and for helpful discussions. This work was supported by the Laboratory-Directed Research and Development program at Pacific Northwest National Laboratory (RLO). Pacific Northwest National Laboratory is a multiprogram national laboratory operated for the U.S. Department of Energy by Battelle Memorial Institute under contract DE-AC06-76RLO 1830.

REFERENCES

Arnold, GE and RL Ornstein. Biopolymers (in press).

Bass, MB, DF Hopkins, WAN Jaquysh, and RL Ornstein. **1992**. Proteins Struct. Funct. Genet. 12:266-277.

Careaga, CL, J Sutherland, J Sabeti, and JJ Falke. **1995**. Biochemistry 34:3048-3055.

Chasman, DL, KM Flaherty, PA Sharp, and RD Kornberg. **1993**. Proc. Natl. Acad. Sci. USA 90:8174-8178.

Coleman, RA, AKP Taggart, LR Benjamin, and BF Pugh. **1995**. J. Biol. Chem. 270:13842-13849.

Cormack, BP and K Struhl. **1992**. Cell 69:685-696.

Gerstein, M, AM Lesk, and C Chothia. **1994**. Biochemistry 33:6739-6749.

Gerstein, M, BF Anderson, GE Norris, EN Baker, AM Lesk, and C Chothia. **1993**. J. Mol. Biol. 234:357-372.

Gibrat, J-F and N Gō. **1990**. Proteins Struct. Funct. Genet. 8:258-279.

Ichiye, T and Karplus, M. **1991**. Proteins Struct. Funct. Genet. 11:205-217.

Kato, K, Y Makino, T Kishimoto, J Yamauchi, S Kato, M Muramatsu, and T Tamura. **1994**. Nucleic Acids Res. 22:1179-1185.

Kim, Y, JH Geiger, S Hahn, and PB Sigler. **1993a**. Nature 365:512-520.

Kim, JL, DB Nikolov, and SK Burley. **1993b**. Nature 365:520-527.

Kim, JL and SK Burley. **1994**. Nature Struct. Biol. 1:638-653.

Li, Y-C and GT Montelione. **1995**. Biochemistry 34:2408-2423.

Mao, B, MR Pear, and JA McCammon. **1982**. J. Biol. Chem. 257:1131-1133.

Meyers, RE and PA Sharp. **1993**. Mol. Cell. Biol. 13:7953-7960.

Miaskiewicz, K and RL Ornstein. **1996**. J. Biomol. Struct. & Dyn. 13:593-600.

Nakajima, N, M Horikoshi, and RG Roeder. **1988**. Mol. Cell. Biol. 8:4028-4040.

Nikolov, DB, S-H Hu, J Lin, A Gasch, A Hoffmann, M Horikoshi, N-H Chua, RG Roeder, and SK Burley. **1992**. Nature 360:40-46.

Nikolov, DB and SK Burley. **1994**. Nat. Struct. Biol. 1:621-637.

Paulsen, MD and RL Ornstein. **1995**. Proteins Struct. Funct. Genet. 21:237-243.

Pearlman, DA, DA Case, JC Caldwell, GL Seibel, UC Singh, P Weiner, and PA Kollman. **1991**. AMBER 4.0, University of California, San Francisco, California.

Perez-Howard, GM, PA Weil, and JM Beechem. **1995**. Biochemistry 34, 8005-8017.

Schultz, MC, RH Reeder, and S Hahn. **1992**. Cell 69:697-702.

Sharff, AJ, LE Rodseth, JC Spurlino, and FA Quiocho. **1992**. Biochemistry 31:10657-10663.

Starr, DB, BC Hoopes, and DK Hawley. **1995**. J. Mol. Biol. 250:434-446.

Thomas, DD, JC Seidel, and J Gergely. **1979**. J. Mol. Biol. 132:257-273.

Tirion, MM and D Ben-Avraham. **1993**. J. Mol. Biol. 230:186-195.

Weiner, SJ, PA Kollman, DT Nguyen, and DA Case. **1986**. J. Comp. Chem. 7:230-252.

Yao, N, S Trakhanov, and FA Quiocho. **1994**. Biochemistry 33:4769-4779.

Yoshimura, H, T Nishio, K Mikashi, K Kinosita, and A Ikegami. **1984**. J. Mol. Biol. 179:453-467.

MOLECULAR DYNAMICS SIMULATIONS ON A ZINC-FINGER PROTEIN: DNA COMPLEX

M.G. Cooney,[1] K. Miaskiewicz,[1] M. Shibata[2] and J.H. Miller[1]

[1]Pacific Northwest National Laboratory, Richland, WA 99352

[2]Roswell Park Cancer Institute, Buffalo, NY 14263

Key Words: *Molecular Dynamics Simulation, Protein Hydration, Transcription Factor-DNA Binding, Zinc Finger Proteins*

ABSTRACT

We have performed 500 picoseconds of molecular dynamics (MD) simulation on an aqueous complex of the erythroid chicken transcription factor GATA-1 with an 8-base-pair segment of its cognate DNA, with nucleotide sequence 5'-AGATAAAC-3', using the AMBER 4.0 package. We began the simulation with the NMR-derived structure in which localized water molecules have been identified near hydrophobic sidechain groups. The terminal base pairs of the DNA were restrained to prevent strand separation. In the course of the simulation the N-terminal arm became more extended, while the β-turn region (amino acids 1-12) expanded and shifted away from the α-helix (amino acids 28-39) of the zinc finger, leaving the cysteines that coordinate the zinc ion more exposed to solvent by the 400th picosecond. The α-helix moved away from its initial close contact with the DNA major groove, leaving new solvent-accessible surfaces and creating a potential space for additional water. The carboxyl terminal arm remained in the DNA minor groove but migrated along the helix, probably reflecting a change in the shape of the minor groove. Most of the methyl groups identified in the literature as near localized water had at least one water molecule within 3.5 Å during most of the MD simulation. Solute-solvent hydrogen bonds persisted to varying degrees during the simulation, and only a few direct protein-DNA hydrogen bonds were found. Our results show the large range of possible conformations that MD may reveal in such complexes that can be difficult to discern from NMR data alone.

INTRODUCTION

The transcription factor GATA-1 is an erythroid-specific protein required for the regulation of erythroid-expressed genes. The GATA-1 factor binds as a monomer to a consensus target hexamer sequence (T/A)GATA(A/G) which occurs in the *cis*-regulatory elements of globin and other erythroid-specific genes. Two zinc-finger regions have been identified in the chicken GATA-1 protein; however, only the carboxyl-terminal one, spanning residues 164-192, is required for DNA binding. The solution structure of the minimal DNA-binding region, residues 158-223, complexed with 8 bp of its cognate DNA, has been previously determined by aqueous 2D NMR methods (Omichinski et al., 1993; Clore et al., 1994). As shown in Figure 1, this structure includes an N-terminal region with two approximately antiparallel β-sheets (residues 5-27), followed by an α-helix (residues 28-38) and a long C-terminal loop (residues 39-51).

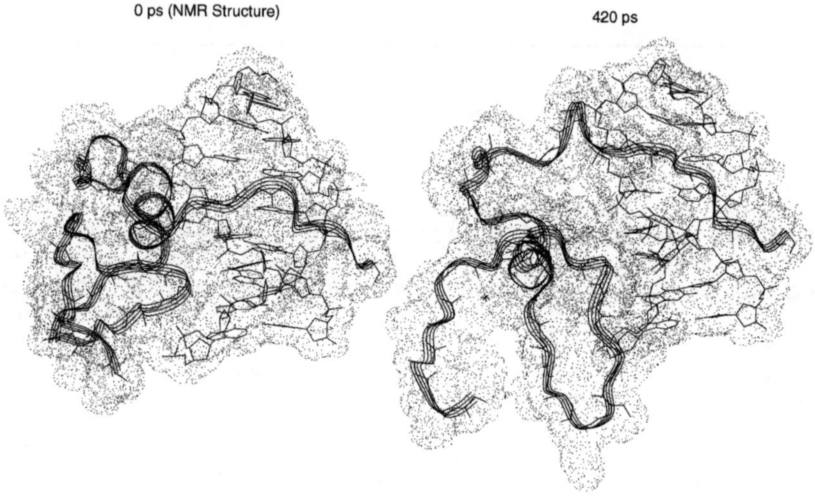

Figure 1. Stick drawings of GATA1:DNA complex with 8 base pairs of its cognate DNA, as deduced from published 2D NMR (NOESY and ROESY) studies by Clore et al. (1994). The DNA duplex sequence is 5'-AGATAAAC-3':5'-GTTTATCT-3'. A ribbon shows the backbone and, for clarity, only the backbone residues of the protein are shown in this figure. Gray dots depict calculated Connolly solvent-accessible surfaces; at 420 psec new cavities are seen in the upper and lower left of the drawing, indicating increased exposed surface area.

We have chosen the solvated NMR structure determined by Clore et al. (1994) as the starting point for 500 psec of molecular dynamics (MD) simulation, including water and counterions. In this aqueous structure, 13 hydrophobic side chains are in close proximity to bound water; the hydration of these groups is probably an enthalpy-driven process, as it should have a large negative entropy. One important goal of the MD simulation is to estimate the average numbers of water molecules associated, over a 500-psec time period, with the hydrophobic residues identified in the NOESY and ROESY studies.

COMPUTATIONAL METHODS

Computations were performed on SGI (Silicon Graphics) and Apollo (Hewlett-Packard) workstations, using the AMBER 4.0 molecular modeling software developed by Peter Kollman's group (University of California, San Francisco). Molecular models were visualized using the Insight-II package from Biosym (San Diego, CA). The aqueous conformation of the chicken GATA1 fragment reported by Clore et al. to span residues 158 to 223, complexed with 8 bp of cognate DNA with sequence [5'-d(AGATAAAC)-3'] : [5'-d(GTTTATCT)-3'], was placed in a water droplet, and Na^{+} and Cl^{-} counterions added. Additionally, a Zn^{++} ion was placed in its coordination site in the zinc-finger motif (residues 164-192 in the full-length protein). The protein sequence was KRAGTV**CSNCQTSTTTLWRRSPMGDPVCNAC** GLYYKLHQVNRPLTMRKDGIQTRNRKVSSKGKKRR, with the zinc finger indicated by bold type, and hydrated nonpolar groups (see below) underlined. The resulting structure was energy-minimized, and the optimized conformation subjected to 10 0.2-psec steps of heating molecular dynamics, followed by 1 psec of equilibration at 300K and 500 psec of production MD at 300K. During that time the terminal base pairs of the DNA were constrained ("belly dynamics") to prevent strand dissociation. Conformations were saved every 7 psec and analyzed with Insight-II modeling software.

RESULTS AND DISCUSSION

Figure 1 shows stick drawings of the GATA1:DNA complex at 0 psec (the NMR-deduced structure), and at 420 psec, a time point near the end of the simulation at which many portions of the GATA1 protein have become markedly disordered. The N-terminal loop, seen at the left of each representation and comprising amino acids 1-27, has greatly expanded, while the α-helix (residues 28-39) has pulled away from the DNA major groove; the N-terminal loop has migrated away from the α-helical part as well. By contrast, the C-terminal arm, residues 40-60, remains bound to the DNA minor groove but has migrated somewhat along the helix axis. In the 420-psec conformation, the two antiparallel β-sheets are evident as a distinct loop that has become more distant from the rest of the complex. This is the largest shift of a single motif to occur in our MD simulation.

The overall DNA helical conformation appears to be preserved, taking into account that the terminal base pairs were constrained to prevent major distortions during the MD simulation. Even after 420 psec of dynamics simulation, the α-helix has maintained its overall conformation, as is seen in both Figures 1 and 2, although it has dissociated considerably from the DNA. Figures 2A and 2B show details of the zinc-finger region of the GATA1 protein; the four cysteines, 7, 10, 28 and 31, have shifted position but remain close enough together to continue to coordinate the zinc ion and thus maintain the structural integrity of this DNA-binding region.

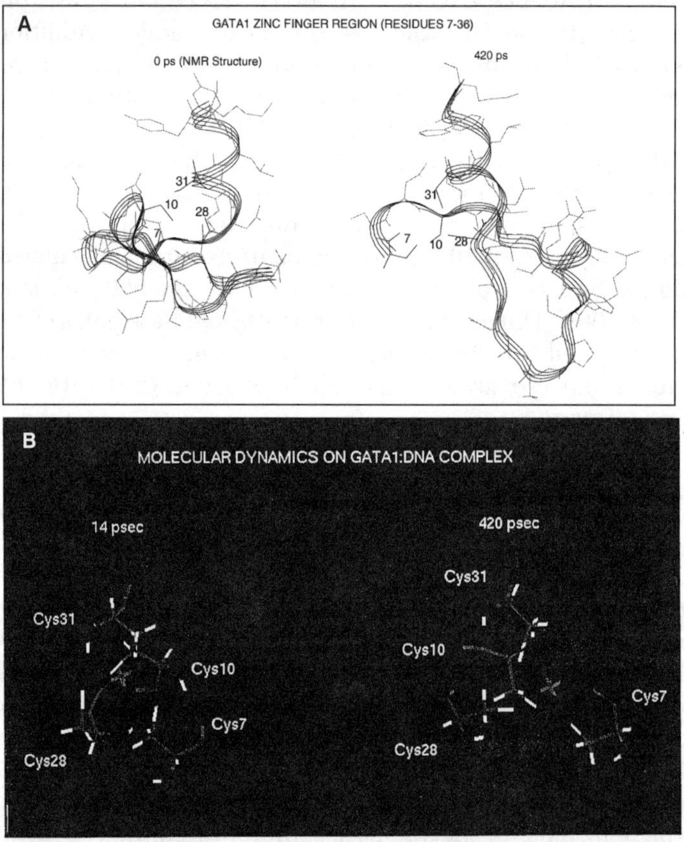

Figure 2. **A**, Detail of the zinc-finger region of the GATA1 protein in the complex at 0 and 420 psec. Side chains and backbone are shown without hydrogens; the four cysteines, 7, 10, 28 and 31, which form an approximately tetrahedral complex with the Zn^{++} ion, are shown in black; other residues are gray. As in Figure 1, a ribbon outlines protein backbone. **B**, Closeup view of the four cysteines at 14 and 420 psec. Though their positions have shifted considerably at the later time, the near-tetrahedral geometry persists, as does coordination of the zinc.

The hydration of GATA1 residues that are normally hydrophobic was observed in the NMR solution structure of the complex by Clore et al (1994). Even the most detailed NMR data in aqueous solution cannot give a precise picture of the differential degree of hydration of various residues. Therefore, we attempted to examine this hydration through MD simulations, which can reveal molecular motions much faster than those on the NMR time scale. To accomplish this we developed a FORTRAN program to calculate the number of solvent water molecules within a specified distance of any desired atom in a given MD structure or "snapshot." The extent of hydration of a methyl group is the average of the hydration of all three of its hydrogens. We chose the cutoff radius as 3.5 Å and plotted hydration vs. time for selected methyl groups, three of which are shown in Figures 3A and 3B. Nonpolar residues with hydrated side chains are Ala_3, Val_6, Leu_{17}, Met_{23}, Ala_{30}, Tyr_{34}, Tyr_{35}, Leu_{37}, Val_{40}, Leu_{44}, Met_{46}, Ile_{51}, and Val_{58}. The dynamic behavior of the two leucines (17 and 37) and two methyls, γ_1 and γ_2, of Val_{40} is particularly interesting: Leu_{37} becomes more hydrated, since it is near a cavity that opens up between the α-helix and the DNA major groove and allows extra water to enter. The hydration dynamics of Leu_{17} appear to mirror approximately those of Leu_{37}. Shifts in solvation patterns over time are evident from the near-reversal of hydration properties of these two Leu residues: Leu_{37} acquires about two more average waters within the 3.5-Å radius examined, while Leu_{17} loses at least one. At about 500 psec, however, the hydration patterns revert nearly to their original state, suggesting the reversibility of the dissociation of DNA from GATA1 protein. The Val_{40} γ_2 methyl shows similar changes to Leu_{37} but at later times (after 350 psec), again as a result of increased solvation in the cavity.

Figure 4 displays bar graphs showing the average number of hydrogen bonds between solute and solvent as a function of residue number for two successive intervals in the MD simulation. In contrast to Figure 3, which shows water molecules within an arbitrary radius (3.5 Å) of hydrophobic sidechain methyl groups, this figure indicates hydrophilic hydrogen bond interactions. The figure shows only H-bonds to the aqueous solvent, rather than direct H-bonds between protein and DNA (only two of which have been found in the NOESY studies). Note that the DNA (residues 62-95) is extensively solvated, as expected, and that part of the α-helix of GATA1 (residues 28-39) exhibits more solvation in the later interval, 205-420 psec, than in the earlier one, 135-205 psec.

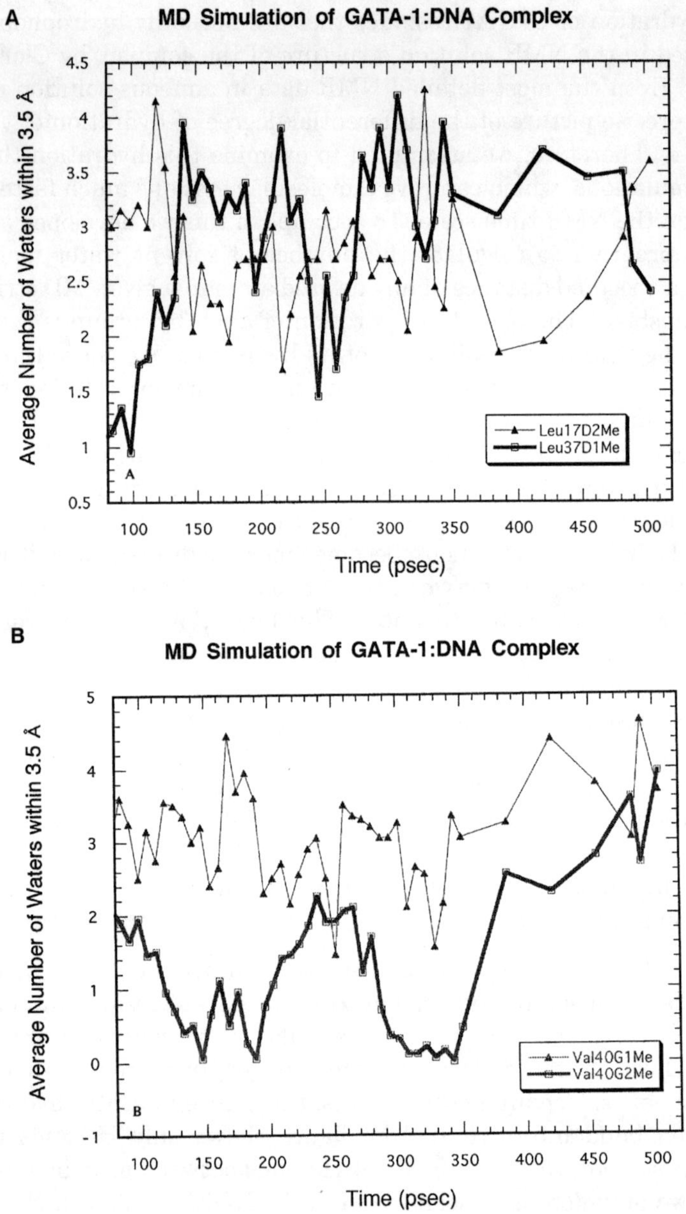

Figure 3. Graph of hydration of selected nonpolar side chains in GATA1, expressed as the number of water molecules within a 3.5-Å radius of each methyl group hydrogen averaged over all three hydrogens. **A**, Leucines 17 and 37; **B**, Valine 40, γ_1 and γ_2 methyls.

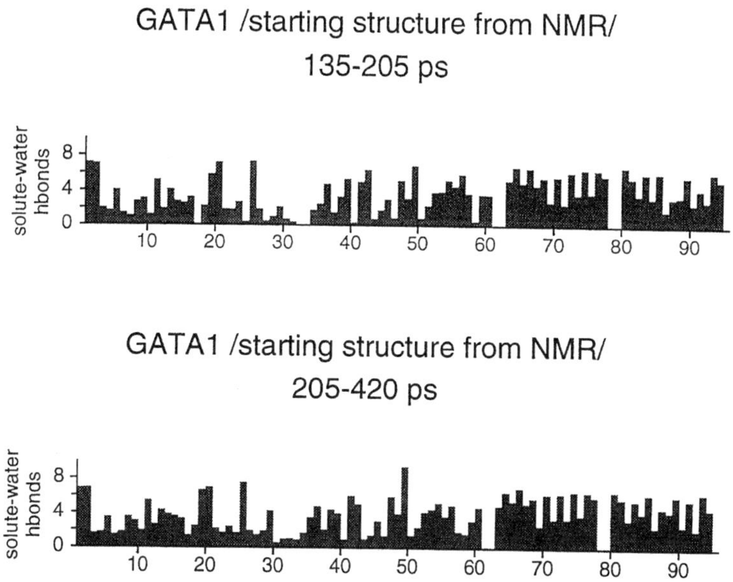

Figure 4. Bar graphs showing average number of hydrogen bonds of polar groups (N-H and O-H hydrogens) on DNA and GATA1 protein with solvent water, as a function of residue number. Stable interval from 135 to 420 psec is depicted, divided into two subintervals.

This result is consistent with the fact that partial dissociation has created a cavity at this location (see also Figure 1).

CONCLUSIONS

The most impressive feature of our MD simulation results is the unexpected flexibility and resulting variety of conformations in the protein part of the GATA1-DNA complex. At least three subsections of GATA1 exhibit novel motions not revealed on the NMR time scale: the β-sheet, region which becomes distorted and distant from the DNA; the α-helix, which pivots away from the DNA major groove; and the C-terminal arm, which slides "up" the DNA axis but remains in the minor groove. These changes persist from about 400 up to at least 500 psec; an even longer simulation (1 nsec or more) would be necessary to see whether the alternate conformations are stable. The second noteworthy characteristic of the GATA1-DNA complex is the presence of water molecules in close proximity to hydrophobic side chains. Despite considerable fluctuation in the average numbers of water molecules within 3.5 Å of these methyl and methylene groups, hydration of these residues generally persists throughout the MD simulation. One methyl, in Leu_{37}, actually becomes more hydrated after 500 psec (see Figure 3B). Since Leu_{37} is part of the α-helix, whose contact with the DNA major groove excludes water in the NMR structure, the increased hydration is a logical result of dissociation of the α-helix from the DNA. In contrast, Leu_{17} shows no net increase in hydration from 0 to 500 psec, although there are short-term transient fluctuations.

In summary, the essential conclusions from our present work are: (1) the GATA1:DNA complex partially dissociates during the MD simulation; (2) much of the α-helix-DNA contact is lost, and the β-loop becomes deformed, creating considerable additional space for solvent water. The opening between the α-helix and the DNA major groove creates a specific cavity for more waters, manifested as increased hydration of some of the side chains in the α-helix; (3) hydration of the methyl and methylene groups generally persists, though some of these, e.g., $Val_{40}\gamma_2$ methyl, have no waters within 3.5 Å for up to 50 psec; (4) longer simulations (at least 750-1000 psec) are necessary to determine whether the changes described are transient or stable (permanent).

ACKNOWLEDGMENTS

This research was supported by Associated Western Universities, Inc., Northwest Division (AWU NW) under Grant DE-FG06-89ER-75522 or DE-FG06-92RL-12451 with the U.S. Department of Energy, and by the Office of Health and Environmental Research (OHER) under contract DE-AC06-76RLO 1830.

REFERENCES

Clore, GM, A Bax, JG Omichinski, and A Gronenborn. **1994**. Localization of bound water in the solution structure of a complex of the erythroid transcription factor GATA-1 with DNA. Structure 2: 89-94.

Omichinski, JG, GM Clore, O Schaad, G Felsenfeld, C Trainor, E Appella, SJ Stahl, and A Gronenborn. **1993**. NMR structure of a specific DNA complex of a Zn-containing DNA binding domain of GATA-1. Science 261:438-446.

PROTEINS: STRUCTURE, FUNCTION AND BIOTECHNOLOGY

THE P450 GENE SUPERFAMILY: STRUCTURAL SIMILARITIES AND FUNCTIONAL DIFFERENCES

J. A. Peterson and S. E. Graham-Lorence

Department of Biochemistry, University of Texas Southwestern Medical Center at Dallas, Dallas, Texas

Key Words: *P450, protein-protein interaction, substrate specificity, mutagenesis*

ABSTRACT

The importance of P450s in bioorganic synthesis, chemical defense mechanisms, and chemical carcinogenesis is well recognized. With the determination of the amino acid sequences of greater than 400 unique P450s in 62 gene families, the relationship between amino acid sequence, protein structure, and biological function has been the focus of intense study. Sequence alignments indicate that there are only three amino acid residues that are identical in all members of this gene superfamily of proteins. The cysteinyl residue in the carboxy termial one-fourth of the molecule provides one of the ligands to the heme iron. The amino acid grouping EXXR, which is amino terminal to the cysteinyl residue, is in the K-helix and is charge-paired on the interior of the molecule. Other residues that are usually conserved are the I-helix threonine, the acid residue amino terminal to the threonine, and several positively charged residues that serve to neutralize the charges of the propionate side chains of the heme. Although there are relatively few conserved residues in the P450 gene superfamily, when the three-dimensional structures of P450cam, P450terp, and P450BM-P are overlaid in space, approximately 50% of the C backbone structure is similar, with an RMS deviation among the three structures of approximately 2 Å. The conserved tertiary structural elements include the four α-helical bundle into which the heme is intercalated, the J-helix, the heme-binding loop containing the conserved cysteine, and strands of β-sheet 1. The remainder of the tertiary structure is not well conserved in three-dimensional space. The implications which the conserved P450 structural fold has for the structure/function of other P450s, and their interaction with substrates and redox partners, will be discussed.

INTRODUCTION

With over 400 proteins in the P450 gene superfamily thus far identified and sequenced from plants, animals, bacteria, and yeast (Nelson et al., 1993), the importance of P450s in reactions other than xenobiotic/drug metabolism (e.g., drug and toxic materials) is beginning to be realized. For example, P450s have been shown to catalyze key reactions in steroidogenesis; in insecticide and herbicide resistance in insects and plants, respectively; in pigment synthesis in flower coloring; and in environmental bioremediation by microorganisms. Gene sequence studies have shown that the proteins which were originally identified as P450 all belong to a unique gene superfamily; however, there are several other proteins, not the least of which is nitric oxide synthase, which meet the original definition of a P450 but which do not belong to the gene superfamily (Nelson et al., 1993). With the solution of the crystal structure of four soluble, bacterial P450s—P450cam (Poulos et al., 1985, 1987), P450BM-P [the heme domain of the fusion protein P450BM-3 (Ravichandran et al., 1993)], P450terp (Hasemann et al., 1994), and P450eryF (Cupp-Vickery and Poulos, 1995), we believe that we can now define the common structure features of this family of proteins.

DISCUSSION

The P450 structural fold (Figure 1) appears to be present in all classes of P450s, i.e., for Class I P450s such as P450cam (CYP101), P450terp (CYP108), and P450scc (CYP11A; cholesterol side-chain cleavage), which require both an iron-sulfur protein (ferredoxin) and an FAD-containing NAD(P)H-ferredoxin reductase for catalysis; for Class II P450s such as mammalian xenobiotic P450s (e.g., CYP1, CYP2, and CYP3 families), steroidogenic, microsomal P450s (e.g., CYP17, 17a-hydroxylase/lyase, and CYP19, P450 aromatase), and the soluble bacterial fatty-acid monoxygenase P450BM-3, all of which require an FAD/FMN-containing NADPH-P450 reductase; and Class III P450s such as allene oxide synthase [CYP74; (Song et al., 1993)] and thromboxane synthase [CYP5; (Yokoyama et al., 1991)], which do not require reduction via a separate protein or protein domain. In the latter class, a representative structure has not yet been determined; however, the structural fold appears to be present, based on sequence alignments and structural features that will be discussed below.

P450cam was the first member of this gene superfamily whose structure was determined by x-ray crystallography (Poulos et al., 1985, 1987). This soluble, bacterial P450 stereoselectively catalyzes the hydroxylation of camphor to 5-exo-hydroxy-camphor (Katagiri et al, 1968; Hedegaard and Gunsalus, 1965). The determination of the atomic structure of P450cam with and without substrate (Poulos et al., 1986, 1987) and with substrate analogs and inhibitors (Raag and

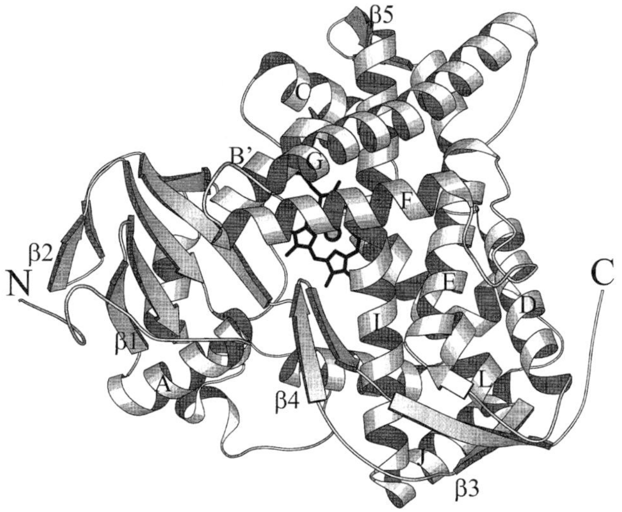

Figure 1. Ribbon drawing of P450BM-P, illustrating the β-sheet-rich domain on the left and the α-helical-rich domain on the right. Helices are indicated by coils, and sheets by arrows.

Poulos, 1991; Poulos and Howard, 1987) has been important in establishing the paradigm for structure/function relationships of P450s. Additionally, site-directed mutagenesis and biophysical techniques applied to P450cam have increased our knowledge of the mechanism of regio- and stereoselectivity of hydroxylation, and of redox-partner binding (Sligar et al., 1991; Paulsen et al., 1993; Loida and Sligar, 1993; Brewer and Peterson, 1988; Hintz and Peterson, 1981). Each of these structures has been informative, and the determination of the structures of other Class I P450s was important (Hasemann et al., 1994); but equally important was the structural determination of P450s from other classes (Ravichandran et al., 1993).

Although we still cannot account for all the variables related to P450 structure/function, we believe that dramatic strides have been made in understanding the overall P450 structure. This has occurred because of the determination of the structures of P450terp (Hasemann et al., 1994), another soluble, bacterial, Class I P450 that monooxygenates α-terpineol, and P450BM-P (Ravichandran et al., 1993), a soluble, bacterial, Class II P450 that monooxygenates long-chain fatty acids. With the addition of these structures, we are now able to differentiate between Class I and Class II P450s, (i.e., the differences in topology between the putative redox-partner binding sites). Also, we have a better understanding of substrate recognition and binding; and, finally, we can identify structural elements highly conserved among P450s, (i.e., the "core structure" of P450s), as well as those elements that are more variable. Because of this, we make certain generalized statements that provide us with a structural definition of P450s. This definition permits us to extend our knowledge to the structurally unknown P450s, thereby enabling us to make certain predictions about their membrane-associated regions, redox-partner binding, and substrate binding (Hasemann et al., 1995).

As seen in Figure 1, P450s are composed of two domains: a portion that is predominantly α-helical, accounting for approximately 70% of the protein, and one that is predominantly β-sheets, accounting for 22% of the protein (Poulos et al., 1987; Hasemann et al., 1994, 1995; Ravichandran et al., 1993). The α-helical domain contains helices B′ through K, helix L, and sheets $\beta 3$, $\beta 4$, and $\beta 5$. The β-sheet domain contains sheets $\beta 1$ and $\beta 2$, and helices A, B, and K′ (Hasemann et al., 1995). The position of the secondary structural elements of P450BM-P and P450cam are indicated in the sequence alignment given in Figure 2.

In comparing sequence alignments of P450s (Hasemann et al., 1995), we have found that only three residues are **absolutely** conserved among this superfamily of proteins: the cysteine in the heme-binding region, which serves as the fifth-coordinating ligand to the heme iron, and a *Glu* and an *Arg* in the K helix that are one helical turn apart, with the *Glu* and *Arg* forming a salt bridge facing into a region termed the "meander" (Ravichandran et al., 1993). Along with the cysteine in the heme-binding region is the highly conserved consensus sequence F(G/S)XGX(H/R)X<u>C</u>XGX(I/L/F)A. Also, there are several residues that are highly but not absolutely conserved, the most notable of which is a *Thr* in the I helix which points into the active site and, just N-terminal to that, a conserved acidic residue (*Glu* or *Asp*), both of which, it has been suggested, play roles in oxygen activation (molecular oxygen bond-splitting) (Raag et al., 1991; Raag and Poulos, 1989). These residues are found in highly conserved sequence (A/G)-G-X-(E/D)-T that is located near the middle of the I helix, directly over pyrrole ring B of the heme; they appear to form a slight bend in the I helix.

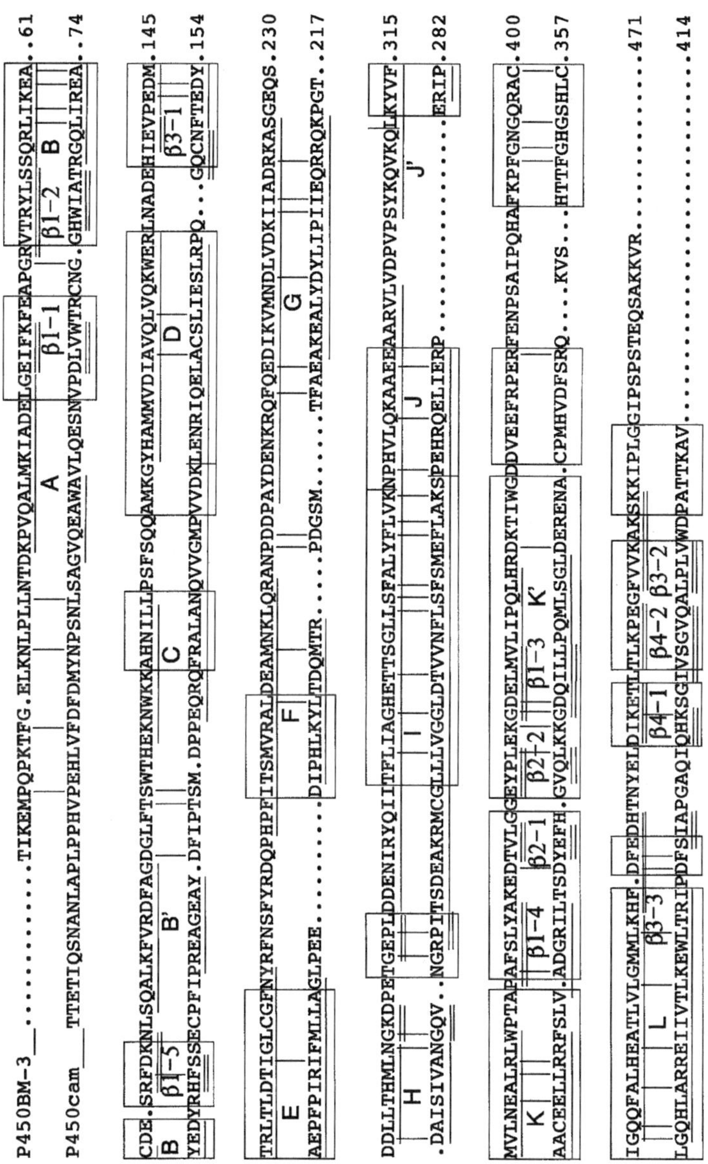

Figure 2. Structural alignment of P450cam and P450BM-P. The α-helices are indicated by the letter and single underline. Strands of β-sheets are indicated by double underline and numerical designation of strand. Structurally conserved elements are boxed, while functionally conserved amino acids are indicated by vertical lines.

The regions of greatest structural difference are those thought to be involved in substrate recognition and binding, and those involved in redox-partner binding. That is, while the overall helices and sheets are generally conserved in these variable regions, the lengths and positions of the α-helices, β-sheets, and loops (or coils) differ among P450s to accommodate the various substrates and different redox partners. For example, the best substrates for P450BM-P are long-chain fatty acids containing more than 12 carbons, while that for P450cam is camphor, a small, bicyclic monoterpene. Thus, it is not surprising that those regions that are at the mouth of the access channel and appear to be involved in substrate recognition and binding, such as helices F and G and the F-G loop, may vary in length and position to adequately accommodate the substrate of choice.

The process of substrate binding in P450s may be more complex than in proteins whose substrates are hydrophilic and whose active sites are solvent-exposed. That is, substrate may be initially recognized and bound on the surface of the molecule; substrate access to the active site will then be down a hydrophobic channel, and substrate orientation in the active site or heme pocket will be accomplished by hydrophobic interaction with amino acid residues (Ravichandran et al., 1993). The structural elements and the residues that are important in substrate recognition and binding can be expected to vary from one P450 to another P450, depending on the particular substrate. As mentioned above, these differences are best demonstrated when comparing P450cam to P450BM-3 and their corresponding substrates, i.e., camphor (a small, bicyclic monoterpene for P450cam), and long-chain fatty acids or eicosanoids for P450PM-3. Thus, to accommodate the large substrates in P450BM-P, the access channel to the active site/heme pocket is cavernous (Ravichandran et al., 1993), whereas, in P450cam, the access channel is almost nonexistent (Poulos et al., 1987).

There are also contrasts in hydroxylation selectivity among P450terp, P450cam, and P450BM-P. P450cam and P450terp are regio- and stereoselective in camphor and α-terpineol hydroxylation, forming 5-*exo*-hydroxy-camphor (Katagiri et al., 1968) and 7-hydroxy-terpineol (Fruetel et al., 1994), respectively. In contrast, P450BM-3 monooxygenates many different compounds, and at several positions, e.g., hydroxylation at ω-1, ω-2 or ω-3 of palmitate (Miura and Fulco, 1975; Boddupalli et al., 1990, 1992) and 18(R)-hydroxylation or 14(S),15(R)-epoxidation of arachidonic acid (manuscript in preparation).

With these differences in specificity in mind, one should not be surprised to find that there are differences in the topology of the structural elements involved in substrate recognition, binding, and orientation. Shown in Figure 3 is a cross-section of the substrate binding pocket in the vicinity of the heme iron on P450cam and P450BM-P. P450cam is seen to be more open directly over the heme iron than is P450BM-P. This openness is a reflection of the structure of the portion of each of the respective substrates that must fit into the active site. In the case of P450cam, the whole camphor molecule must be accommodated in the active site. In the case of P450BM-P, only the methyl end of the fatty acid must slip past the F87 residing over the heme ring so that the appropriate carbon atom is positioned correctly with respect to the activated oxygen. Dynamic motion in the active site of P450BM-P must occur, as indicated by both molecular dynamics and the product distribution from arachidonic acid metabolism. The 14,15-olefin of arachidonic acid can be oxidized to the epoxide only if F87 is rotated away from the heme ring.

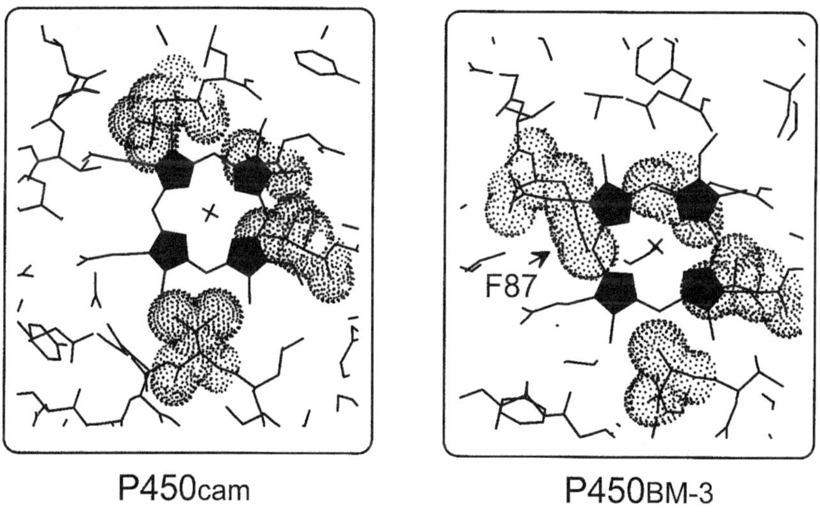

Figure 3. Van der Waals surfaces in vicinity of heme of P450cam and P450BM-P. 8-Å slab of residues on distal side of heme, showing shape of "substrate binding site."

In summary, understanding the structural features of P450s as determined by x-ray crystallography is important in rational redesign of these enzymes for biotechnological applications. However, understanding the atomic structure of the protein crystal is only the starting point for explaining the dynamics of substrate binding and protein/protein interaction in these extremely important enzyme systems. We must understand the dynamic motions of these proteins before we can rationally redesign them.

ACKNOWLEDGMENT

This research was supported in part by research grants from the National Institutes of Health GM43479 and GM50858 to JAP.

REFERENCES

Boddupalli, SS, RW Estabrook, and JA Peterson. **1990**. Fatty acid monooxygenation by cytochrome P-450BM-3. J. Biol. Chem. 265:4233-4239.

Boddupalli, SS, BC Pramanik, CA Slaughter, RW Estabrook, and JA Peterson. **1992**. Fatty acid monooxygenation by P450BM-3: Product identification and proposed mechanisms for the sequential hydroxylation reactions. Arch. Biochem. Biophys. 292:20-28.

Brewer, CB, and JA Peterson. **1988**. Single turnover kinetics of the reaction between oxycytochrome P-450cam and reduced putidaredoxin. J. Biol. Chem. 263:791-798.

Cupp-Vickery, JR, and TL Poulos. **1995**. Structure of cytochrome P450eryF involved in erythromycin biosynthesis. Nature Struct. Biol. 2:144-153.

Fruetel, JA, RL Mackman, JA Peterson, and PR Ortiz de Montellano. **1994**. Relationship of active site topology to substrate specificity for cytochrome P450terp (CYP108). J. Biol. Chem. 269:28815-28821.

Hasemann, CA, KG Ravichandran, JA Peterson, and J. Deisenhofer. **1994**. Crystal structure and refinement of cytochrome P450terp at 2.3 Å resolution. J. Mol. Biol. 236:1169-1185.

Hasemann, CA, RG Kurumbail, SS Boddupalli, JA Peterson, and J Deisenhofer. **1995**. Structure and function of cytochromes P450: a comparative analysis of three crystal structures. Structure 3:41-62.

Hedegaard, J, and IC Gunsalus. **1965**. Mixed function oxidation. IV. An induced methylene hydroxylase in camphor oxidation. J. Biol. Chem. 240:4038-4043.

Hintz, MJ, and JA Peterson. **1981**. The kinetics of reduction of cytochrome P-450cam by reduced putidaredoxin. J. Biol. Chem. 256:6721-6728.

Katagiri, M, BN Ganguli, and IC Gunsalus. **1968**. A soluble cytochrome P-450 functional in methylene hydroxylation. J. Biol. Chem. 243:3543-3546.

Loida, PJ, and SG Sligar. **1993**. Engineering cytochrome P-450cam to increase the stereospecificity and coupling of aliphatic hydroxylation. Protein Eng. 6:207-212.

Miura, Y, and AF Fulco. **1975**. Omega-1, Omega-2, and Omega-3 hydroxylation of long-chain fatty acids, amides and alcohols by a soluble enzyme system of *Bacillus megaterium*. Biochim. Biophys. Acta 388:305-317.

Nelson, DR, T Kalmataki, DJ Waxman, FP Guengerich, RW Estabrook, R Feyereisen, FJ Gonzalez, MJ Coon, IC Gunsalus, O Gotoh, K Okuda, and DW Nebert. **1993**. The P450 superfamily: update on new sequences, gene mapping, accession numbers, early trivial names of enzymes, and nomenclature. DNA Cell Biol. 12:1-51.

Paulsen, MD, D Filipovic, SG Sligar, and RL Ornstein. **1993**. Controlling the regiospecificity and coupling of cytochrome P450cam: T 185F mutant increases coupling and abolishes 3-hydroxynorcamphor product. Protein Sci. 2:357-365.

Poulos, TL, BC Finzel, IC Gunsalus, GC Wagner, and J Kraut. **1985**. The 2.6-Å crystal structure of *Pseudomonas putida* cytochrome P-450. J. Biol. Chem. 260:16122-16130.

Poulos, TL, BC Finzel, and AJ Howard. **1986**. Crystal structure of substrate-free *Pseudomonas putida* cytochrome P-450. Biochemistry 25:5314-5322.

Poulos, TL, BC Finzel, and AJ Howard. **1987**. High-resolution crystal structure of cytochrome P450cam. J. Mol. Biol. 195:687-700.

Poulos, TL, and AJ Howard. **1987**. Crystal structures of metyrapone- and phenylimidazole-inhibited complexes of cytochrome P-450cam. Biochemistry 26:165-8174.

Raag, R, SA Martinis, SG Sligar, and TL Poulos. **1991**. Crystal structure of the cytochrome P-450CAM active site mutant Thr252Ala. Biochemistry 30:11420-11429.

Raag, R, and TL Poulos. **1989**. Crystal structure of the carbon monoxide-substrate-cytochrome P-450CAM ternary complex. Biochemistry 28:7586-7592.

Raag, R, and TL Poulos. **1991**. Crystal structures of cytochrome P-450CAM complex with camphane, thiocamphor, and adamantane: Factors controlling P-450 substrate hydroxylation. Biochemistry 30:2674-2684.

Ravichandran, KG, SS Boddupalli, CA Hasemann, JA Peterson, and J Deisenhofer. **1993**. Crystal structure of hemoprotein domain of P450BM-3, a prototype of microsomal P450's. Science 261:731-736.

Sligar, SG, D Filipovic, and PS Stayton. **1991**. Mutagenesis of cytochromes P450cam and b5. Meth. Enzymol. 206:31-49.

Song, WC, CD Funk, and AR Brash. **1993**. Molecular cloning of an allene oxide synthase: a cytochrome P450 specialized for the metabolism of fatty acid hydroperoxides. Proc. Natl. Acad. Sci. USA 90:8519-8523.

Yokoyama, C, A Miyata, H Ihara, V Ullrich, and T Tanabe. **1991**. Molecular cloning of human platelet thromboxane A synthase. Biochem. Biophys. Res. Commun. 178:1479-1484.

THE ROLE OF AROMATIC CONTACTS IN BUILDING STREPTAVIDIN'S LARGE ACTIVATION BARRIER TO BIOTIN DISSOCIATION

P.S. Stayton, L.A. Klumb, and A. Chilkoti

Center for Bioengineering, University of Washington,
Seattle, Washington

Key Words: *Structure-based drug design, activation barrier, high affinity, site directed mutagenesis*

ABSTRACT

The molecular origins of the high-affinity streptavidin-biotin interactions have been investigated by elucidating the contributions of tryptophan side chains to the large dissociation activation barrier. We have gained insight into the underlying energetic contributions of these residues and to the related structure of the transition state through site-directed mutagenesis and biophysical studies. Equilibrium and kinetic analysis has yielded the thermodynamic and activation parameters for wild-type streptavidin and for each of the tryptophan mutants. The alterations in the activation enthalpies and entropies demonstrate a unique role for each of the three tryptophans at positions 79, 108, and 120. *Trp* 79 plays a key role in controlling the entropic costs of biotin immobilization in the bound state; *Trp* 120 is necessary for the large equilibrium enthalpy associated with biotin binding; and *Trp* 108 is required to manage the entropy of the transition state.

INTRODUCTION

Insight into the molecular mechanisms controlling high-affinity interactions and slow protein-ligand dissociation kinetics is of general significance to biology and to the field of structure-based drug design, where these features generally represent the target goals for therapeutic development. The dissociation process is of particular importance, as protein-small molecule on-rates typically fall within two orders of magnitude of the diffusion limit, whereas the corresponding off-rates vary over a large range that can be conservatively estimated at 10^{12}. Relatedly, from a drug development standpoint, it is much more straightforward to design interactions from the bound state than to design features that would increase second-order on-rates significantly. It is interesting to note, though, how little is known about the "simple" protein-ligand reaction coordinate. In many respects, we have a much clearer understanding of the protein folding transition state and reaction coordinate energetics. Similarly, site-directed mutagenesis approaches have provided considerable insight into the mechanisms by which enzymes control the catalytic activation barriers associated with substrate turnover. There has been remarkably little dissection of the activation barriers governing dissociation kinetics, however. From the perspective of slow-dissociation kinetics, the problem for the protein is lowering the bound-state energy level and raising the transition-state barrier, the inverse of our classical picture of enzyme-mediated substrate catalysis. In addition, by analogy to this description of the enzyme reaction coordinate, the protein must achieve this differential recognition with a chemically identical ligand, rather than with the structurally altered transition-state substrate. As a starting point in efforts to better elucidate the dissociation reaction coordinate, we have utilized the streptavidin-biotin model system to elucidate the molecular origins of the high-affinity and slow-dissociation rate. We summarize here studies of the streptavidin aromatic contacts and their contribution to building the large-dissociation activation barrier.

THE STREPTAVIDIN-BIOTIN DISSOCIATION COORDINATE

The differential recognition of substrate vs. transition-state structures by enzymes is now well-established as a catalytic strategy. Concerted site-directed mutagenesis and enzymatic studies have shown that specific active-site side chains preferentially bind the transition state in relation to the starting substrate, and thus lower the catalytic activation barrier by stabilizing the transition state. Detailed mapping of this nature at active sites can reconstruct transition-state structures, providing molecular snapshots along the catalytic reaction coordinate. While this strategy has proven invaluable in the field of enzymology, it is interesting to note that the molecular nature of the transition state for protein-ligand dissociation remains remarkably poorly characterized in terms of structure and activation barrier relationships. As a first approach to characterizing the protein-ligand dissociation coordinate, we have thus applied a model that is analogous to the enzymatic model. Our reference state is the unbound state, for which the system free energy is assumed to be unaltered by site-directed mutagenesis (Figure 1). This assumption then allows us to assign variations in the experimentally determined thermodynamic free energy and enthalpy to the ground-state (ligand-bound) system.

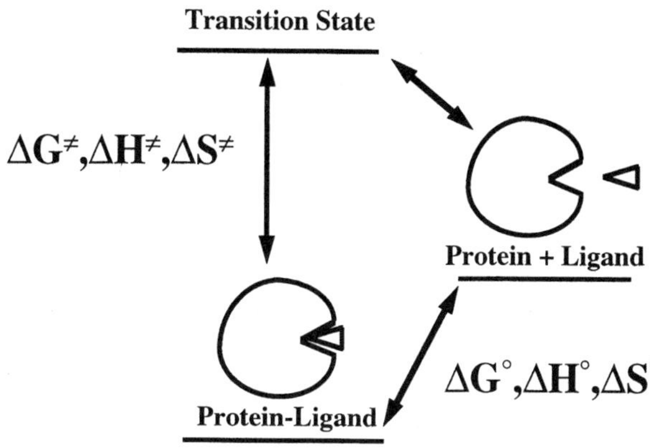

Figure 1. Schematic protein-ligand reaction coordinate. Measurement of the equilibrium free energy and enthalpy determines the level of the bound states for the mutant and wild-type proteins, under the assumption that the unbound state is unaltered by mutagenesis. Transition-state levels for the mutants relative to wild-type can then be determined by measuring temperature dependence of dissociation rate. The relative role of the side-chain contacts in the ground state versus the transition state can thus be determined by comparing how these energetics are changed by side-chain alterations.

In common with other high affinity systems, the exceptionally tight binding observed in the streptavidin/avidin interaction with biotin ($K_a = 10^{14} - 10^{15}\ M^{-1}$) is controlled by the slow dissociation kinetics of the protein-ligand complex. The x-ray crystal structure of the streptavidin tetramer complexed with biotin reveals an extensive set of hydrogen-bonding and van der Waals interactions between protein and ligand, and comparison with the biotin-free structure suggests that biotin association is also accompanied by the ordering of two surface loops (Weber et al., 1989; Hendrickson et al., 1989). These structural motifs have been observed in numerous protein-ligand systems, and the involvement of aromatic side chains is particularly widespread in high-affinity complexes. The elucidation of the structure-function relationships underlying the biotin dissociation kinetics are thus of broad interest as a paradigm for tight-binding protein-ligand interactions. Computational (Miyamoto et al., 1993) and site-directed mutagenesis studies (Chilkoti et al., 1995) have suggested that the large free energy decrease associated with biotin binding is largely attributable to hydrophobic and van der Waals interactions between biotin and four *Trp* residues in the binding site of streptavidin. Three *Trp* residues line the biotin-binding pocket, *Trp* 79, *Trp* 92, and *Trp* 108, while *Trp* 120 is contributed by an adjacent dyad-related subunit. We have initially probed the structural and energetic contributions of the tryptophans at positions 79, 108, and 120 with conservative alterations of tryptophan side chains to phenylalanine (Figure 2).

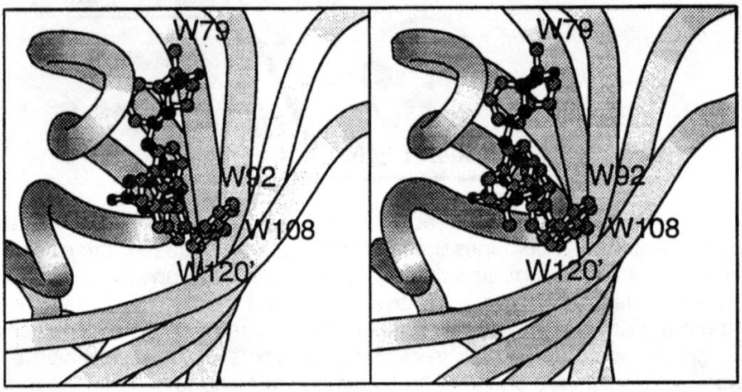

Figure 2. Tryptophan contacts within the biotin-binding pocket. Three tryptophans, at positions 79, 92, and 108, directly interact with biotin in the complex. The fourth tryptophan, *Trp* 120, is contributed by a dyad-related subunit.

Equilibrium characterization of the tryptophan mutants has yielded estimates of $\Delta\Delta G°$ relative to wild-type streptavidin, and direct quantitation of the $\Delta H°$ for biotin association (Chilkoti and Stayton, 1995). These results are summarized in Table 1. The W108F and W79F mutations result in small alterations of $\Delta\Delta G°$, with a relative ordering of W108F ($\Delta\Delta G°$ = 0.5 kcal/mol) < W79F ($\Delta\Delta G°$ = 0.8 kcal/mol). The W120F mutation results in a larger perturbation of $\Delta\Delta G°$, which is largely a consequence of enthalpic reductions, as this mutation also exhibits a loss of binding enthalpy of + 5.1 kcal/mol. The W108F mutant displays a small decrease in binding enthalpy, and the W79F actually displays an increase in binding enthalpy. This was unexpected, as biotin-binding is enthalpically driven and we had assumed streptavidin would thus be enthalpically optimized, given its high binding efficiency. An explanation for this effect is given in a later section. These equilibrium measurements allow us to set the $\Delta G°$, $\Delta H°$, and $\Delta S°$ levels of the bound-state system for each of these mutants, which is a necessary step in defining the dissociation coordinate.

Table 1. Summary of Equilibrium and Kinetic Characterization of *Trp* Mutants.

Protein	$\Delta\Delta G°$ (kcal/ml)	$\Delta H°$ (kcal/mol)	$k_{off}(S^{-1})$
WT	0	–24.5	5.4×10^{-6}
W79F	0.8	–26.0	3.6×10^{-5}
W120F	2.7	–19.4	3.7×10^{-4}
W108F	0.5	–23.5	9.6×10^{-5}

In order to define the activation barrier to dissociation, the off-rate of the biotin-streptavidin complex was determined at temperatures ranging from 10°C to 35°C (Chilkoti and Stayton, 1995). The activation thermodynamic parameters, $\Delta G^{\neq}$, $\Delta H^{\neq}$, and $\Delta S^{\neq}$, were determined by transition-state analysis of the dependence of the dissociation rate constant upon temperature, which is described by the Eyring equation:

$$k_{off} = \frac{k_B T}{h} \exp(-\Delta G^{\neq}/RT), \qquad (1)$$

where k_B, h, and R are the Boltzmann, Planck, and universal gas constants, $\Delta G^{\neq}$ is the activation energy of dissociation of the protein-ligand complex, and T is the absolute temperature. The transmission coefficient k (not shown) on the righthand side of equation (1) has been assumed to be unity. The activation enthalpy and entropy can be determined by partitioning $\Delta G^{\neq} = \Delta H^{\neq} - T\Delta S^{\neq}$ into a linearized form of equation (1):

$$\ln(k_{off}/T) = -\Delta H^{\neq}/R\,(1/T) + \text{Constant}. \qquad (2)$$

A linear least-squares fit of a plot of $\ln(k_{off}/T)$ versus 1/T provided the activation enthalpy, and the entropy of activation could then be determined by summation. The activation parameters derived from these analyses are compiled in Table 2.

It can be seen that the activation barrier for native streptavidin is built through a large activation enthalpy ($\Delta H^{\neq}$ = 29 kcal/mol), and that progress to the transition state is actually accompanied by a favorable activation entropy ($T\Delta S^{\neq}$ = 3.9 kcal/mol at 25°C). Our initial interpretation of these results is a model where the $\Delta H^{\neq}$ is generated by a loss of binding contacts and an endothermic conformational change that is necessary to allow the dissociation of biotin. The positive activation entropy would then be generated by an increase in conformational entropy of both biotin and the side-chain residues that have lost contact with the ligand. An additional potential source of positive activation entropy is the disordering of the flexible loop at residues 45 to 50, which undergoes an order to disorder transition upon the release of biotin. In order to test this model, we performed the activation parameter analysis of the tryptophan mutants. The W108F mutant exhibited large alterations in activation enthalpy and entropy, despite being the mutant which displays the smallest perturbation in equilibrium binding affinity. The W120F mutant exhibited a large change in activation enthalpy, and an activation entropy that is nearly within the experimental error of the native streptavidin value. The W79F also displayed a large alteration of activation enthalpy and entropy.

Table 2. Eyring Activation Analysis of *Trp* Mutants at 25°C

Protein	ΔΔG≠ (kcal/ml)	ΔH ≠ (kcal/mol)	TΔS ≠ (kcal/mol)
WT	24.4	43.0	7.6
W79F	23.3	34.9	11.6
W120F	21.9	28.5	6.6
W108F	22.7	36.5	13.8

With the combined equilibrium and activation analysis, a mapping of the ground- and transition-state energetics can be conducted to clarify the role of these tryptophan side chains. Free energy maps of the dissociation coordinate are shown in Figure 3. The most striking feature is the clear definition of *Trp* 108's role in the transition state. Altering this side chain to *Phe* has little effect on the ground-state energetics but induces large effects on the energetics of the transition state. Similarly, the large effects of the W120F alteration on affinity and off-rate are clearly dominated by an equilibrium alteration of binding enthalpy. The W79F mutation results in a free energy perturbation of the ground state, and transition states, with a partitioning of enthalpic alterations between the ground and transition states. We have interpreted these results as suggesting that *Trp* 108 and *Trp* 79 maintain an important contact with biotin in the transition state, whereas *Trp* 120 has largely lost its interaction with biotin. This mechanistic model (Figure 4) is consistent with the need for the protein to "open" a dissociation pathway, and is also consistent with the interpretation of the large positive activation enthalpy as arising from a concerted loss of binding contacts and conformational change.

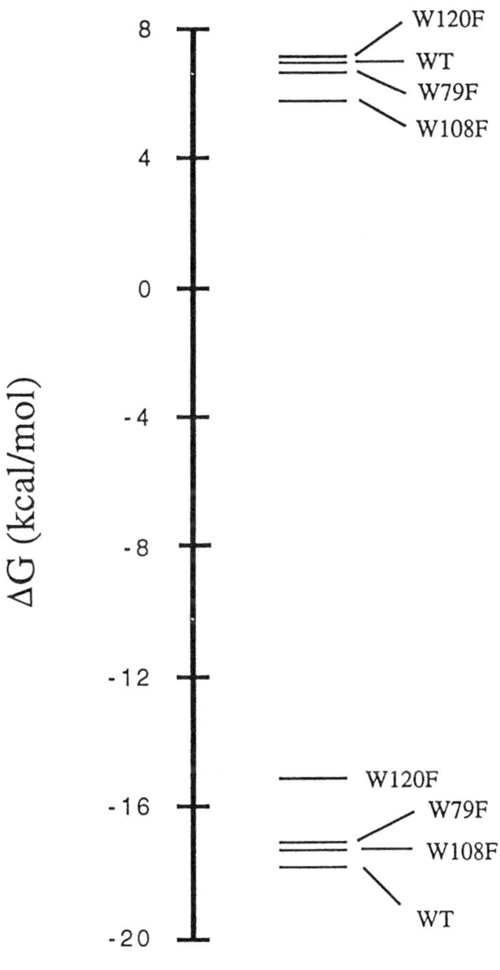

Figure 3. Transition state mapping of the tryptophan free energy ($\Delta G^{\neq}$) contributions.

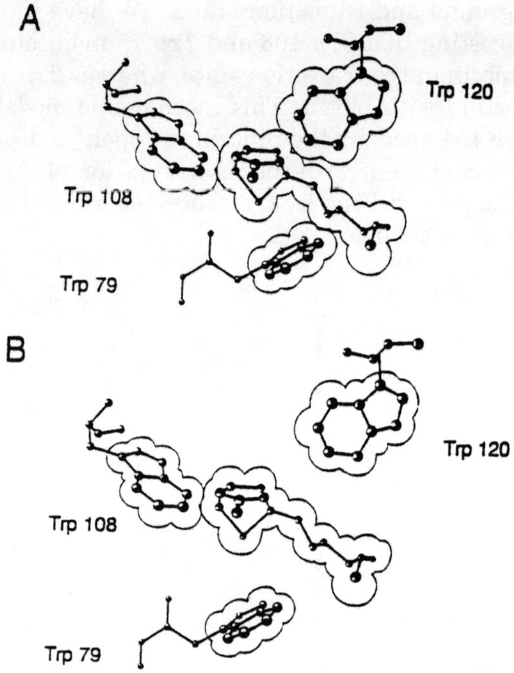

Figure 4. Schematic model for the dissociation coordinate transition state. **A**, Ground-state structure is known from crystallographic studies. The three tryptophans are in close contact with the ligand. **B**, Activation analysis and transition state mapping studies suggest that *Trp* 120 and *Trp* 79 have lost significant contact with biotin in the transition state. Results also suggest that *Trp* 108 maintains a key contact and role in management of transition-state energetics.

SUMMARY OF ENERGETIC ROLES OF TRYPTOPHAN CONTACTS

Transition state mapping suggests that *Trp* 108 remains an important contact to biotin in the transition state. The *Phe* mutant actually has a more positive activation enthalpy than wild-type, which would raise the activation barrier, but this effect is overcompensated by a corresponding increase in the favorable activation entropy. It would thus appear that the *Phe* side chain retains less favorable enthalpic interaction with biotin in the transition state, and this loss of interaction results in an increase in protein and/or ligand conformational entropy. The *Trp* 108 side chain is structurally constrained via a hydrogen-bonding interaction between the indole nitrogen and the side-chain carboxylate of *Asp* 128. This interaction is lost with the *Phe* mutant, and thus the increase in side-chain entropy associated with loss of binding interactions could be an important component of the increased activation entropy. Weber has previously suggested that the fixation of binding-site side chains via hydrogen bonds would reduce the entropic losses associated with biotin binding. The mutagenesis results are in interesting agreement with this hypothesis, although they clearly suggest that tryptophan has been chosen at position 108 for its ability to manage the entropy of the transition state, rather than the ground state.

The role of *Trp* 79 also appears to be entropy management, but this residue is key to the ground state. The W79F alteration results in a more negative binding enthalpy, which was unexpected, given the high efficiency of the streptavidin-biotin interaction. This increase in binding enthalpy, however, is accompanied by an overcompensating reduction in equilibrium entropy. The tryptophan side chain has thus been chosen over phenylalanine because it optimizes the balance of enthalpy and entropy associated with biotin binding. *Trp* 79 is in direct contact with the propionic acid tail of biotin, which is the most flexible portion of the ligand. This portion of the ligand can thus be expected to contribute the most unfavorable conformational entropy loss upon going from the free state to the bound state. We have thus hypothesized that nature has sacrificed potential binding enthalpy in choosing tryptophan over phenylalanine at this position in order to allow a weaker interaction that preserves some of the conformational entropy of the propionic acid tail. At present, it is unclear why having the smaller phenylalanine side chain at position 79 results in a more negative binding enthalpy and greater loss of binding entropy.

The largest perturbation of binding energetics was observed for *Trp* 120. This residue is the only biotin-binding contact that is contributed by an adjacent subunit, and is located on a loop that partially caps the biotin binding site. The isothermal binding titration results demonstrated that a large reduction in binding enthalpy is the underlying alteration that leads to the reduction in equilibrium binding affinity and the increased dissociation rate. The transition-state analysis suggests that *Trp* 120 has lost contact with biotin in the transition state, which is consistent with the need to open a dissociation pathway for biotin by moving the loop containing this residue. The crystallographic

B-factors for *Trp* 120 and the loop display relatively high values, which is consistent with the need for mobility at this position.

REFERENCES

Chilkoti, A and PS Stayton. **1995**. Molecular origins of the slow streptavidin-biotin dissociation kinetics. *J. Am. Chem. Soc.* **117**, 10622-10628.

Chilkoti, A, PH Tan, and PS Stayton. **1995**. "Site-directed mutagenesis studies of the high affinity streptavidin-biotin complex: contributions of tryptophan residues." *Proc. Natl. Acad. Sci. USA* **92**, 1754-1758.

Hendrickson, WA, A Pahler, JL Smith, Y Satow, YS Meritt, and RP Phizackerley. **1989**. "Crystal structure of core streptavidin determined from multiwavelength anomalous diffraction of synchrotron radiation." *Proc. Natl. Acad. Sci. USA* 86, 2190-2194.

Miyamoto, S and P Kollman. **1993a**. "Absolute and relative binding free energy calculations of the interaction of biotin and its analogs with streptavidin using molecular dynamics/free energy perturbation approaches." *Proteins* 16, 226-245.

Miyamoto, S and P Kollman. **1993b**. Proc. Natl. Acad. Sci. USA 90, 8402-8406.

Weber, PC, DH Ohlendorf, JJ Wendoloski, and FR Salemme. **1989**. "Structural origins of high-affinity biotin binding to streptavidin." *Science* 243, 85-88.

IMPACT OF CARBON TETRACHLORIDE AND OTHER SOLVENTS ON ENZYME INTRAMOLECULAR HYDROGEN BONDING: SUBTILISIN

R. Ornstein* and Y.-J. Zheng

Environmental Molecular Sciences Laboratory,
Pacific Northwest National Laboratory, Richland, Washington

Key Words: *Carbon tetrachloride, hydrogen bonding, subtilisin*

ABSTRACT

Remediation and monitoring of subsurface contamination has proven to be problematic, thus cost-effective, enzyme-based alternatives may prove useful in dealing with heavily halogenated substances in the deep subsurface. At the U.S. Department of Energy's Hanford Reservation in the State of Washington, carbon tetrachloride ($CC1_4$) is the most serious recalcitrant, subsurface, chlorinated hydrocarbon. A hallmark of enzyme structure is the detailed pattern of intramolecular hydrogen bonding. We compare the hydrogen bonding pattern obtained from molecular dynamics simulation on the enzyme Carlsberg subtilisin in $CC1_4$, in water, and in two other solvents.

INTRODUCTION

Halogenated solvents derived from industrial wastes, agricultural practices, waste products of energy production, and national defense activities have entered the environment through routine disposal practices and accidents. Contamination of water and soil sediments by heavily chlorinated and synthetic hydrocarbons is a serious environmental problem. An Environmental Protection Agency (EPA) listing of 129 priority pollutants contained 62 halogenated organic compounds, while 27 of 65 compound classes on a list of toxic pollutants were halogenated hydrocarbons (Keith and Telliard, 1979). The extent an nature of known chemical contamination on U.S. Department of Energy (DOE) lands was summarized in a 1992 DOE report (Riley et al., 1992). Heav-

*To whom correspondence should be addressed: Email: rl_ornstein@pnl.gov; fax 509-375-6904.

ily halogenated hydrocarbons (mostly one- to three-carbon-containing solvents) are the most prevalent DOE contaminant class in ground water and second (or third) most prevalent class in soils (Riley et al., 1992).

Some wastes at DOE sites are stored in tanks or in the ground in a retrievable form, awaiting additional treatment before permanent subsurface disposal (U.S. DOE, 1987). However, most wastes have been disposed to the ground surface, ponds, cribs, basins, pits, piles, injection wells, and landfills, leading to the subsurface contamination. Subsurface contamination is also the result of leaks from process sewer lines, fuel and hazardous-waste underground storage tanks, and breached drums of buried chemicals and wastes (Riley et al., 1992).

Remediation and monitoring of subsurface contamination has proved to be problematic, thus cost-effective and novel biotechnological (enzyme-based) alternatives may prove useful in dealing with heavily halogenated substances in the deep subsurface (Brockman and Ornstein, 1991; Ornstein, 1991a,b, 1994). Toward this goal, it is necessary to first gain a fundamental understanding of the possible perturbations to enzyme structure and dynamics induced by nonaqueous environments. It may then be possible to rationally redesign a given enzyme in order to reverse undesirable structural and/or dynamic perturbations that limit their functionality in a particular nonaqueous environment.

Several advantages of carrying out enzymatic reactions in nonaqueous environments have been identified. These include higher solubility of substrate in organic solvent, increased thermostability of enzymes (Zaks and Klibanov, 1984), altered substrate specificity (Zaks and Klibanov, 1988; Gololobov et al., 1992; Tawaki and Klibanov, 1992) and regiospecificity (Riva et al., 1988; Rubio et al., 1991), improved enzyme recovery from organic solvent, and the ability to carry out reactions which are impossible in water due to either kinetic or thermodynamic factors (Klibanov, 1986).

Recently, x-ray crystallographic and NMR techniques have been used to study protein structure and dynamics in organic solvents. Crystallographic work by Ringe and Klibanov and coworkers (Fitzpatrick et al., 1993, 1994) on subtilisin Carlsberg with bound acetonitrile and the work by Farber and coworkers (Yennawar et al., 1994, 1995) on chymotrypsin with bound hexane suggest that there are small structural changes compared to the crystal structures of these two proteins from aqueous medium.

In contrast, NMR studies seem to suggest considerable changes in structure and dynamics of proteins in organic solvents (Wu and Gorenstein, 1993; Desai and Klibanov, 1995). Hartsough and Merz (1992, 1993) employed computer simulation techniques to examine the effect of chloroform on the small protein BPTI and found only subtle structural changes compared to the structure in aqueous solution. Recently, we compared the protein dynamics of a protease enzyme, subtilisin Carlsberg, in aqueous solution and in acetonitrile (Zheng and Ornstein, 1996a), and in dimethyl sulfoxide (DMSO) (Zheng and Ornstein, 1996b). We found that subtilisin in acetonitrile and DMSO shows considerable flexibility. Although this conclusion is in agreement with the 2D NMR study of cytochrome C in tetrahydrofuran (THR) (Wu and Gorenstein, 1993), it is clearly in conflict with previous studies in nonpolar medium. Since acetonitrile and DMSO can act as hydrogen bond acceptors, and since they have relatively large dielectric constants (greater than 36 and 46, respectively), they are expected to behave differently from a nonpolar organic solvent.

At the DOE Hanford Reservation in Washington state, carbon tetrachloride ($CC1_4$) is the most serious recalcitrant subsurface chlorinated hydrocarbon (Riley et al., 1992). $CC1_4$ is a nonpolar solvent and has very low dielectric constant (2.238), with no hydrogen bonding capability. We recently completed a detailed molecular dynamics simulation study on subtilisin in $CC1_4$, to address the question of how protein structure and dynamics are affected by solvent properties such as hydrogen bonding capacity and dielectric constant (Zheng and Ornstein, 1996c). We briefly compare our subtilisin simulations in $CC1_4$, acetonitrile, DMSO, and water with respect to the impact on protein intramolecular hydrogen bonding interactions.

METHODS

All calculations were carried out using the AMBER 4.0 program (Pearlman et al., 1991). The all-atom AMBER force field was used for proteins (Weiner et al., 1986). For simulations in water, $CC1_4$, and DMSO, starting coordinates for all heavy atoms of subtilisin, three metal ions, and crystal waters were obtained from the crystal structure determined in aqueous solution (Fitzpatrick et al., 1993). For the simulation in acetonitrile, starting coordinates for all heavy atoms of subtilisin, crystal acetonitriles, one metal ion, and crystal waters were taken from the crystal structure with diffused in acetonitrile (Fitzpatrick et al., 1993). The TIP3P model was used for water (Jorgensen, 1981; Jorgensen et al., 1983), while OPLS parameters were used for acetonitrile (Jorgensen and Briggs, 1988) and DMSO (Jorgensen, unpublished data). For $CC1_4$, Kollman's parameters were used (DeBolt and Kollman, 1990). For simulations in water and acetonitrile, the protein was solvated by a CAP of solvent molecules (Zheng and Ornstein, 1996a). The simulations in $CC1_4$ (Zheng and Ornstein, 1996c) and DMSO (Zheng and Ornstein, 1996b) were carried out in a box with periodic boundary conditions at 298 K and 1 atm. The duration of the simulations was 450, 480, 714, and 745 ps in water in acetonitrile, in $CC1_4$, and in DMSO, respectively.

RESULTS AND DISCUSSION

Figure 1 shows the total number of intramolecular protein hydrogen bonds during the simulations for subtilisin in $CC1_4$ and water. There are more intramolecular protein hydrogen bonds in the protein in $CC1_4$

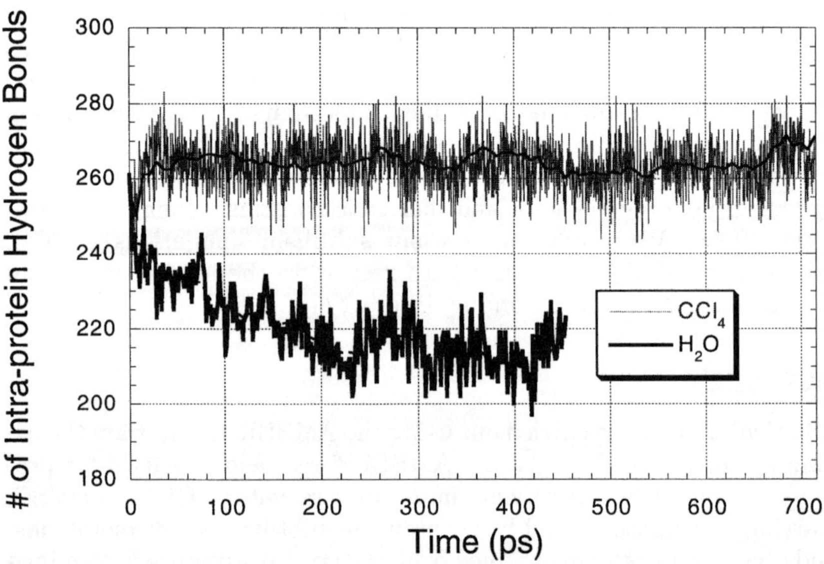

Figure 1. Calculated total number of intraprotein hydrogen bonds in different solvents.

than in water. The number of calculated intramolecular protein hydrogen bonds in each simulation (over the last 300 ps) is 220 ± 10, 249 ± 7, 260 ± 8, and 263 ± 7 in water, DMSO, acetonitrile, and $CC1_4$, respectively. The number of intramolecular protein hydrogen bonds is highest in $CC1_4$, which has the lowest dielectric constant. This probably explains some of the unusual properties observed for proteins and enzymes in nonaqueous environments; for example, reduced flexibility and increased thermostability. The calculated root-mean square deviation (RMSD) for each individual secondary structural element of subtilisin in each of the four solvent simulations are summarized in Table 1. As can be seen, the secondary structures are intact during all four simulations; thus the bulk of the hydrogen bond losses (in water) or gains (in other solvents) involve the side chains.

We observed that a hydrophilic solvent such as DMSO is capable of stripping water molecules away from the protein surface (Zheng and Ornstein, 1996b), while a hydrophobic solvent such as $CC1_4$ is incapable of doing so (Zheng and Ornstein, 1996c). This is in agreement with experimental observations and is related to the hydrophobicity of the solvent (Wasacz et al., 1987; Gorman and Dordick 1992).

The effect of solvent on peptide hydrogen bonding interaction has been investigated using N-methylacetamide (NMA) (Klotz and Franzen, 1960). The equilibrium constants for the dimerization of NMA were determined in three solvents: $CC1_4$, dioxane, and water. The equilibrium constants at 25°C in the three solvents are 5.2, 0.55, and 0.005 M^{-1}, respectively (Klotz and Franzen, 1960). Thus, hydrogen bonds involving protein donor and acceptor sites will afford greater stabilization in nonpolar, hydrophobic solvents, such as $CC1_4$.

In a solvent with little or no hydrogen bonding capability, the hydrogen bonding contribution to the free energy of protein folding will be an even more important determinant than in aqueous solution. Normally, removing protein-solvent hydrogen bonds would increase the free energies of both folded and unfolded states. In aqueous solution, a properly folded protein is believed to have more intramolecular hydrogen bonds than the unfolded form. In nonaqueous solution, an increase in total number of intramolecular protein hydrogen bonds could stabilize either an active or inactive conformation. Understanding these interactions is important in designing new biotechnological uses for proteins in nonaqueous environments.

Table 1. The calculated root-mean square deviation (RMSD) (Å) of C_α and standard deviation for each individual secondary structural element of subtilisin in four different solvents.

Element	Residues	Water	DMSO	Acetonitrile	$CC1_4$
H1	6-11	0.28 ± 0.10	0.22 ± 0.07	0.69 ± 0.21	0.26±0.07
H2	12-19	0.39 ± 0.11	0.36 ± 0.08	0.39 ± 0.14	0.79 ± 0.12
H3	63-74	0.33 ± 0.08	0.36 ± 0.09	0.61 ± 0.12	0.40 ± 0.09
H4	103-117	0.37 ± 0.08	0.44 ± 0.12	0.52 ± 0.17	0.98 ± 0.16
H5	132-146	0.43 ± 0.10	0.40 ± 0.08	0.73 ± 0.34	0.39 ± 0.07
H6	220-238	0.64 ± 0.25	0.65 ± 0.16	0.55 ± 0.10	0.56 ± 0.11
H7	242-253	0.49 ± 0.13	0.52 ± 0.13	0.54 ± 0.09	0.53 ± 0.08
H8	269-274	0.27 ± 0.09	0.26 ± 0.09	0.34 ± 0.08	0.26 ± 0.09
Sheet[a]		0.73 ± 0.12	0.84 ± 0.11	1.05 ± 0.18	0.99 ± 0.18

[a]The sheet consists of residues 43-50, 89-95, 26-32, 120-124, 148-153, 175-180, 197-220, 265-268.

ACKNOWLEDGMENTS

This work was supported by the Laboratory-Directed Research and development program at the Pacific Northwest National Laboratory as part of the Microbial Biotechnology Initiative (RLO). Pacific Northwest National Laboratory is a multiprogram laboratory operated for the U.S. Department of Energy by Battelle Memorial Institute under Contract DE-AC06-76RLO 1830. Y.-J. Zheng is an AWU-NW-DOE Post-doctoral Fellow.

REFERENCES

Brockman, FJ and RL Ornstein. **1991**. In Hazardous Materials Control/Superfund '91, HMCRI, Greenbelt, Maryland, pp. 264-266.

Debolt, SE and PA Kollman. **1990**. J. Am. Chem. Soc. 112:7515.

Desai, UR and AM Klibanov. 1995. J. Am. Chem. Soc. 117:3940.

Fitzpatrick, PA, ACU Steinmetz, D Ringe and AM Klibanov. **1993**. Proc. Natl. Acad. Sci. USA 90:8653.

Fitzpatrick, PA, D Ringe and AM Klibanov. **1994**. Biochem. Biophys. Res. Commun. 198:675.

Gololobon, MY, TL Voyushina, VM Stepanov, and P Adlercreutz. **1992**. FEBS Lett. 307:309.

Gorman, LA and JS Dordick. **1992**. Biotechnol. Bioeng. 39:392.

Hartsough, DS and KM Merz, Jr. 1992. J. Am. Chem. Soc. 114:10113.

Hartsough, DS and KM Merz, Jr. 1993. J. Am. Chem. Soc. 115:6529.

Jorgensen, WL. **1981**. J. Am. Chem. Soc. 103:335.

Jorgensen, WL, and JM Briggs. **1988**. Mol. Phys. 63:547.

Jorgensen, WL, J Chandrasekhar, JD Madura, RW Impey and ML Klein. **1983**. J. Chem. Phys. 79:926.

Keith, LH and WA Telliards. **1979**. Environ. Sci. & Technol. 13, 416.

Klibanov, AM. **1986**. CHEMTECH 16:354.

Klibanov, AM. **1990**. Acc. Chem. Res. 23:114.

Klotz, IM and JS Franzen. **1960**. J. Am. Chem. Soc. 88:5241.

Ornstein, RL. **1991a**. In On Site Bioreclamation: Processes for Xenobiotic and Hydrocarbon Treatment, Hinchee, R. E. and Olfenbuttel, R. F., Eds., Butterworth-Heinemann, Boston, Massachusetts, pp. 509-514.

Ornstein, RL. **1991b**. In Proceedings of National Research & Development Conference on the Control of Hazardous Materials, HMCRI, Greenbelt, Maryland, pp. 314-319.

Ornstein, RL. **1994**. In Structural Biology: State of the Art 1993, Sarma, R. H. and Sarma, M. H., Eds., Adenine Press, Albany, New York, vol. 1, pp 59-76.

Pearlman, DA, DA Case, JC Caldwell, GL Seibel, C Singh, P Weiner and PA Kollman. **1991**. AMBER 4.0, University of California, San Francisco, California.

Riley, RG, JM Zachara and FJ Wobber. **1992**. Chemical Contaminants on DOE Lands and Selection of Contaminant Mixture for Subsurface Science Research. U.S. Department of Energy Report.

Riva, S, J Chopineau, APG Kieboom and AM Klibanov. **1988**. J. Am. Chem. Soc. 110:584.

Rubio, E, A Fernandez-Mayorales and AM Klibanov. **1991**. J. Am Chem. Soc. 113:695.

Tawaki, S and AM Klibanov. **1992**. J. Am. Chem. Soc. 114:1882.

U.S. Department of Energy. **1987**. Integrated Data Base for 1987: Spent Fuel and Radioactive Waste Inventories, Projections and Characteristics, DOE/RW-0006, Rev. 3.

Wasacz, FM, JM Olinger, RJ Jakobsen. **1987**. Biochemistry 28:1464.

Weiner, SJ, PA Kollman, DT Nguyen and DA Case. **1986**. J. Comput. Chem. 7:230.

Wu, J and DG Gorenstein. 1993. J. Am. Chem. Soc. 115:6843.

Yennawar, NH, HP Yennawar and GK Farber. **1994**. Biochemistry 33:7326.

Yennawar, NH, HP Yennawar and GK Farber. **1995**. J. Am. Chem. Soc.

Zaks, A and AM Klibanov. **1984**. Science 224:1249.

Zaks, A and AM Klibanov. **1986**. J. Am. Chem. Soc. 108:2767.

Zaks, A and AM Klibanov. **1988**. J. Biol. Chem. 263:3194.

Zheng, Y-J and RL Ornstein. **1996a**. Biopolymers 38:791.

Zheng, Y-J and RL Ornstein. **1996b**. J. Am. Chem. Soc. 118:(in press).

Zheng, Y-J and RL Ornstein. **1996c**. Prot. Eng. 9:(in press).

USING FREE-ENERGY CALCULATIONS TO OPTIMIZE CHLOROALKANE FORCE-FIELD PARAMETERS

M. D. Paulsen

Environmental Molecular Sciences Laboratory,
Pacific Northwest National Laboratory, Richland, Washington

Keywords: *Molecular dynamics, free energy of solvation, biodehalogenation, chloroalkanes, force field parameterization*

ABSTRACT

Lennard-Jones parameters for chloroalkanes were optimized in order to reproduce experimental absolute-hydration free-energy data. Absolute-solvation free energies were calculated using the recently developed separation-shifted scaling method in conjunction with thermodynamic integration calculations. Coefficients for the repulsive interaction of carbon, hydrogen, and chlorine with extended simple point charge water were optimized to simultaneously improve the agreement between calculated and experimental absolute-hydration free energies for both substituted and unsubstituted short-chain alkanes.

INTRODUCTION

Computer simulations are being widely used to examine a variety of chemical and biochemical phenomena. In particular, the use of computer simulations to evaluate the free-energy differences for processes such as solvation, substrate binding, and host-guest complexation has become widespread (Jorgensen, 1989; Straatsma and McCammon, 1991; Kollman, 1993). To date, the vast majority of such studies have employed molecular mechanics force fields rather than quantum mechanical descriptions for both solutes and solvents. Thus the utility of such simulations will be critically dependent on the parameterization of the force field. Several force fields appropriate for organic or biochemical systems have been developed (Brooks et al., 1983; Weiner et al., 1986; van Gunsteren and Berendsen, 1987; Jorgensen and Tirado-Rives, 1988; Cornell et al., 1995). These force fields have been optimized to reproduce a variety of data including crystallographic and spectroscopic data, *ab initio* calculations, and properties of pure liquids such as densities and heats of vaporization. However, in studies of solvation effects, if both solvent and solute have been parameterized to reproduce properties of the respective pure substances, the force field may not reproduce solution properties nearly as well. A recent study of alkane hydration using the TIP3P water model showed relatively poor agreement between calculated and experimental values for the hydration free energy of ethane (Sun and Kollman, 1995). Similar discrepancies have been noted for solvation free energies of simple alkanes in simulations using the SPC/E water model (T. P. Straatsma, personal communication). Since many of the processes of interest involved the transfer of organic compounds from aqueous solution to, for instance, an enzyme active site, it is important that the interactions of simple organics with water be well reproduced in any force field.

Increases in computer speed and improvements in the algorithms for free energy determinations now permit solvation free energies to be used directly in the parameterization of molecular mechanics force fields. We present here an optimization of Lennard-Jones repulsive parameters for unsubstituted short-chain alkanes and chloroalkanes which result in improved agreement between calculated and experimental hydration free energies.

METHODS

Free energy differences can be evaluated from molecular dynamics simulations using the technique of thermodynamic integration. The two states A and B, for which the free energy difference is to be computed, are coupled by a control variable λ, which is slowly varied during the course of the molecular dynamics trajectory. For solvation free energies, state A is a gas-phase solute and a pure solvent, which do not interact with each other, while state B is a solution. The free energy difference between states A and B, where state A is given by $\lambda = 0$ and state B by $\lambda = 1$, is then given by

$$\Delta G = \int_0^1 \left\langle \frac{\partial H(\lambda)}{\partial \lambda} \right\rangle d\lambda \approx \sum_i \left\langle \frac{\partial H(\lambda)}{\partial \lambda} \right\rangle \Delta\lambda \cdot \tag{1}$$

This expression requires that the derivative of the Hamiltonian with respect to the control variable λ be evaluated at each time step. An ensemble average is then constructed for each discrete value of λ used in the trajectory. The interactions between solvent and solute are modeled by a Lennard-Jones potential in combination with a set of atom-centered partial charges. A linear coupling of the LJ term as a function of λ would give the following expression:

$$V_{LJ}(\lambda) = (1-\lambda)\left[\frac{A_a}{r^{12}} - \frac{B_a}{r^6}\right] + \lambda\left[\frac{A_b}{r^{12}} - \frac{B_b}{r^6}\right]. \tag{2}$$

However, since, in state A the solute and solvent do not interact, the second term in the equation is zero. With this type of λ dependence for the Hamiltonian, there is very poor sampling for small values of r except when $\lambda = 1$. Unfortunately, this region of conformational space makes large contributions to the calculated free energy, which means that it will be difficult to obtain well-converged free energies. An alternative method of coupling the Hamiltonians that describe states A and B has been proposed by Zacharias et al. (1994) and is given by

$$V_{LJ} = (1-\lambda)\left[\frac{A_a}{(r^2+\delta\lambda)^6} - \frac{B_a}{(r^2+\delta\lambda)^3}\right]. \tag{3}$$

In this expression, the atomic separation, r, is shifted by a λ-dependent constant which permits efficient sampling at small values of r, resulting in significantly improved convergence for calculated free energies. This expression has the correct limiting behavior at $\lambda = 0$ and $\lambda = 1$, and the potential and its derivative are finite and continuous everywhere except for $\lambda = 0$ and $r = 0$.

The ARGOS simulation package was used for all simulations (Straatsma and McCammon, 1990). The SPC/E model was used to describe water molecules (Berendsen et al., 1987). Initially, solute molecules were represented using the AMBER-OPLS force field (Tirado Rives and Jorgensen, 1990). All simulations were performed at constant temperature and pressure with a temperature of 298 K and a pressure of 1×10^5 Pa. The pressure and temperature of the system were maintained by coupling to an external bath (Berendsen et al., 1984). All simulations were performed on a periodic system containing the solute and 410 water molecules in a box 2.4 nm on a side. The cutoff radius for all nonbonded interactions was set to 1.1 nm. All bonds involving hydrogen were constrained using SHAKE (Ryckaert et al., 1977), and a 2.0-fs time step was used for all simulations. Multiconfiguration thermodynamic integration, as implemented in the ARGOS simulation package, was used to evaluate the hydration free energies (Straatsma and McCammon, 1991). All thermodynamic integration calculations were performed using 21 equally spaced steps in λ, with 750 steps of equilibration and 6000 steps of data-gathering for a total simulation time of 283.5 ps. The recently developed separation-shifted scaling method was used in all thermodynamic integration calculations (Zacharias et al., 1994).

RESULTS

Initially, absolute solvation free energies were determined for ethane, propane, and butane using published Optimized Parameters for Liquid Simulations (OPLS) nonbonded parameters. The results are summarized in Table 1. For all three alkanes, the calculated absolute solvation free energy is substantially too positive. Although the calculated absolute solvation free energies for all three compounds are in poor agreement with experiments, the resulting relative free energies of solvation are in agreement with the experimental data to within the statistical uncertainty of the calculations.

Table 1. Alkane free energy of hydration

	OPLS[a] (kJ/mol)	Revised (kJ/mol)	Experimental[b] (kJ/mol)
Ethane	14.24 ± 1.46	7.38 ± 1.26	7.66
Propane	15.11 ± 1.66	9.37 ± 1.29	8.18
Butane	17.44 ± 1.66	6.43 ± 1.64	8.70

[a]Optimized parameters for liquid simulations.
[b]Cabani et al., 1981

To characterize the sensitivity of the calculated solvation free energy to the choice of nonbonded parameters, the derivative of the energy with respect to each of the nonbonded force-field parameters was determined. This analysis indicated that the calculated free energies were insensitive to changes in the partial charges assigned to the hydrogen and carbon atoms. Considering the nonpolar nature of these solutes, such a result was expected. For more polar molecules, the calculated solvation free energy will be much more sensitive to small variations in the assignment of partial charges. Thus, optimizing the agreement between calculated and experimental solvation free energies by scaling the partial charges of the solute was not pursued. However, the calculated solvation free energy for these solutes is sensitive to the choice of Lennard-Jones parameters, as shown by the derivatives summarized in Table 2. The derivatives for the repulsive Lennard-Jones term were then used to develop revised force-field parameters which would improve the agreement between calculated and experimental solvation free energies.

Table 2. Energy derivatives with respect to C_{12} nonbonded parameters for unsubstituted alkanes

	Ethane	Propane	Butane
$(\partial U/\partial C_{12})C_o$ (nm^{-12})	3.0240×10^6	3.6069×10^6	4.0548×10^6
$(\partial U/\partial C_{12})H_o$ (nm^{-12})	5.5221×10^7	6.6847×10^7	7.4369×10^7

Since the derivatives are strictly appropriate only at the value of the parameter for which they were calculated, it was necessary to repeat the procedure for each revision of the nonbonded parameters. After three rounds of optimization, the agreement between calculated and experimental solvation free energies for ethane, propane, and butane was substantially improved. Solvation free energies using the revised parameters are summarized in Table 1. In contrast to the calculation with the original OPLS parameters, the calculated and experimental solvation free energies for ethane and propane now agree to within the statistical uncertainty of the calculated values. The difference between calculated and experimental solvation free energies for butane is still slightly larger than the statistical uncertainty of the calculation, but it is significantly smaller than with the unoptimized parameters. Additional improvement in the calculated value for butane might be possible if the nonbonded parameters for methyl and methylene carbons are optimized separately. In Figure 1, the interaction energy between an SPC/E water oxygen and an alkane carbon is plotted as a function of interatomic distance for both the original and revised carbon nonbonded parameters. The position of the minimum in the interaction potential has decreased from 0.37 nm to 0.36 nm, and the well depth has increased by 25%.

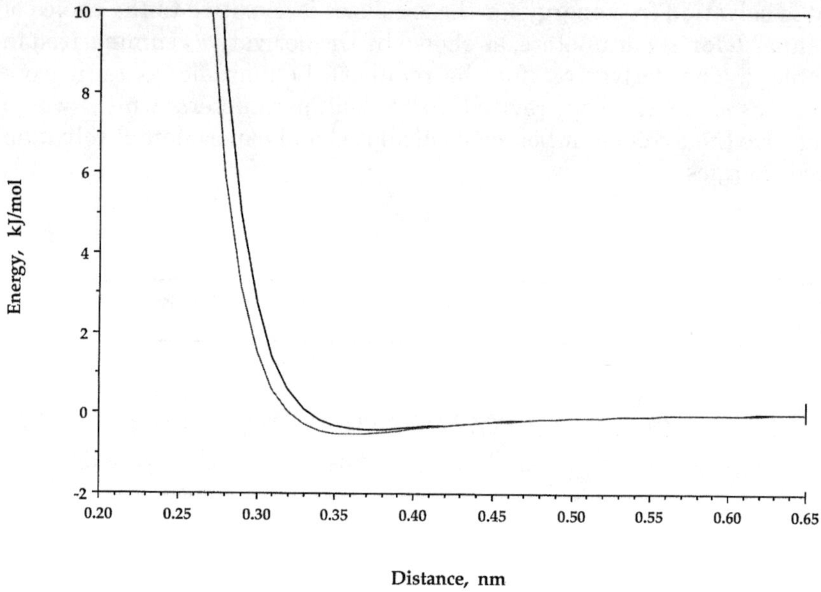

Figure 1. The Lennard-Jones interaction potential for carbon and water oxygen as a function of interatomic distance for both the original parameters (heavier line) and the revised parameters (lighter line).

After optimizing nonbonded parameters for unsubstituted alkanes, these parameters were used to calculate solvation free energies for hexachloroethane (HCA) and 1,1,2,2- tetrachloroethane (TetCA). The results of these calculations are shown in Table 3. Even with the modified carbon and hydrogen parameters, the solvation free energies for HCA and TetCA are substantially too positive and, for both compounds, have the wrong sign. However, this discrepancy is not apparent in the calculated relative solvation free energy where, as with the unsubstituted alkanes, there is reasonable agreement between the calculated and experimental values.

As with the unsubstituted alkanes, the calculated solvation free energies for HCA and TetCA are not particularly sensitive to modest scaling

Table 3. Chloroalkane free energy of hydration

	Initial (kJ/mol)	Revised (kJ/mol)	Experimental[(a)] (kJ/mol)
Hexachloroethane	9.59 ± 2.37	-4.89 ± 1.97	-5.88
1,1,2,2-tetrachloroethane	5.29 ± 1.86	-8.729 ± 1.81	-9.87

[(a)]Cabani et al., 1981.

of the partial charges. Thus, improved agreement between calculated and experimental solvation free energies again required modification of the Lennard-Jones parameters. For both solutes, the ensemble average of the derivative of the energy with respect to the chlorine-water oxygen r^{12} parameter was determined. These derivatives are summarized in Table 4. Using this information, revised chlorine repulsive parameters were determined. After two rounds of optimization, the agreement between calculated and experimental free energies of solvation was substantially improved. For both solutes, the calculated and experimental values differ by less than the statistical uncertainty of the calculation. In Figure 2, the interaction energy between an SPC/E water oxygen and chlorine is plotted as a function of interatomic distance for both the original and revised chlorine nonbonded parameters. With the revised parameters, the position of the minimum in the interaction potential has decreased from 0.37 nm to 0.35 nm and the well depth has been increased by 25%.

Table 4. Energy derivatives with respect to C_{12} nonbonded parameters for chloroalkanes

	Hexachloroethane	1,1,2,2-tetrachloroethane
$(\partial U/\partial C_{12})Cl_o$ (nm^{-12})	9.3377×10^6	6.3339×10^6

Additional calculations are currently under way to test the ability of the revised nonbonded parameters to reproduce the solvation free energy for a range of chlorinated alkanes not included in the parameterization described here. The performance of the revised force field in simulations of the pure liquid chloroalkanes is also being investigated. Once a set of parameters is validated for the entire range of chloroethanes, investigations of the interactions of these solutes with dehalogenating enzymes such as a haloalkane dehalogenase or cytochrome P450 will be possible.

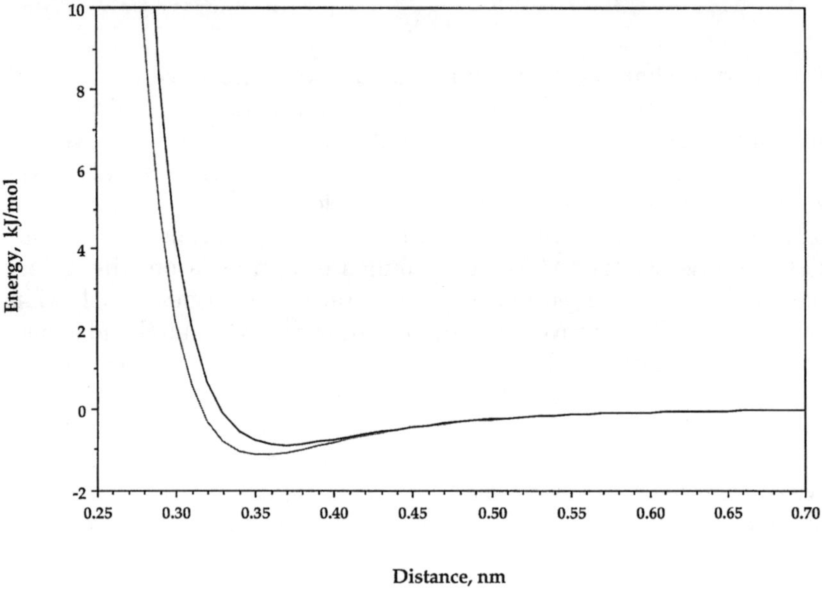

Figure 2. The Lennard-Jones interaction potential for chlorine and water oxygen as a function of interatomic distance for both the original parameters (heavier line) and the revised parameters (lighter line).

ACKNOWLEDGMENTS

The author thanks Dr. Tjerk Straatsma for many useful discussions and for advice in using the ARGOS software package. Pacific Northwest National Laboratory is a multiprogram national laboratory, operated for the U.S. Department of Energy by Battelle Memorial Institute under Contract DE-AC06-76RLO 1830. This work was supported by a grant (KP0402) from the Health Effects and Life Sciences Research

Division within the Office of Health and Environmental Research of the Office of Energy Research of the U.S. Department of Energy.

REFERENCES

Berendsen, HJC, JR Grigera, and TP Straatsma. **1987**. The missing term in effective pair potentials. J. Phys. Chem. 91:6269-6271.

Berendsen, HJC, JPM Postma, WFv Gunsteren, A DiNola, and JR Haak. **1984**. Molecular Dynamics with Coupling to an External Bath. J. Chem. Phys. 81:3684-3690.

Brooks, BR, RE Brucoleri, BD Olafson, DJ Slater, S Swaminathan, and M Karplus. 1983. CHARMM: A program for macromolecular energy, minimization, and dynamics calculations. J. Comp. Chem. 4:187-217.

Cabani, S, P Gianni, V Mollica, and L Lepori. **1981**. Group Contributions to the thermodynamic properties of nonionic organic solutes in dilute aqueous solution. J. Solution Chem. 10:563-595.

Cornell, WD, P Cieplak, CI Bayly, IR Gould, KM Merz Jr, DM Ferguson, DC Spellmeyer, T Fox, JW Caldwell, and PA Kollmam. **1995**. A second generation force field for the simulation of proteins, nucleic acids, and organic molecules. J. Am. Chem. Soc. 117:5179-5197.

Jorgensen, WL. **1989**. Free Energy Calculations: a breakthrough for modeling organic chemistry in solution. Accounts Chem. Res. 22:184-189.

Jorgensen, WL and J Tirado-Rives. 1988. The OPLS potential functions for proteins, energy minimization for crystals of cyclic peptides and crambin. J. Am. Chem. Soc. 110:1657-1666.

Kollman, P. **1993**. Free Energy Calculations: Application to Chemical and Biochemical Phenomena. Chem. Rev. 93:2395-2417.

Ryckaert, J, G Ciccotti, HJC Berendsen. **1977**. Numerical integration of the cartesian equations of motion of a system with constraints. J. Comput. Phys. 23:327-341.

Straatsma, TP and JA McCammon. **1990**. ARGOS, A vectorized general molecular dynamics program. J. Comp. Chem. 11:943-951.

Straatsma, TP and JA McCammon. **1991**. Multiconfiguration Thermodynamic Integration. J. Chem. Phys. 95:1175-1188.

Straatsma, TP and JA McCammon. 1991. Theoretical Calculations of Relative Affinities of Binding. Methods in Enzymology. San Diego, Academic Press. 497-511.

Sun, Y. PA and Kollman. **1995**. Hydrophobic Solvation of Methane and Nonbonded Parameters of the TIP3P Water Model. J. Comp. Chem 16:1164-1169.

Tirado Rives, J. and WL Jorgensen. **1990**. Molecular Dynamics of Proteins with the OPLS potential Functions: Simulation of the third domain of silver pheasant ovomucoid in water. J. Am. Chem. Soc. 112:2773-2781.

van Gunsteren, WF and HJC Berendsen. 1987. Groningen Molecular Simulations (GROMOS) Library Manual. Biomos, Groningen, The Netherlands.

Weiner, SJ, PA Kollman, DT Nguyen, and DA Case. **1986**. An all atom force field for simulations of proteins and nucleic acids. J. Comp. Chem. 7:230-252.

Zacharias, M, TP Straatsma, and JA McCammon. **1994**. Separation-Shifted Scaling, a New Scaling Method for Lennard-Jones Interactions in Thermodynamic Integration. J. Chem. Phys. 100:9025-9031.

METAL-DEPENDENT SELF-ASSEMBLY OF PROTEIN TUBES, CABLES, AND SHEETS

P. Schurke, C. M. Jochheim, M. J. Dabrowski, and W. M. Atkins

University of Washington, Department of Medicinal Chemistry, Seattle, Washington

Key Words: *Supramolecular self-assembly/protein-protein interactions/ protein-metal binding sites*

ABSTRACT

Amino acid side chains from *His*-4, *Met*-8 and *His*-12 from adjacent GS dodecamers in a docked complex provide an apparent binuclear metal binding site which mediates formation of protein tubules. Replacement of *Met*-8 and *His*-12 with cysteine results in mutant proteins with altered metal ion specificity for the stacking reaction. Studies with model peptides demonstrate the feasibility of such an intermolecular metal-binding site. Additionally, when this metal-binding site on the surface of the GS dodecamer was destroyed by site-directed mutagenesis, a metal-dependent lateral aggregation occurred, which generated sheets of hexagonally packed GS dodecamers.

INTRODUCTION

Molecules which self-assemble into supramolecular aggregates are desirable as components in nanofabrication strategies and nanoengineering applications (Jorgensen, 1993; Whitesides, et al., 1995). Proteins possess stringent molecular recognition and regulatory properties which make them ideal candidates for applications in "advanced materials" which require self-organization. In addition, many proteins naturally exhibit self-assembly properties which yield aggregates with useful shapes, sizes, and symmetry. Paradoxically, a potential limitation of proteins is their inherent specificity, which usually allows for generation of a single supramolecular aggregate from any individual protein. Proteins with the ability to assemble into several supramolecular aggregates, with different quaternary structures, would provide a versatile component for nanofabrication technologies. Redesign of proteins which possess self-assembly behavior offers one approach for generating additional components with novel supramolecular architecture.

Bacterial glutamine synthetases (GS) are dodecamers of identical subunits formed from two, face-to-face hexameric rings. The x-ray structures of the *Salmonella typhimurium* GS in the absence (Almassey et al., 1986) and presence of ligands (Shwu-Huey et al., 1993) clearly indicate a highly symmetrical ring structure resulting from head-to-tail arrangement of subunits within each hexameric ring. Several decades ago, it was reported that some bacterial GS assemble into protein tubules in the presence of ~10 μM to 100 μM Zn^{2+} (Miller et al., 1974). Moreover, optical image analysis of Co^{2+} -precipitated GS tubules indicated that these "stacked" dodecameric complexes associate laterally to generate three- and seven-stranded cables (Frey et al., 1975). An understanding of the molecular mechanisms of these protein self-assembly properties will accelerate efforts to engineer proteins with novel and useful aggregation properties. Therefore, we have constructed a model to guide our understanding of the molecular mechanism of GS stacking. We have challenged this model using site-directed mutagenesis of *Escherichia coli* GS and refined the model using biophysical characterization of peptides which mimic the GS metal-binding site.

METHODS

Site-directed mutant proteins were obtained as previously described (Dabrowski et al., 1994). Electron microscopy was performed on a JEOL microscope at 80 keV, 50,000–60,000 × magnification, with uranyl acetate as negative stain. Kinetics of GS stacking were determined by 90° light-scattering measurements, at high dilution (0.03 μM GS dodecamers) in 50 mM HEPES, 100 mM KCl, 1.0 mM $MnCl_2$ at 5°C. Light-scattering data were collected at the low extent of reaction, which was within the first 5 minutes after addition of the indicated metal ion, and fitted to a bimolecular reaction scheme (Yanchunas et al., 1994). MALDI-MS was performed with a Finnigan Lasermat at 20 keV. The matrix which afforded the most efficient desorption, 2,5-dihydroxybenzoic acid, as observed by others (Hutchens et al., 1992), was used for all samples. Samples were 50 pmoles peptide/mL in 55% H_20/45% trifluoroethanol. When present, metal ions were 0.15 mM, added as the acetate salt. Electrospray mass spectra were obtained on a Fisons VG Quatro II triple quadrupole instrument. Samples were as described above, in 50% methanol/49% H_2O/1% trifluoroethanol.

RESULTS

A molecular model for the dodecamer-dodecamer interface when GS molecules stack was generated from the available x-ray structure of the *S. typhimurium* GS (Dabrowski et al., 1994). Briefly, when two dodecamers are stacked and rotated ~10° with respect to each other about the 6-fold axis of symmetry, the N-terminal helices from each subunit form the "spokes of a wheel," with the helical axes emanating from the 6-fold axis of the dodecamer. The helices from one dodecamer are interdigitated between the helices from the adjacent, "docked," dodecamer. Each helix provides potential metal-ligating amino acid side chains at positions 4, 8, and 12, resulting in an interdodecameric metal-binding site (Figure 1). Notably, this model predicts that either a mononuclear metal binding site exists between *His*-4 and *Met*-8, or between *Met*-8 and *His*-12, or that the site binds two metal ions at each of these locations.

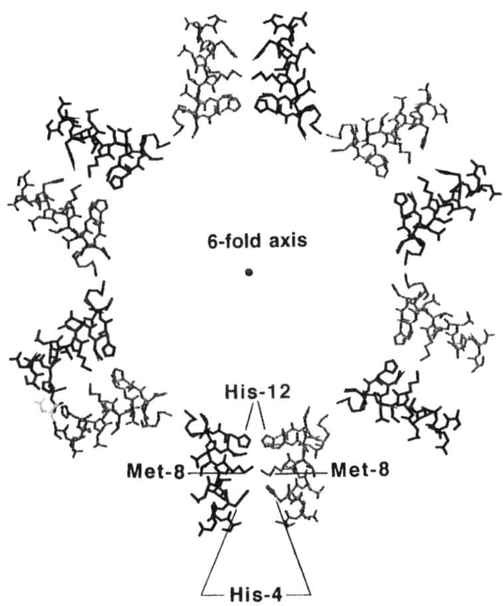

Figure 1. Interface between two dodecamers in a docked complex. The N-terminal helices, residues 1-13, of each subunit lie perpendicular to the 6-fold axis of symmetry through the dodecameric rings, and form interdigitated "spokes" of a wheel. Alpha-carbon traces for helices from both dodecamers are shown, with helices from one dodecamer in black and the other dodecamer in gray. Side chains are included for two pairs of helices. The side chains from *His*-4, *Met*-8, and *His*-12 generate six intermolecular metal-binding sites between the dodecamers.

This model was tested by constructing site-directed mutants, with aliphatic amino acid side chains replacing *His*-4, *Met*-8, or *His*-12. The rates of stacking of wild-type and GS mutants were obtained by 90° light-scattering measurements, at the low extent of reaction and at high dilution (0.03-μM dodecamers). Under these conditions, the kinetics closely approximate a simple bimolecular reaction scheme, and bimolecular rate constants are obtained, as described previously (Yanchunas et al., 1994). The H12L and H4A mutants stack 20-fold and 12-fold (respectively) more slowly than the wild-type protein with Zn^{2+}; they are similarly impaired when other metals are used to initiate stacking. The M8L mutant protein exhibits a 3-fold reduction in rate constant, compared to the wild type. These results are summarized in Figure 2.

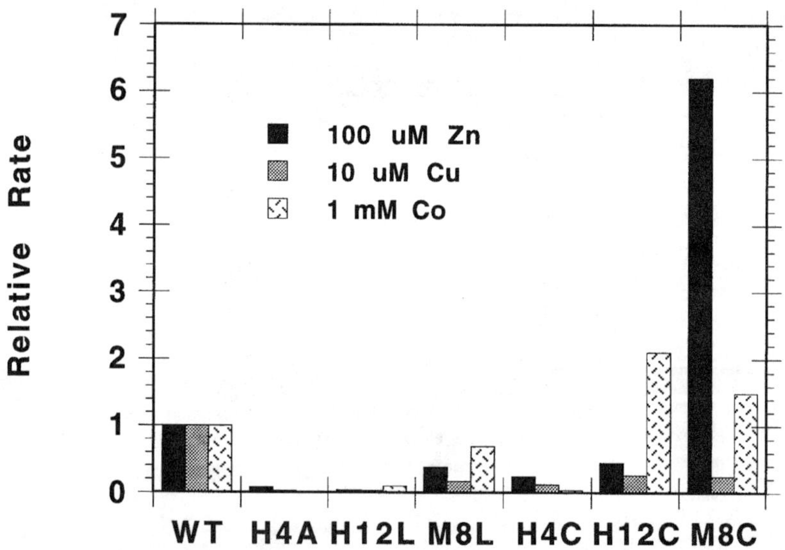

Figure 2. Metal ion specificity of wild-type and mutant stacking reactions. The wild-type glutamine synthetase (GS) "stacks," with nearly identical bimolecular rate constants in the presence of 10 μM Cu^{2+}, 100 μM Zn^{2+}, or 1 mM Co^{2+}. The site-directed mutants H4A and H12L stack with dramatically reduced rate constants. The rate constants for the M8L mutant are also reduced, to a lesser extent. When cysteine is substituted at these positions, stacking behavior is restored and, in some cases, exceeds that of the wild-type GS. Relative rates are based on time-dependent changes in intensity of scattered light after addition of the indicated metal ion. Conditions as described in Methods.

In principle, replacement of *His*-4, *Met*-8, or *His*-12 with cysteine will provide potential metal ligands from the thiol function, and these mutations would be expected to afford proteins which stack efficiently. This is observed experimentally and, in fact, the M8C mutant stacks 6-fold and 1.5-fold faster than the wild-type protein with Zn^{2+} and Co^{2+}, respectively (Figure 2). Moreover, the M8C mutant demonstrates a 25-fold increase in the specificity for Zn^{2+}-dependent stacking versus Cu^{2+}, in marked contrast to the wild-type GS, which stacks most efficiently with Cu^{2+} ions. Together, these results indicate that the N-terminal helices of the GS subunits are critical for metal-dependent formation of GS tubes, and that metal ligands are provided by *His*-4, *Met*-8, and *His*-12 at the dodecamer-dodecamer interface of a stacked complex. However, these experiments do not directly demonstrate whether two metal ions bind at each helical dimer, or whether one metal ion binds to either end of the helical pairs. In order to distinguish between these possibilities, model peptides have been studied by circular dichroism (CD), MALDI-TOF mass spectrometry (MS), and electrospray ionization mass spectrometry.

The model peptides contain variations of the sequence found in the N-terminal helices of wild-type GS. The sequences were N-AEH$\underline{\text{A}}$LTML-NEHE$\underline{\text{A}}$-C (peptide 1) and N-AEH$\underline{\text{A}}$LT$\underline{\text{C}}$LNEHE$\underline{\text{A}}$-C (peptide 2). The underlined alanines in both sequences replace valines found in the native sequence and were incorporated in order to maximize helical content while minimizing the tendency to form coiled-coils in the absence of metal ions (Zhou et al., 1992). The cysleine at position 7 in peptide 2 is analogous to the M8C mutant GS. In 55% water/45% trifluoroethanol (v/v), the peptides exist as a mixture of helix and random coil. Upon addition of a 4-fold excess of Cu^{2+}, the peptides become predominantly α-helix, based on CD spectroscopy (data not shown). These experiments establish conditions which yield α-helices ligated to metal ions. Peptide 1 and peptide 2 were >60% and >75% helical in the presence of excess Cu^{2+}.

It was necessary to determine whether metal ions stabilized helical dimers in solution, rather than simply binding to monomeric helical species. This was determined by MALDI-TOF-MS. Solutions of the peptides (500 μM) in 55% H_2O/45% TFE were made in 1.5 mM Zn^{2+} or Cu^{2+} and examined by MALDI-MS. Samples containing Zn^{2+} yielded poor spectra, presumably due to inefficient desorption. Samples containing Cu^{2+}, however, clearly demonstrated a metal-dependent formation of peptide dimers, including (2M), (2M + Cu^{2+}) and (2M + 2Cu^{2+}) species, where M is the molecular ion of the peptide (Figure 3).

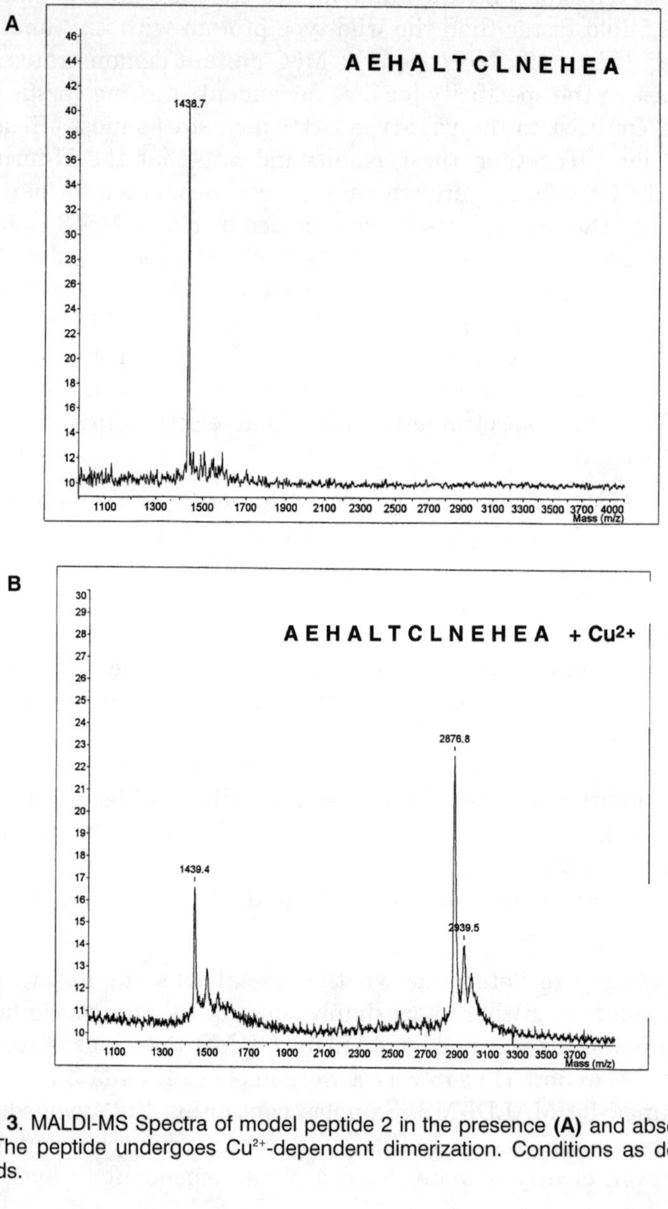

Figure 3. MALDI-MS Spectra of model peptide 2 in the presence **(A)** and absence **(B)** of Cu^{2+}. The peptide undergoes Cu^{2+}-dependent dimerization. Conditions as described in Methods.

In addition to the MALDI-MS experiments, electrospray ionization MS was used to corroborate the metal-dependent formation of peptide dimers. Results are shown for peptide 2 in the presence of Zn^{2+} (Figure 4). The ions corresponding to $(2M + 2Zn + 2H)^{6+}$, $(2M + 2Zn + H)^{5+}$, and $(2M + 2Zn)^{4+}$ at m/z 500.4, 600.3, and 750.9, respectively, indicate the presence of peptide dimers with 2 Zn^{2+} ions bound. Notably, no ions are observed which correspond to the peptide dimer with a single zinc ion bound, $(2M + Zn + 3H)^{5+}$, $(2M + Zn + 2H)^{4+}$, or $(2M + Zn + 1H)^{3+}$, at m/z 588.0, 734.8, or 979.3. Together, these results indicate that peptide dimer formation occurs in the presence of two metal ions, and that the dimers are present exclusively with two Zn^{2+} ions bound.

FORMATION OF GS MONOLAYERS

Proteins which form more than one type of supramolecular aggregate will provide versatile components for nanofabrication. It has been observed previously (Yanchunas et al., 1994) that side-to-side aggregation of GS tubules is enhanced in the presence of high concentrations (~3 mM to 5 mM) of Zn^{2+} or Co^{2+}. Therefore, we hypothesized that an interdodecameric metal binding site also existed on the edges of the GS structure. We reasoned that mutant proteins in which the metal-binding sites have been destroyed between *His*-4, *Met*-8, and *His*-12 should retain the ability to aggregate laterally in the presence of Co^{2+} or Zn^{2+}. In order to test this hypothesis, the H12D mutant, which demonstrates no stacking behavior, was exposed to varying concentrations of Co^{2+}, and the light-scattering intensity was measured as described in Methods. At 0.3 mM Co^{2+}, the H12D mutant demonstrated no time-dependent increase in light-scattering intensity. Under these conditions, the wild-type GS stacked and exhibited a rapid increase in light-scattering intensity. Upon addition of 3 mM Co^{2+} to the H12D mutant, there was a gradual increase in light scattering, indicating a metal-dependent aggregation. Examination of the sample by electron microscopy indicated that the mutant GS dodecamers were mainly aggregated in islands of hexagonally packed protein monolayers, in marked contrast to the wild-type GS, which exists as laterally associated tubules (Figure 5). The largest islands observed were ~900Å x 1200Å. These results indicate that there is an intermolecular metal-binding site on the lateral edges of the *E. coli* GS dodecamers which drives this self-assembly process. Future experiments will attempt to identify and modify this site for generation of extended protein monolayers from hexagonally packed GS dodecamers.

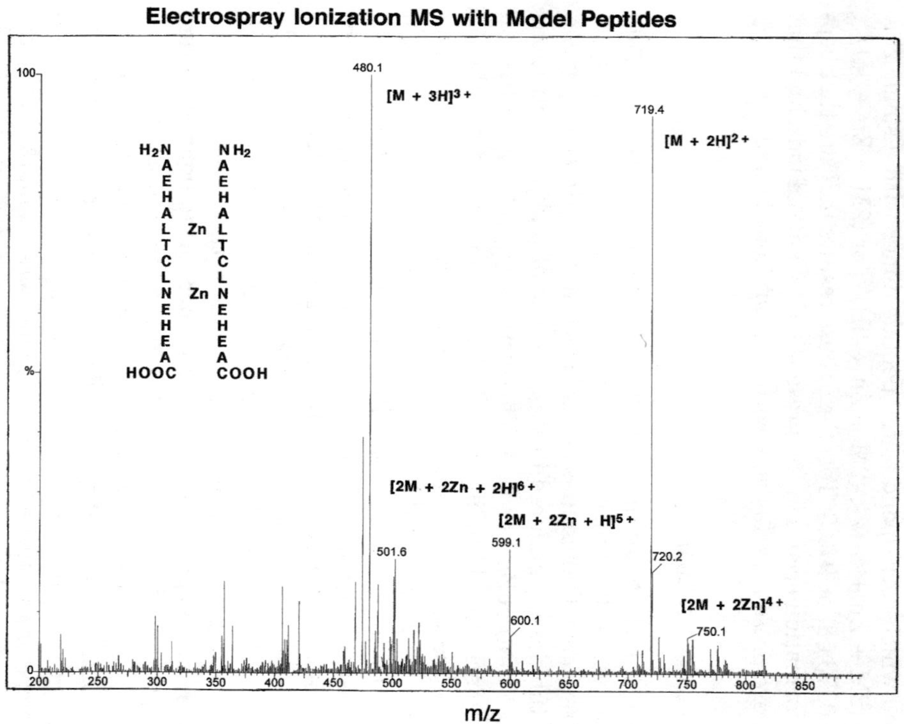

Figure 4. Electrospray ionization mass spectroscopy (MS) of peptide 2 in the presence of Zn^{2+}. In addition to the free monomeric peptide, $[M+3H]^{3+}$ and $[M+2H]^{2+}$, the predominant species is the dimer with two Zn^{2+} ions bound, $[2M+2Zn+2H]^{6+}$, $[2M+2Zn+H]^{5+}$, $[2M+2Zn]^{4+}$.

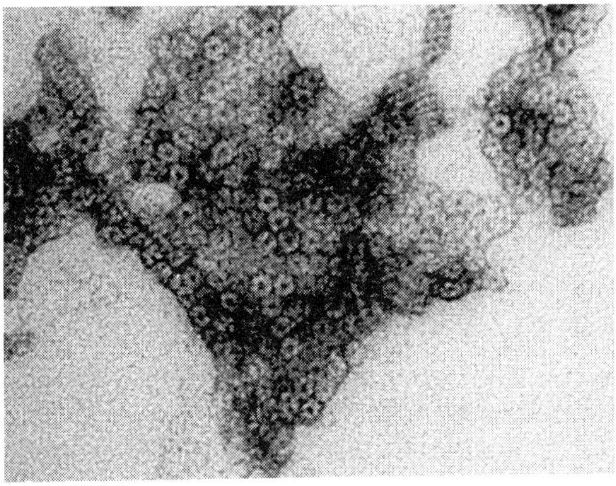

Figure 5. Electron micrographs of glutamine synthetase (GS) sheets. The H12D mutant forms hexagonally packed GS monolayers in the presence of excess Co^{2+}. After addition of 3 mM Co^{2+} to a solution of 0.03-mM H12D GS, the sample was negatively stained with 1% uranyl acetate on a Formvar-coated grid, air-dried, and mounted. Microscopy was at 80KeV, 70,000× magnification before photographic enlargement.

ACKNOWLEDGMENT

This work was supported by NSF (#MCB9305202) and the Whitaker Foundation (#93-0272).

REFERENCES

Almassey, RJ, CA Janson, R Hamlin, N-H Xuong, and D Eisenberg. **1986**. Novel subunit-subunit interactions in the structure of glutamine synthetase. Nature 323:304-309.

Dabrowski, MD, J Yanchunas, Jr., BC Villafranca, EC Dietze, P Schurke, and WM Atkins. **1994**. Supramolecular self-assembly of glutamine synthetase: Mutagenesis of a novel intermolecular metal binding site required for dodecamer stacking. Biochemistry 33:14957-14964.

Frey, TG, D Eisenberg, and FA Eiserling. **1975**. Glutamine synthetase forms three- and seven-stranded helical cables. Proc. Natl. Acad. Sci. USA 72:3402-3406.

Hutchens, TW, RW Nelson, MH Allen, CM Li, and TT Yip. **1992**. Peptide-metal ion interactions in solution: Detection by laser desorption time-of-flight mass spectrometry and electrospray ionization mass spectrometry. Biol. Mass Spectrom. 21:151-159.

Jorgensen, WL. **1993**. Supramolecular chemistry. Proc. Natl. Acad. Sci. USA 90:1635-1636.

Miller, RE, E Shelton, and ER Stadtman. **1974**. Zinc-induced paracrystalline aggregation of glutamine synthetase. Arch. Biochem. Biophys. 163:155-171.

Shwu-Huey, L, C Pan, and D Eisenberg. **1993**. Feedback inhibition of fully unadenylylated glutamine synthetase from *Salmonella typhimurium* by glycine, alanine, and serine. Proc. Natl. Acad. Sci. USA 90:4996-5000.

Whitesides, GM, EE Simanek, JP Mathias, CT Seto, DN Chin, M Mammen, and DM Gordon. **1995**. Noncovalent synthesis: Using physical organic chemistry to make aggregates. Accounts Chem. Res. 28:37-44.

Yanchunas, J Jr., MD Dabrowski, P Schurke, and WM Atkins. **1994**. Supramolecular self-assembly of *Escherichia coli* glutamine synthetase: Characterization of dodecamer stacking and high order association. Biochemistry 33:14957-14964.

Zhou, NE, BY Zhu, CM Kay, and RS Hodges. **1992**. The two-stranded α-helical coiled-coil is an ideal model for studying protein stability and subunit interactions. Biopolymers 32:419-426.

MOLECULAR DYNAMICS SIMULATION STUDIES OF ELECTRON TRANSFER PROPERTIES OF IRON-SULFUR PROTEINS

T. Ichiye, R. B. Yelle, J. B. Koerner, P. D. Swartz, and B. W. Beck

Department of Biochemistry/Biophysics, Washington State University, Pullman, Washington

Key Words: *Rubredoxin, redox potential, electron transfer*

ABSTRACT

Molecular dynamics simulations of the oxidized and reduced forms of several iron-sulfur proteins are being used to study the electron transfer properties, including redox potentials and electron transfer rates, of these proteins. In particular, deviations from Marcus theory due to the nonlinear response of the environment will be discussed.

INTRODUCTION

The iron-sulfur proteins are ubiquitous electron transfer proteins that are involved in many biological processes, such as photosynthesis, respiration, and enzymatic reactions (Cammack, 1992; Zeng et al., in press). One of the most intriguing questions about electron transfer proteins is how they affect the electron transfer properties (i.e., redox potentials and electron transfer rates) of their redox sites. For instance, homologous rubredoxins with the same redox site exhibit different redox potentials, from about -60 to 40 mV (Moura et al., 1979; Zeng et al., in press). Analogs of the rubredoxin redox site generally have redox potentials that are considerably lower: as low as about -800 mV (Lane et al., 1977). Since the conformational changes of the redox sites of the iron-sulfur proteins are small (Backes et al., 1991), attention has focused on the protein matrix itself (Sweeney and Rabinowitz, 1980). Our work has examined the microscopic electrostatic environment of the redox site. Here, we describe how this work demonstrates that the structural origins of observed redox potentials in iron-sulfur proteins are mainly due to a combination of the electrostatic contributions of the protein backbone and polar side chains as well as the solvent. These interactions are difficult to predict from structural data alone because they depend dynamically on the distance and orientation of the polar group.

The studies presented here mainly concentrate on the rubredoxins (Figure 1), which are small proteins (generally about 6 kDa), usually with a single [Fe] redox site consisting of a single iron tetrahedrally bound to four cysteinyl sulfurs. We have also studied high potential iron-sulfur proteins (HiPIP), which are also relatively small (about 9 kDa), with a [4Fe-4S] redox site consisting of four irons and four inorganic sulfurs arranged in a cubane that is ligated to the protein via cysteines. Our calculations used the molecular mechanics program CHARMM 22g2 (Brooks et al., 1983), using parameters from CHARMM 19 plus additional parameters for TIP3P water (Jorgensen et al., 1983) and the iron-sulfur site (Yelle et al., 1995).

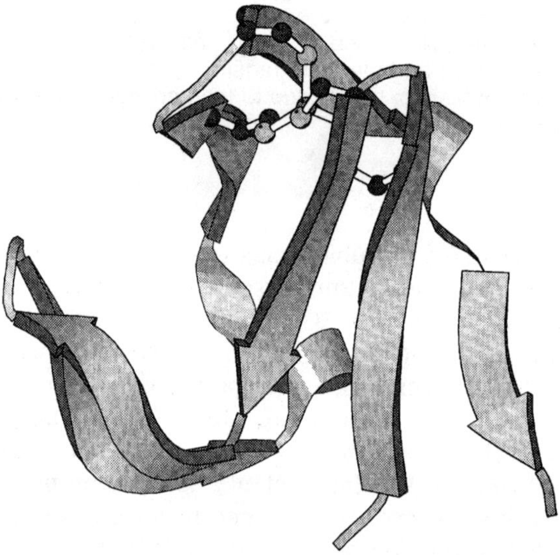

Figure 1. Structure of *Clostridium pasteurianum* rubredoxin (Watenpaugh et al., 1980) with redox site represented in ball-and-stick form. Figure generated using MOLSCRIPT (Kraulis, 1991), by B. W. Beck.

First, a study of the electrostatic potentials at the redox site ϕ in crystal structures of four homologous rubredoxins and four homologous HiPIP (Swartz et al., in press) has shown a rather surprising correlation of ϕ due to just the backbone and polar side chains with the redox potentials of a given set of homologous proteins. More importantly, a key residue was identified that separates the rubredoxins into two groups differing in redox potential by 50 mV, which also holds when a total of eight single [Fe] rubredoxins with both sequence and redox potential data collected by Zeng et al. (in press) are examined. Also, of the four HiPIP studied, the two with unusually low redox potentials (by about 200 mV) can be differentiated from the other two in that ϕ is more negative due to the arrangement of the backbone and several polar side chains in a core region around the redox site, rather than in just one key residue.

Furthermore, solvation does not appear to play a major role in the principal differences observed in these two sets of homologous proteins. For the rubredoxins studied here, this is because they are structurally very similar. For the HiPIP studied here, this is because, of the two with the higher redox potentials one is significantly larger and the other significantly smaller than the two with unusually low redox potentials. Thus, they have the least and most solvated redox sites, respectively, so solvation cannot be the root of the low redox potentials. It was also shown by molecular dynamics simulations of pure water around frozen crystal structures that charged side chains (all of which occur at the surface in the proteins studied) have very little effect, in large part due to dielectric shielding by the water. Thus, these studies indicate the importance of the microscopic environment created by the protein, which includes the nearby backbone and polar side chains but not the more distant surface-charged side chains in the proteins studied here.

A second set of studies involved molecular dynamics simulations of the iron-sulfur protein rubredoxin from *Clostridium pasteurianum* (Cp) (Yelle et al., 1995) and, more recently, that from the hyperthermophilic *Pyrococcus furiosus* (Pf) (Swartz and Ichiye, submitted), in the oxidized and reduced forms. In both cases, good agreement between both forms and crystal data has been obtained, with rms deviations of backbone atoms of about 1.1 to 1.5 Å, which was due in part to the use of explicit solvent and counterions to balance these highly charged proteins. These simulations further confirm the lack of importance of charged side chains, since the presence of counterions more fully screens their charge. However, the reduced form at room temperature exhibits an unexpected structural change. The redox site becomes much more solvent-accessible, so that water enters a channel between the surface and the site (Figure 2) but with little actual structural rearrangement (the rms deviation of backbone atoms between the oxidized and reduced forms is 0.77 Å in Cp rubredoxin). The water is able to enter through a “gate”: a valine side chain lying across the redox site that alternates between opening and closing the path inward in both the oxidized and reduced state. Since the effective charge of the redox site increases in magnitude from -1 to -2 in going from the oxidized to the reduced state, the water is drawn in toward the redox site in the reduced state due to a greater attraction between the two.

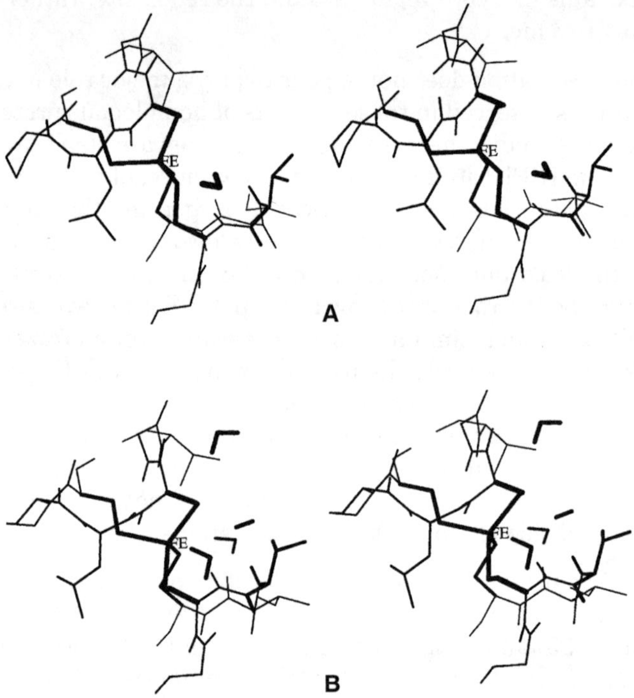

Figure 2. Frame from simulation of oxidized form **(A)** and reduced form **(B)**, showing redox site and waters within 6 Å of iron. Note difference in structure of Val 8 side chain between the two simulations. Picture generated using MOLSCRIPT (Kraulis, 1991); figure adapted from Yelle et al., 1995.

There are several indirect pieces of evidence for the increased solvent accessibility of the redox site upon change in redox state in the rubredoxins. Most interestingly, simulations of Pf rubredoxin at high temperatures (90°C) show no significant water penetration in either redox state (Swartz and Ichiye, submitted), and the calculated change in ϕ of -314 to -247 mV between low and high temperatures is in accord with the unusual temperature dependence of the redox potential, which gives rise to a drop of -180 mV over the same temperature range (Adams, 1992). In fact, the change in ϕ must be due to the difference in water penetration of the reduced form at low and high temperatures, since the change is in the wrong direction if the water penetration is neglected. Moreover, the change in solvent accessibility can account for mutational studies of Cp rubredoxin in which two nonpolar residues very near the redox site were mutated to both a positive *Arg* and a negative *Asp*, in all cases resulting in an increase in redox potential (Zeng et al., in press). Since the mutated residues were at the same positions as those that were shown to be responsible for the solvent accessibility in the simulations of reduced Cp and Pf rubredoxin, changes in solvent accessibility rather than a direct electrostatic effect can explain this observation. In essence, since the charged side chains tend to be more solvated than the nonpolar side chains of the wild type and thus increase exposure of the redox site to solvent, there is an increase in solvent accessibility for a charged side chain regardless of the sign of the charge. These mutations are currently under study in our group. Furthermore, water penetration can also explain the dramatic difference between the redox potentials of rubredoxin and its analogs, and we are also currently studying this question. Unfortunately, there is, so far, no clear experimental evidence for whether or not this water penetration occurs in rubredoxin. No high-resolution crystal exists for reduced Cp rubredoxin, and no such change is seen between the oxidized and reduced crystal structures of Pf rubredoxin (Day et al., 1992). However, such a change may be prevented by crystal packing forces, since crystals of the reduced form are obtained by reducing crystals of the oxidized form at cryotemperatures (Rees, personal communication). Also, a solution-state NMR study of a zinc-substituted Pf rubredoxin did not specifically look for water (Blake et al., 1992). However, there is evidence for solvent in the redox site of other electron transfer proteins, such as the resonance Raman studies of ferredoxin (Backes et al., 1991) and the NMR studies of cytochrome *c* (Qi et al., 1994). The latter is of particular interest because the crystal studies did not show a change in the water location and number upon change in redox state as was seen in the NMR solution-state study, lending support to the idea that crystal packing tends to inhibit the change.

In light of these simulations, it appears that solvation can also play an important role in determining redox properties. In fact, although solvent accessibility does not appear to be the major factor giving rise to the unusually low redox potentials of two of the HiPIP studied, it could explain the remaining differences between the calculated ϕ and the experimental data for the two HiPIP with higher redox potentials. Moreover, resonance Raman studies indicate that of two types of [4Fe-4S]-containing proteins, the redox sites of ferredoxin are accessible to solvent while that of the HiPIP is not, which has led to the proposition that solvent effects lead to the differing redox couples of these two proteins (Backes et al., 1991).

The degree of water penetration observed in our simulations of rubredoxin has also brought into question the accuracy of the potential function for the redox site. The original parameters (Yelle et al., 1995) included electrostatic potential (ESP) "partial" charges from an X- electronic structure calculation (Noodleman et al., 1985), which determined electron populations by methods not geared for optimizing the charges for electrostatic calculations. This set of charges had a slight negative charge on the iron, which seems unphysical. In addition, the van der Waals parameters were based on ionic radii for Fe(II) and Fe(III), which are the formal charges of the iron in the redox site but which do not take into account that the properties of this iron are considerably altered by the cysteinyl ligands.

Originally, it was thought that the details of the potential would make little difference, since no protein atom or solvent had been observed close to the iron of the redox site in the crystal. Also, energy minimization studies indicated that the protein structure was relatively insensitive to the exact charge distribution (Shenoy and Ichiye, 1993). However, based on the simulation results showing water penetration, *ab initio* calculations (unrestricted Hartree-Fock) using the GAMESS (Schmidt et al., 1990) and Gaussian 92 (Frisch et al., 1992) programs were performed on a redox site analog by itself and when interacting with a water molecule (Koerner and Ichiye, unpublished results). New ESP charges for the isolated complex were calculated by the CHELPG method (Breneman and Wiberg, 1990), which fits the charges to the electrostatic potential due to the electron density from the *ab initio* wavefunctions. The latter exhibited considerable changes from the old charges, including a more reasonable positive partial charge for the iron.

The optimized configurations for the redox site and a water molecule in different orientations showed that the distance of closest approach of water to the iron was too small, as expected. Thus, a new nonbonded interaction potential for classical simulations was calculated by fitting the interaction energy as a function of separation generated from *ab initio* calculations of the redox site with water, methane, and ammonia. The new potential has a radius for the iron much larger than that previously used, and the attractive potential was seen to decay according to $1/r^4$, indicating that it was due to the interaction of a charge and an induced dipole. Preliminary simulations with the new parameters indicate that there is still enhanced solvent accessibility upon reduction and, in fact, the use of the old parameters was fortuitous, since the overestimated solvent accessibility brought the phenomenon to our attention.

The question of water penetration really becomes important only if it makes a significant contribution to the redox potential and/or the electron transfer rates. Our temperature-dependent studies of Pf and the mutational studies of Zeng et al. (in press) indicate that water penetration is important in redox potentials, as is the magnitude of the water contribution (17 kcal/mole), although this is an overestimate, since the water is too close, based on our electronic structure calculations. A study was also made of the contribution of this water to the activation energy due to environmental reorganization for electron exchange between two rubredoxins (Yelle and Ichiye, submitted) by analyzing the same simulations of Cp rubredoxin (Yelle et al., 1995). Not surprisingly, this water makes a significant contribution of about 4 kcal/mole to the barrier, shown by calculations of the activation energy $\Delta G\ddagger$ with and without the water (Table 1). This water penetration gives rise to a nonlinear response of the environment, since it enters through a "gate," leading to a deviation from Marcus theory. It can be seen from Table 1 that the "exact," Marcus, and Gaussian values should all be identical if the response is linear. It is somewhat surprising that the deviation was only about 2 kcal/mole, even in this extreme case. In addition, the contributions of each component are additive within the Marcus approximation; the breakdown shows that the protein contributes relatively little to the activation barrier, and the major component is the solvent. These observations lead to a model for controlling specific versus nonspecific electron transfer (Yelle and Ichiye, submitted), based on the idea that specific binding of a physiological electron transfer partner would tend to prevent the change in solvent accessibility of the redox site upon change in redox state, thus reducing the barrier to electron transfer (Figure 3). Therefore, an electron transfer protein with a solvent-accessible redox site would preferentially react with a specifically bound partner.

Table 1. Activation free energies (kcal/mole) for the self-exchange reaction of rubredoxin, excerpted from Yelle and Ichiye (in press).

	ΔG‡, with Penetrating Water	ΔG‡, without Penetrating Water
"Exact"	21.3	17.2
Marcus theory	20.8	16.4
Backbone and polar side chains	2.7	2.7
Charged side chains and counterions	-1.6	-1.6
Total solvent	19.7	15.3
Gaussian approximation	23.9	15.5

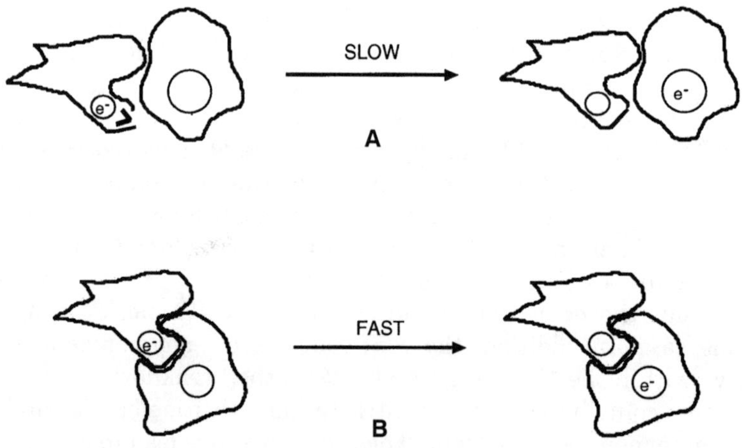

Figure 3. Proposed mechanism for control of electron transfer (Yelle and Ichiye, in press): **A**, nonspecific binding, **B** specific binding.

In summary, the structural origins of the observed differences in redox potentials appear to be a combination of the electrostatics due to the backbone, polar side chains, and solvent. These are complex phenomena that cannot be readily evaluated by simple inspection of the three-dimensional structures and can be of varying importance in explaining different data. Moreover, the change in solvent accessibility of the redox site upon change in redox state leads to large changes in redox potentials and electron transfer rates, which has led to a model for controlling specific versus nonspecific electron transfer.

ACKNOWLEDGMENTS

This work was supported by a grant from the National Institutes of Health (1R29 GM45303). We also acknowledge the Pittsburgh Supercomputer Center, the Maui High Performance Computing Center, and the VADMS center at Washington State University for computational resources.

REFERENCES

Adams, M. **1992**. Novel iron-sulfur centers in metalloenzymes and redox proteins from extremely thermophilic bacteria. Adv. Inorg. Chem. 38:341-396.

Backes, G, Y Mino, TM Loehr, TE Meyer, MA Cusanovich, WV Sweeny, ET Adman, and J Sanders-Loehr. **1991**. The environment of Fe_4S_4 clusters in ferredoxins and high-potential iron proteins. New information from x-ray crystallography and resonance Raman spectroscopy. J. Am. Chem. Soc. 113:2055-2064.

Blake, PR, J-B Park, ZH Zhou, DR Hare, MWW Adams, and MF Summers. **1992**. Solution-state structure by NMR of zinc-substituted rubredoxin from the marine hyperthermophilic archaebacterium *Pyrococcus furiosus*. Protein Sci. 1:1508-1521.

Breneman, CM, and KB Wiberg. **1990**. Determining atom-centered monopoles from molecular electrostatic potentials. The need for high sampling density in formamide conformational analysis. J. Comp. Chem. 11:361-373.

Brooks, B, RE Bruccoleri, BD Olafson, DJ States, S Swaminathan, and M Karplus. **1983**. CHARMM: A program for macromolecular energy, minimization, and dynamics calculations. J. Comp. Chem. 4:187-217.

Cammack, R. **1992**. Iron-sulfur cluster in enzymes: themes and variations, pp. 281-322. In: *Iron-Sulfur Proteins*. Academic Press, Inc., San Diego, CA.

Day, MW, BT Hsu, L Joshua-Tor, J-B Park, ZH Zhou, MWW Adams, and DC Rees. **1992**. X-ray crystal structures of the oxidized and reduced forms of the rubredoxin from the marine hyperthermophilic archebacterium *Pyrococcus furiosus*. Protein Sci. 1:1494-1507.

Frisch, MJ, GW Trucks, M Head-Gordon, PMW Gill, MW Wong, JB Foresman, BG Johnson, HB Schlegel, MA Robb, ES Replogle, R Gomperts, JL Andres, K Raghavachari, JS Binkley, C Gonzalez, RL Martin, DJ Fox, DJ Defrees, J Baker, JJP Stewart, and JA Pople. **1992**. Gaussian 92, Revision A; Gaussian Inc.; Pittsburgh, PA.

Jorgensen, WL, J Chandrasekhar, JD Madura, RW Impey, and ML Klein. **1983**. Comparison of simple potential functions for simulating liquid water. J. Chem. Phys. 79:926-935.

Kraulis, PJ. **1991**. MOLSCRIPT: a program to produce both detailed and schematic plots of protein structures. J. Appl. Crystallogr. 24:946-950.

Lane, RW, JA Ibers, RB Frankel, GC Papaefthymiou, and RH Holm. **1977**. Synthetic analogues of the active sites of iron-sulfur proteins. 14. I. Synthesis, properties, and structures of bis(*o*-xylyl-α,α'-dithiolato) ferrate (II,III)anions, analogues of oxidized and reduced rubredoxin sites. J. Am. Chem. Soc. 99:84-98.

Moura, I, JG Moura, MH Santos, AV Xavier, and J LeGall. **1979**. Redox studies on rubredoxin from sulphate and sulphur reducing bacteria. FEBS 107:419-421.

Noodleman, L, J Norman, JG, JH Osborne, A Aizman, and DA Case. **1985**. Models for ferredoxins: Electronic structures of iron-sulfur clusters with one, two, and four iron atoms. J. Am. Chem. Soc. 107:3418-3426.

Qi, PX, JL Urbauer, EJ Fuentes, MF Leopold, and JA Wand. **1994**. Structural water in oxidized and reduced horse cytochrome c. Struct. Biol. 1:378-382.

Schmidt, MW, KK Baldridge, JA Boatz, JH Jensen, S Koseki, MW Gordon, KA Nguyen, TL Windus, and ST Elbert. **1990**. GAMESS (General Atomic and Molecular Electronic Structure System). QCPE Bull. 10:52.

Shenoy, VS, and T Ichiye. **1993**. Influence of protein flexibility on the redox potential of rubredoxin: Energy minimization studies. Proteins Struct. Funct. Genet. 17:152-160.

Swartz, PD, and T Ichiye. The temperature dependence of the redox potential of rubredoxin from *Pyrococcus furiosus*: A molecular dynamics study. Biochemistry.: submitted.

Swartz, PD, BW Beck, and T Ichiye. Structural origins of redox potential in iron-sulfur proteins: Electrostatic potentials of crystal structures. Biophys. J. (in press).

Sweeney, WV, and JC Rabinowitz. **1980**. Proteins containing 4Fe-4S clusters: an overview. Ann. Rev. Biochem. 49:139-161.

Watenpaugh, K, LC Sieker, and LH Jensen. **1980**. Crystallographic refinement of rubredoxin at 1.2 Å resolution. J. Mol. Biol. 138:615-633.

Yelle, RB, and T Ichiye. Calculation of electron transfer activation energies from molecular dynamics simulations of rubredoxin: Role of water in electron transfer proteins. Biochemistry: submitted.

Yelle, RB, NS Park, and T Ichiye. **1995**. Molecular dynamics simulations of rubredoxin from *Clostridium pasteurianum*: Changes in structure and electrostatic potential during redox reactions. Proteins Struct. Funct. Genet. 22:154-167.

Zeng, Q, ET Smith, DM Kurtz, and RA Scott. Protein determinants of metal site reduction potentials. Site directed mutagenesis studies of *Clostridium pasteurianum*: rubredoxin. Inorg. Chim. Acta (in press).

MOLECULAR DYNAMICS STUDY OF HALOALKANE DEHALOGENASE; IMPLICATIONS FOR SOLVENT CHANNEL ACCESS TO THE ACTIVE SITE

G. E. Arnold and R. L. Ornstein

Environmental Molecular Sciences Laboratory, Pacific Northwest Laboratory, Richland, Washington

Key Words: *Protein dynamics, crystal packing forces, aqueous simulation, dehalogenation*

ABSTRACT

Xanthobacter grows on 1,2-dichloroethane as a sole carbon and energy source. The haloalkane dehalogenase enzyme catalyzes the first metabolic step in the 1,2-dichloroethane *Xanthobacter* degradation pathway. The enzyme mechanism is very similar to that of a serine protease except that an aspartic acid has replaced the nucleophilic serine in the catalytic triad. Several crystal structures of this 310 amino acid enzyme have been solved. The x-ray structures where either the substrate is bound or a covalent alkyl intermediate has been trapped indicate that the active site pocket is occluded from the surface of the enzyme. The underlying dynamics operative in controlling the access and exit channel(s), for substrate and product, are not well understood. To better understand this process, molecular dynamics (MD) simulations of the fully solvated enzyme were calculated using the following three crystal structures as starting points: (1) substrate free simulation, using PDB entry 2HAD (HAD); (2) substrate bound simulation, using PDB entry 2DHC (DHC); and (3) the trapped covalent intermediate alkyl intermediate simulation, using PDB entry 2DHD (DHD). Temporal analysis of the trajectories versus potential energy, hydrogen bond stability, root mean square deviation from the starting crystal structure, atomic fluctuations, and protein radius of gyration indicated that the simulations were both stable and representative of the average experimental observables. However, significant differences were found between the experimental B factor values and those calculated from the MD trajectories for active site residues *Glu* 56, *Phe* 164, *Trp* 175 and *Asp* 260. It has been suggest that the dynamic motions of *Glu* 56 may possibly control access of the hydrolytic water to the active site in the final enzymatic step. The simulations indicated a heterogeneity in sidechain position of this residue and approaching that of two equal populations. *Asp* 260 is located near the entrance of the substrate binding pocket and, in all three trajectories, its sidechain atoms rotated out from the binding pocket to the surface of the protein. These dynamic motions may be part of the mechanism facilitating access of substrate to the active site and need to be studied further.

INTRODUCTION

Synthetic compounds are becoming an increasingly significant source of carbon in the environ ment. Herbicides, pesticides, plasticizers, and solvents are but a few. Many of these compounds possess carbon-halogen bonds or novel structural conformations which have not been observed in the environment previous to their introduction by man. Most of these chemicals are recalcitrant to both biological and physical mineralization and, as a result, have accumulated in various food chains and geological sites. The presence of these xenobiotic compounds in the environment poses a potential threat to both the environment and the health of human society.

Bioremediation offers a very elegant approach to transform contaminants into innocuous byproducts. There are only a few known cases of microorganisms that have developed the intrinsic capability to naturally attenuate xenobiotic species. Unfortunately, the number of halogenated materials that cannot be naturally degraded, on time scales of relevance to human beings, remains significant. In order to utilize indigenous microorganisms as effective biodegradative agents, we must ultimately understand the molecular-level chemistry of the key enzymes of their degradation pathways. Specifically, we must first understand the structure/function mechanisms of those enzymes capable of metabolizing carbon-halogen bonds by nonreductive dehalogenation.

Several halogenated hydrocarbons have been identified at various Department of Energy (DOE) sites; in particular, 1,2-dichloroethane (1,2-DCE) has been targeted as a priority cleanup pollutant. The bacterium *Xanthobacter autotrophicus* is capable of growing on 1,2-DCE as its sole carbon and energy source, and a key enzyme, haloalkane dehalogenase, has been identified as the catalyst in the first degradation step in the 1,2-DCE metabolic pathway (Janssen et al., 1989). This enzyme converts 1-haloalkanes to their corresponding alcohols, with concomitant release of halide ion. The enzyme reaction is thought to be a two-step process in which (1) substrate initially undergoes nucleophilic attack by an aspartic acid residue, forming a covalent ester intermediate, followed by (2) hydrolysis of the ester by an activated water molecule. Many of the details underlying its catalytic activity are poorly understood, both in terms of the chemical intermediates, transition-state species along the reaction coordinate, and the specific details of how the various residues in the active site contribute to the catalysis. Both the affinity of the enzyme toward its substrate and its turnover number are quite low (K_m = 1.1 mM and K_{cat} = 6/sec). Thus, haloalkane dehalogenase presents a challenging protein engineering objective of the design of an efficient detoxifying agent with improved catalytic activity.

In order to better understand the underlying dynamics of the enzyme's active site residues, an all-atom molecular mechanics force field has been used to calculate three different MD simulations of haloalkane dehalogenase, using crystal structures of substrate free enzyme, substrate bound enzyme, and the covalent ester intermediate enzyme (Verschueren et al., 1993a). These simulations provide fundamental information regarding the mobility of the substrate in the active site, the dynamics of key catalytic residues, and regio-specific kinetic data.

METHODS

Protein Models

The haloalkane dehalogenase coordinates were obtained from the Protein Data Bank (PDB) at Brookhaven National Laboratory (Bernstein et al., 1977). PDB coordinate entries 2HAD (substrate free enzyme) 2DHC (substrate bound enzyme) and 2DHD (alkyl intermediate enzyme) were each used to construct the starting protein conformations for the three different trajectories (Verschueren et al., 1993b).

A 10-Å layer of water (constructed from a pre-equilibrated grid of bulk water) was placed around the protein molecules, using the soak command in the Insight II program (Biosym Technologies). Solvent molecules which were in van der Waals contact with atoms of the molecules being solvated were removed. The 2HAD coordinate set contained 203 crystallographically resolved water molecules, and construction of the 10-Å solvent layer added an additional 3474, bringing the aggregate number of atoms in the system to 15,307. The DHC and DHD models included 122 and 151 crystallographically resolved water molecules respectively, and 3367 and 3368 bulk water molecules were added to the respective systems. Hydrogen atoms for both protein and water molecules were explicitly modeled. The coordinates of the added hydrogen atoms were generated according to idealized bond lengths and valence angles, using the Insight II software package (Biosym Technologies). Crystallographic and bulk water molecules were modeled using the CVFF water model (Dauber-Osguthorpe et al., 1988). Side-chain polar hydrogens and water hydrogens were repositioned so as to maximize intramolecular hydrogen bonding, using a previously described automated method (Bass et al., 1992).

All acidic, basic, and terminal protein residues were fully charged. At pH 8.2 protein models HAD and DHC carried a net charge of +17, and protein model DHD carried a net charge of +18, as determined by the program network (Bass et al., 1992). In all cases, histidine 289 was modeled in its charge neutral state with a proton placed in the delta position. Counter ions were added to each system, as previously described (Arnold and Ornstein, 1994), so that each system had a net electric charge of zero.

Minimizations

Atomic coordinate energy minimizations and MD trajectories were calculated using the Discover (version 2.9) simulation software package (Biosym Technologies). To remove artifacts introduced by the addition of explicit hydrogens, the systems were initially energy-minimized, with the positions of all heavy atoms fixed using the methods of steepest descents for 500 steps followed by 500 steps of conjugate gradients minimization. Incorrectly placed water molecules and/or counterions can lead to spurious protein motions during minimization. To avoid this potential problem during minimization, 10 ps of MD was run prior to minimization with the entire protein molecule constrained, thus allowing for the reorientation and relaxation of water dipoles. Next, all constraints were removed from the system and the protein/water/counterion coordinates were minimized for 500 steps, using the steepest descents minimization. The protein/water/counterion coordinates were then further minimized using conjugate gradients until all gradient partial derivatives were less than 1.0 kcal/mole-Å.

MD Simulations

Each MD trajectory was run for 300 ps and calculated using the leapfrog algorithm with a 1-fs time step (Verlet, 1967). Constant temperature during the simulations was maintained by weak coupling to a thermal bath (Berendsen et al., 1984). Atomic velocities were initially assigned using a Maxwellian distribution at 50 K. The systems were then slowly warmed to 300 K, using a graded approach to the target temperature (time constant of 2.0 ps) so that the target temperature was achieved after 10 ps. Following the 10-ps warm-up phase, the time constant was adjusted to 0.1 ps, and dynamics was continued at 300 K for an additional 490 ps.

During the simulation, a half-harmonic restraining force was applied to the water molecules at a distance of greater than 35.0 Å from the alpha carbon atom of residue 54. The restraining force consisted of a force constant of 0.5 kcal mol^{-1} Å^{-2}, with an upper limit of 100.0 kcal mol^{-1} Å^{-2}. No cross-terms were used in the energy expression, and a simple harmonic potential was chosen for the bond-stretching terms. A constant dielectric of 1.0 was used. Both the Coulombic and van der Waals potentials were calculated using the method of dual cutoffs, where cutoff distances of 15.0 and 11.0 Å were used. The use of dual cutoffs is computationally efficient and results in a trajectory that is superior to one using a single, short cutoff (Arnold and Ornstein, 1994). No switching function was used to truncate the nonbonded interactions; the use of a switching function has been previously shown to produce spurious trajectories (Loncharich and Brooks, 1989; Ding et al., 1992; Guenot and Kollman, 1993; Arnold and Ornstein, 1994). The trajectories were calculated using a group-based, nonbonded cutoff, and the nonbonded pair list was updated every 20 time steps. The parameters used in the force field were those of the CVFF force field (Hagler, 1985; Dauber-Osguthorpe et al., 1988).

RESULTS AND DISCUSSION

Several properties that assess the stability and convergence characteristics of the MD trajectories are collected in Figure 1. The incipient potential energies of the three protein models were all quite low (Figure 1A), indicating that the starting structures were all sufficiently well minimized. The initial potential energy of the three trajectories rapidly increased as kinetic energy was added to the systems during the thermalization phase. The trajectories were observed to be energetically converged after ~50 ps of MD simulation. The root mean square deviation (RMSD) of the MD structures with respect to the crystal structures is plotted as a function of time in Figure 1B. The stability of the trajectories is evident from the RMSD values, which appear to level off after ~50 ps and plateau to a value of ~1.5 Å over the final 200 ps of the simulations. Similar characteristics were observed in the time evolution of the MD structures' radius of gyration (Figure 1C). Following an initial increase during the thermalization phase of the simulations, the radius-of-gyration remained stable throughout the remainder of the trajectories. The three crystal structures have identical radius-of-gyration values of 17.9 Å; the average radius-of-gyration values calculated over the final 250 ps of the respective simulations were 18.40 Å ± 0.04 (HAD), 18.50 Å ± 0.05 (DHC), and 18.52 Å ± 0.03 (DHD). These data indicate that no decrease in protein packing density had occurred during the time course of the trajectories. Collectively, this information demonstrates that the simulations are convergent and that the crystal structure geometries are reasonably well maintained, indicating that the trajectories are well stabilized and warrant further analysis.

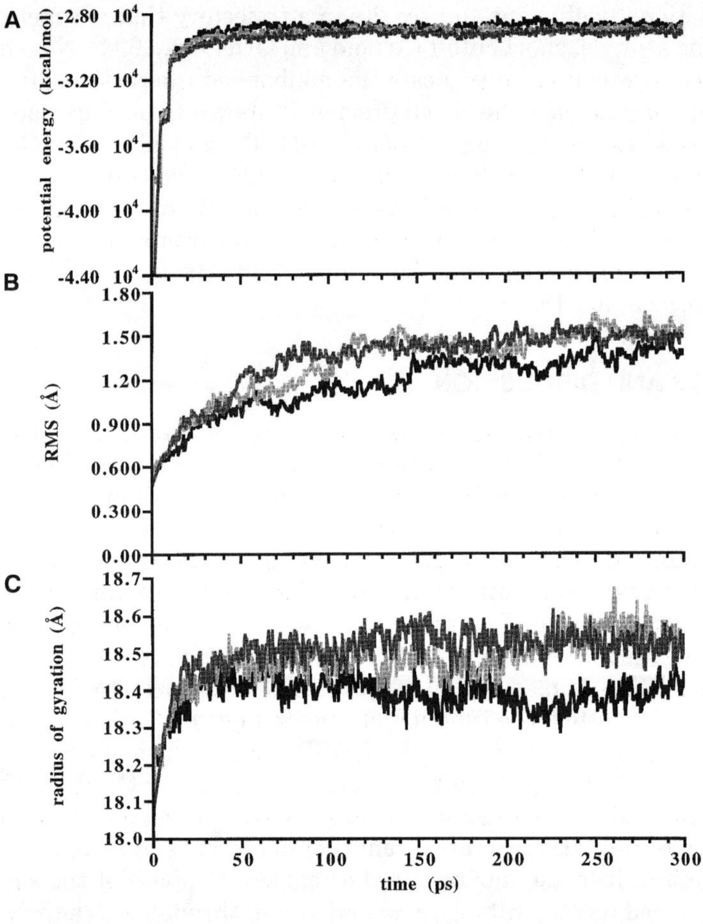

Figure 1. Time evolution convergence profiles for 300 ps of molecular dynamics (MD) simula tion on haloalkane dehalogenase: **A**, calculated potential energies; **B**, root mean square deviation of backbone atoms relative only to crystal structures; **C**, protein radius of gyration. Gray scale of the traces corresponds to: black, substrate-free MD simulation (HAD); dark gray, substrate-bound MD simulation (DHC); and light gray, covalent ester intermediate simulation (DHD).

Atomic temperature factor values are a measure of thermal mobility of the individual atoms comprising the protein molecule. Figure 2 compares the B-factor values determined from the x-ray crystal structures against those calculated from the MD simulations; the B-factor data have been averaged on a per-residue basis and plotted as a function of residue number. The simulated B-factor values were calculated using the expression $B = (8/3)(\pi^2) <u^2>$, where u represents atomic displacements from the mean geometric atomic position. There is, in general, good qualitative agreement between the experimental and simulation values, as demonstrated by the coincident peaks and valleys in both plots. The simulation values are, by and large, greater than the experimental values, and the average values for the simulated B-factors are 17.8 Å^2 (HAD), 21.3 Å^2 (DHC), and 17.9 Å^2 (DHC), compared to 14.3 Å^2 (HAD), 9.4 Å^2 (DHC), and 12.1 Å^2 (DHC) for the crystal structures. Regions of high B-factor values in the simulations generally occur in loop and turn structures between regions of a-helix and b-strand structures. Differences observed in B factor values, between simulation and experiment, are particularly pronounced outside regions of core secondary structure and this is a manifestation of comparing two different environments. The simulation paradigm is that of a protein molecule free in solution and not a protein molecule in a crystalline lattice, hence these particular regions should exhibit greater mobility in the solution environment compared to that in the constrained crystal environment.

In the protein simulations, there is an observed strong correlation between residue mobility (B-factor) and corresponding solvent accessibility, with correlation values of 0.77 (HAD), 0.68 (DHC), and 0.77 (DHD). This trend is evident in Table 1, where B-factor and solvent accessible surface values are compared for the protein secondary structure elements. In general, there is good agreement between the simulation values and those determined from the crystal structure for both secondary structure mobility and the corresponding solvent accessible surface; however, for the mobility parameters there are some noteworthy exceptions, particularly in the cap region of the protein. In the cap domain of the protein crystal, there are several intermolecular contact points (residues *Arg*167, *Ala*169, *Ser*183, *Arg*186, *Gln*189, and *Arg*193) which contribute to the structural rigidity of this region. In contrast, the mobility of this region is not impeded by near-neighbor contacts in the protein simulation. This absence of intermolecular contacts may be why this region exhibits significantly greater mobility in the

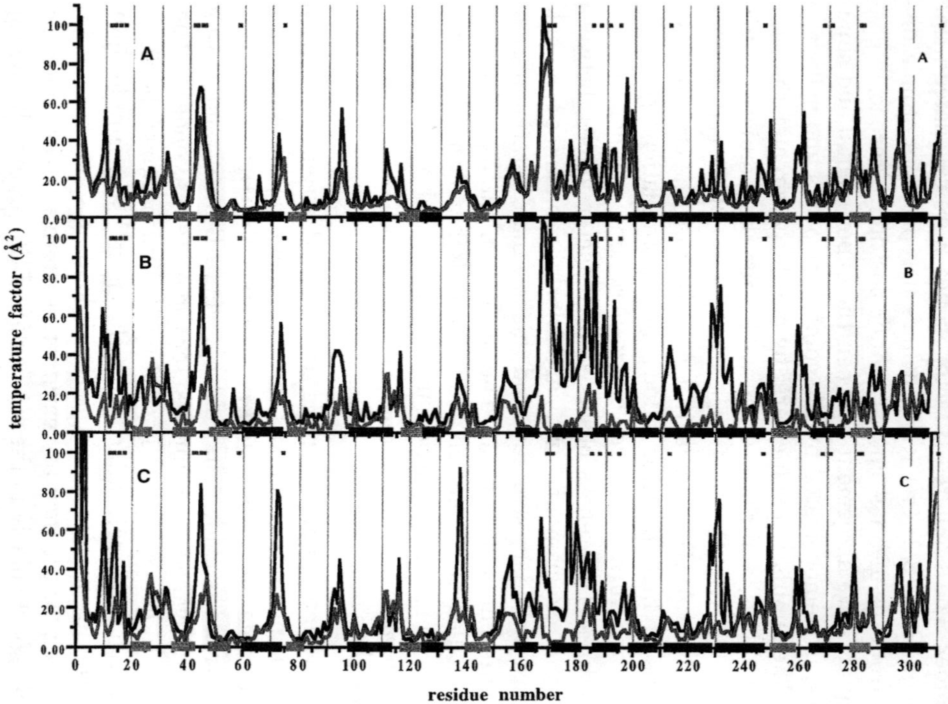

Figure 2. Comparison of all-atom atomic temperature factors between x-ray crystal structures (gray trace) and those calculated from MD simulations (black trace): **A**, substrate-free MD simulation (HAD); **B**, substrate-bound MD simulation (DHC); and **C**, covalent ester intermediate simulation (DHD). Solid bars along abscissa demarcate positions of secondary structure elements and color-coded, black bars (helices) and gray bars (β-strands). Squares at top of plots demarcate crystal contact points.

Table 1. B-factor and solvent accessible surface area values of secondary structure units for the cyrstal structures and molecular dynamics simulations.

Secondary Structure Unit[b]	B-Factors ($Å^2$)[a]						Solvent Accessible Surface Area ($Å^2$)[a]					
	Crystal Structure			Simulation			Crystal Structure			Simulation		
	HAD[c]	DHC[c]	DHD[c]	HAD[c]	DHC[c]	DHD[c]	HAD[c]	DHC[c]	DHD[c]	HAD[c]	DHC[c]	DHD[c]
β1	57.1	35.5	46.6	75.0	100.8	75.7	270.5	270.3	262.3	277.5	250.5	266.3
β2	48.2	30.4	33.6	56.4	92.7	60.1	93.3	93.3	95.2	101.7	134.9	93.9
β3	39.0	16.5	27.9	36.3	43.4	36.5	39.2	43.9	41.5	74.8	79.7	79.0
H_1	140.9	74.9	112.4	137.5	137.8	184.1	334.7	340.6	339.1	395.7	331.6	347.6
β4	51.3	27.7	38.0	30.2	34.2	26.8	7.8	9.0	8.8	19.5	15.3	17.1
H_2	197.7	129.6	175.2	175.9	181.8	161.8	448.9	460.7	449.2	565.3	533.8	528.2
β5	46.7	18.2	35.0	21.2	25.3	20.4	0.8	0.5	0.9	0.8	4.1	1.5
H_3	52.1	17.9	33.4	35.7	57.8	46.0	9.3	10.5	0.6	13.1	58.5	1.7
β6	85.4	32.9	64.9	50.4	53.6	48.6	86.1	92.2	82.8	92.4	118.1	92.3
H_4	59.2	17.6	39.4	98.8	88.8	118.0	203.4	212.1	201.2	190.2	205.7	188.7
H_5	85.9	38.7	53.1	193.8	357.4	336.4	345.7	344.8	338.2	479.8	360.6	619.2
H_6	83.4	29.2	61.1	179.6	263.9	147.8	446.0	444.4	446.4	412.0	468.2	538.2
H_7	81.3	51.6	68.3	102.6	89.3	92.1	273.2	277.7	275.5	292.0	252.5	241.9
H_8	166.7	71.6	119.0	196.4	328.5	180.1	360.4	357.9	356.2	432.2	384.0	410.1
H_9	230.2	145.5	192.4	245.8	359.9	330.7	713.8	719.7	721.2	835.2	862.7	868.3
β7	90.9	35.3	63.2	62.2	49.2	57.0	54.1	58.0	58.1	74.5	75.0	73.9
H_{10}	132.3	44.0	80.0	137.2	144.1	102.2	463.8	464.1	471.3	361.7	467.0	461.9
β8	97.5	67.8	82.8	140.7	80.7	91.9	309.5	334.6	322.3	329.7	296.3	315.9
H_{11}	304.6	289.8	299.6	314.5	227.7	309.3	604.1	605.9	612.8	681.8	698.7	720.7

[a]Values are summation of all residues contained in secondary structure unit.
[b]Secondary structure units as defined in the Brookhaven Protein Data Bank entry, 2HAD.
[c]HAD, substrate-free structure; DHC, substrate-bound structure; DHD, covalent ester intermediate structure.

MD simulations than that observed in the crystal structures. Increased structural mobility is particularly evident for helices H_5 and H_6, situated at the top of the active site cavity. Outside of the cap region, helices H_1, H_9, and H_{10} were all found to have greater mobility in the simulation models as compared to the crystal structure. Here, too, intermolecular contact points exist in the crystal structure and most likely restrict the mobility of these helices in the crystal lattice.

A feature common to all three simulations was the high degree of mobility observed for the catalytic triad residues *Asp*260 and *His*289 (Table 2). In all simulations, the *His*289 imidazole ring was modeled as a net neutral species, with a proton placed at the delta nitrogen position. As a result there were no ionic interactions with *Asp*260. Within the first 50 ps of all three trajector ies, the *Asp*260 side chain was observed to rotate away from the active site toward the surface of the protein, into the solvent. This motion was a result of rotations about the and backbone dihedral angles from starting geometries of roughly -77 and 125 to mean geometries of ~-124 and ~-78 determined over the final 200 ps of the trajectories. Following the rotation of *Asp*260 into the solvent, the mobility of *His*289 greatly increased, and its side chain rotated between its initial position and the position formerly occupied by *Asp*260 in the HAD and DHD trajectories (Figure 3). Similar dynamics were observed in the DHC trajectory for the first ~125 ps; however, during the final ~175 ps, *His*289 moved toward *Phe*290 exposing a solvent channel to the catalytic cavity (Figure 4). The access channel extends from the solvent, through the space left vacant from the rotations of residues *Asp*260 and *His*289, to the active site.

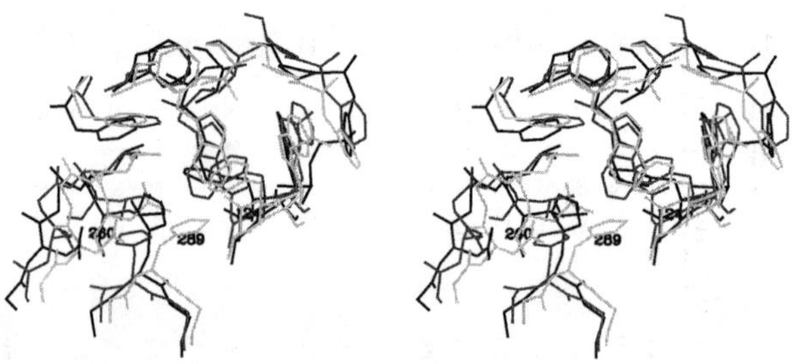

Figure 3. Stereo view of active site residue positions for: **A**, substrate-bound crystal structure, light gray residues; **B**, substrate-free MD simulation (DHC), dark gray residues; and **C**, substrate- bound MD simulation (DHC), black residues. In all three MD simulations there were observed backbone dihedral angle transitions for *Asp*260, which moved the *Asp*260 side chain out into the solvent. The imidazole side chain of *His*289 could then rotate either to fill the *Asp*260 void or rotate to a position proximate to *Phe*290, exposing an access channel.

Table 2. Active site residue B-factor and solvent accessible surface area values for crystal strucutres and molecular dynamics simulations.

Active Site Residue	B-Factors (Å^2)						Solvent Accessible Surface Area (Å^2)					
	Crystal Structure			Simulation			Crystal Structure			Simulation		
	HAD	DHC[a]	[a]DHD[a]	HAD[a]	DHC[a]	DHD[a]	HAD[a]	DHC[a]	DHD[a]	HAD[a]	DHC[a]	DHD[a]
E56	6.13	4.12	4.75	9.06	22.5	8.32	1.20	0.90	0.20	4.30	5.00	14.0
D124	6.72	3.21	10.0 [b]	8.08	6.13	8.31 [b]	5.40	4.50	0.20 [b]	6.10	36.3	1.00 [b]
W125	6.18	2.26	2.51	5.44	11.7	6.96	1.90	2.10	0.40	5.80	11.5	0.70
F128	7.72	2.71	3.70	4.59	9.78	9.64	2.00	3.90	0.00	0.70	10.7	0.00
F164	9.27	1.09	5.30	14.6	18.8	20.5	10.8	11.6	8.40	6.00	25.5	5.00
F172	8.13	4.37	3.48	16.1	32.6	20.5	4.80	2.80	0.50	14.7	12.8	15.1
W175	9.08	2.08	2.98	11.8	24.6	14.0	4.20	3.10	0.00	4.20	14.7	12.0
F222	8.70	4.85	4.81	10.5	24.7	7.03	0.50	0.50	0.60	3.60	0.00	0.10
P223	8.98	4.71	6.05	12.9	20.9	8.81	0.60	0.40	0.30	1.90	6.10	7.60
V226	9.37	4.30	5.29	14.7	23.5	16.6	5.20	5.50	0.50	4.10	15.3	1.90
D260	8.55	2.16	6.34	27.8	41.6	20.3	11.2	12.1	11.5	60.6	101	77.1
L262	12.1	3.99	7.55	12.9	35.2	18.3	3.90	6.80	1.70	6.60	28.4	5.90
L263	10.8	3.79	9.69	13.3	12.6	18.4	1.00	1.10	0.00	0.10	23.1	4.00
H289	6.83	1.48	2.63	15.8	33.6	8.77	3.70	3.10	3.80	0.10	26.9	2.00

[a]Had, substrate-free structure; DHC, substrate-bound structure; and DHD, covalent ester intermediate structure.
[b]Reported values are for the *Asp*124 chloroethyl ester.

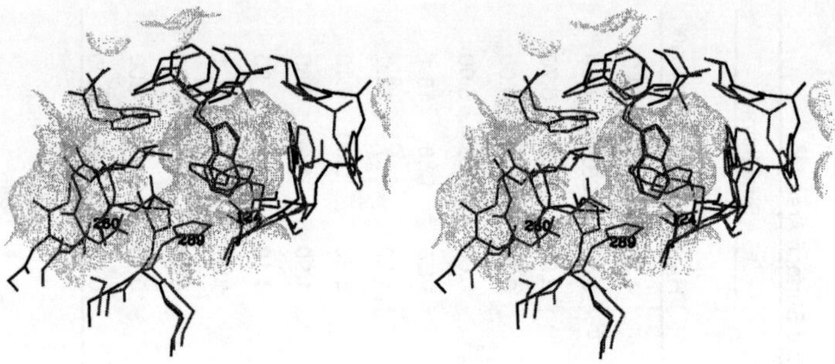

Figure 4. Stereo view of active site residue positions for substrate-bound crystal structure (gray residues) and substrate-free MD simulation (black residues). The access channel, formed as a result of transition of *Asp*260 and *His*289 from their initial crystal structure atomic positions, is depicted by shaded surface. Shaded surface was generated using a rolling probe with a 1.4-Å radius.

The observed active site residue B-factor and accessible surface values for the substrate free trajectory (HAD) corresponded well to those determined from the crystal structure (except for *Asp*260 and *His*289). And the mobility parameter values of *Asp*124 (HAD and DHC trajectories) and the chloroethyl aspartate124 (DHD trajectory) were consistently low for all three trajectories in comparison to the other active site residues. However, for the covalent intermediate MD trajectory (DHD) these values were generally observed to be greater than those determined from the crystal structure, and for the substrate-bound MD trajectory (DHC) these values were approximately a factor of 3 to 5 greater for most active site residues. Thus, the introduction of either the 1,2-dichloroethane substrate or the chloroethyl ester moiety into the catalytic cavity greatly altered the dynamics of the surrounding active site residues. Both residue mobility and solvent accessible surface measures were significantly greater for the DHC simulation compared to those for the DHD simulation. The large increase in accessible surface area for the DHD trajectory resulted from the formation of the solvent channel. The larger B-factor values observed for the DHC trajectory were probably caused by the mobility of the 1,2-dichloroethane substrate in the catalytic cavity and were a result of the corresponding adjustments in residue sidechain position necessary to accommodate substrate movement.

The mobility of the substrate in the active site was interpreted from distance changes of 1,2-dichloroethane relative to *Asp*124. These data are shown in Figure 5. Distances were meas ured as a function of time

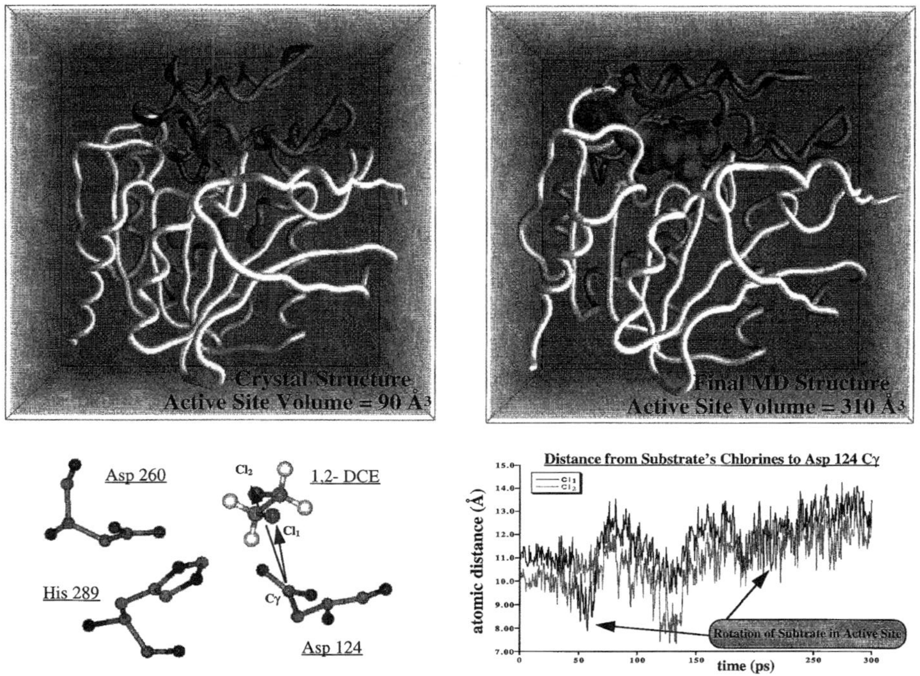

Figure 5. Comparison of final molecular dynamics (MD) structure (right panel) to that of the crystal structure (left panel) of the substrate-bound structures (DHC). Protein is shown in a "worm" representation, with main body of protein shown in white and CAP domain shown in dark gray. Gray surface depicts active site volume, and catalytic triad residues are shown in a wire representation. Ball-and-stick drawing at lower left depicts catalytic triad and shows distances measured from the *Asp*124 gamma carbon to the substrate's chlorine atoms. Graph at bottom right is a time evolution plot of distances measured from *Asp*124 gamma carbon to the 1,2-dichloroethane chlorine atoms for the substrate-bound MD simulation (DHC).

from the *Asp*124 gamma carbon to the substrate's chlorine atoms (Figure 5). The observed changes in distance resulted from a rotation of the substrate in the active site. The tumbling motion consisted of a rotation of the substrate molecule about the C-C carbon atom substrate axis, the first of which occurred at ~60 ps. This motion interchanges the positions of the substrate's carbon atoms relative to the *Asp*124 carboxylic oxygen atoms. In addition, the rotation caused the initial Cl_2 atomic position to move away from *Trp*125 and fill the void space between *Phe*164 and *Phe*172.

The large amount of mobility observed for the substrate in the catalytic cavity over the time course of the DHC MD simulation indicates that the substrate is loosely bound in the active site (Figure 5). This phenomenon can be directly attributed to expansion of the active site volume over the time course of the simulation, and this expansion is primarily due to the movement of the CAP domain away from the rest of the protein. This observed expansion was most likely due to the solution structure paradigm used in the MD calculations. In this model no intermolecular crystal contact points were present to impede the dynamics of the CAP domain. The observed large degree of mobility of the CAP domain in the MD simulations could allow for a change in protein conformation that would facilitate channel access to the active site.

ACKNOWLEDGMENTS

Pacific Northwest Laboratory is operated for the U.S. Department of Energy by Battelle Memorial Institute under Contract DE-AC06-76RLO 1830. This work was supported by a grant KP0402 from the Health Effects and Life Sciences Research Division of the Office of Health and Environmental Research of the U.S. Department of Energy (RLO).

REFERENCES

Arnold, GE and RL Ornstein. **1994**. An evaluation of implicit and explicit solvent model sys tems for the molecular dynamics simulation of bacteriophage T4 lysozyme. Proteins Struct. Funct. Genet. 18:19-33.

Bass, MB, DF Hopkins, WAN Jaquysh, and RL Ornstein. **1992**. A method for determining the positions of polar hydrogens added to a protein structure that maximizes protein hydrogen bonding. Proteins Struct. Funct. Genet. 12:266-277.

Berendsen, HJC, JPM Postma, WFv Gunsteren, A DiNola, and JR Haak. **1984**. Molecular dynamics with coupling to an external bath. J. Chem. Phys. 81:3684-3690.

Bernstein, FC, TF Koetzle, GJB Williams, EFJ Meyer, MD Brice, JR Rodgers, O Kennard, T Shimanouchi, and M Tasumi. **1977**. The protein data bank: A computer-based archive file for macromolecular structures. J. Mol. Biol. 112:535-542.

Dauber-Osguthorpe, P, VA Roberts, DJ Osguthorpe, J Wolff, M Genest, and AT Hagler. **1988**. Structure and energetics of ligand binding to proteins: *Escherichia coli* dihydrofolate reductase- trimethoprim, a drug-receptor system. Proteins Struct. Funct. Genet. 4:31-47.

Ding, H, N Karasawa, and WA Goddard. **1992**. Optimal spline cutoffs for Coulomb and van der Waals interactions. Chem. Phys. Lett. 193:197-201.

Guenot, J and PA Kollman. **1993**. Conformational and energetic effects of truncating nonbonded interactions in an aqueous protein dynamics simulation. J. Comp. Chem. 14:295-311.

Hagler, AT, VJ Hruby, and J Meienhofer, (eds.). **1985**. In: *The Peptides*, pp. 213-299. Academic Press, New York.

Janssen, DB, F Pries, J Van der Ploeg, B Kazemier, P Terpstra, and B Witholt. **1989**. Cloning of 1,2-dichloroethane degradation gens of *Xanthobacter autotrophicus* GJ10 and expression and sequencing of the *dhl*A gene. J. Bacteriol. 171:6791-6799.

Loncharich, RJ, and BR Brooks. **1989**. The effects of truncating long-range forces on protein dynamics. Proteins Struct. Funct. Genetics. 6:32-45.

Verlet, L. **1967**. Computer experiments on classical fluids. I. Thermodynamical properties of Lennard-Jones molecules. Phys. Rev. 159:98-103.

Verschueren, KHG, J Kingma, HJ Rozeboom, KH Kalk, DB Janssen, and BW Dijkstra. **1993a**. Crystallographic and fluorescence studies of the interaction of haloalkane dehalogenase with halide ions. Studies with halide compounds reveal a halide binding site in the active site. *Biochemistry* 32:9031-9037.

Verschueren, KHG, F Seljee, HJ Rozeboom, KH Kalk, and BW Dijkstra. **1993b**. Crystallographic analysis of the catalytic mechanism of haloalkane dehalogenase. *Nature* 363:693-698.

THE X-RAY CRYSTAL STRUCTURE OF A HIGHLY DESICCATED PROTEIN

J. A. Bell

Department of Chemistry, Rensselaer Polytechnic Institute, Troy, New York

Key Words: *X-ray, dehydration, molecular dynamics, crystallization, ribonuclease*

ABSTRACT

Crystals of ribonuclease A in space group $P2_12_12_1$ were grown from aqueous methanol and dried in the air for several days. Crystals thus treated continue to diffract x rays to better than 2.0 Å resolution. The more commonly analyzed crystal form of ribonuclease A (space group $P2_1$) does not diffract x rays at high resolution when so treated. The $P2_12_12_1$ crystals were then hydrated to various levels by sealing them in glass capillaries containing either saturated salt solutions of known vapor pressure, or dry calcium sulfate. Thus, the effect of dehydration on protein structure could be monitored by x-ray diffraction at near-atomic resolution. Protein in crystals rehydrated at high partial pressure of water has a structure very similar to that in crystals that were never desolvated. As vapor pressure decreased, the unit cell volume decreased; many small conformational changes were observed; and certain regions of the protein became disordered, suggesting large increases in backbone mobility.

INTRODUCTION

Almost all protein structures currently known at or near atomic resolution are highly hydrated. Because of the unusual stability of the modification IX crystal form of bovine pancreatic ribonuclease (King et al., 1962), the structure of this protein can be studied at various levels of reduced hydration, even when desiccated with anhydrous calcium sulfate. (For some other cases, where crystal structure information is available for proteins under reduced hydration, see Kodandapani et al., 1990; Kachalova et al., 1991; Fitzpatrick et al., 1994; Yennawar et al., 1995.)

The structure of this crystal form of ribonuclease in 60% aqueous methanol has been deposited with the Protein Data Bank (Abola et al., 1987) and given the identifier 1BEL (M. H. Dung and J. A. Bell, in preparation). Preliminary information regarding the partially refined x-ray structure of calcium-sulfate-desiccated crystals is reported here, together with some information about the crystals at intermediate hydration.

Dehydration of proteins occurs during freeze-drying of protein pharmaceuticals now produced through biotechnology. Rehydration and recovery of activity is a problem that will be better understood when the structural effects of protein dehydration are understood. This project will also provide important experimental data, in which solvent water is not a consideration, to compare with molecular mechanics calculations.

A growing area of biotechnology involves the use of enzymes in organic solvents (Klibanov, 1989). In aid of efforts to design enzymes for these unusual environments, this project will provide the details of the structure of an enzyme significantly depleted of water, first in air, and later in anhydrous organic solvents.

MATERIALS AND METHODS

Diffraction data were collected using an R-axis image plate system mounted on a Rigaku rotating anode x-ray generator RU-200 (50 mA, 100 kV) with a graphite monochromator. Data collection statistics are summarized in Table 1.

Crystals were grown from approximately 56% aqueous methanol solutions of bovine pancreatic ribonuclease A (Sigma) and were dried by evaporation of solvent in the air for at least 48 hr before being sealed in a glass capillary with anhydrous calcium sulfate or with a saturated salt solution to maintain a fixed vapor pressure of water (Rockland, 1960). Data for the desiccated crystal structure were collected 8 days after the capillary was sealed. Figure 1 shows the physical arrangement of desiccant and crystal.

Table 1. Data collection statistics.

Sample	Rmerge, %	Resolution range, Å	Completeness $I/\sigma(I) > 2$, %	Mosaicity, degrees
60% Methanol	8.3	∞ –1.6	82	0.25
Dried, $CaSO_4$	4.4	∞ –2.1[a]	74	1.0

[a]Data set includes some reflections to 1.9 Å.

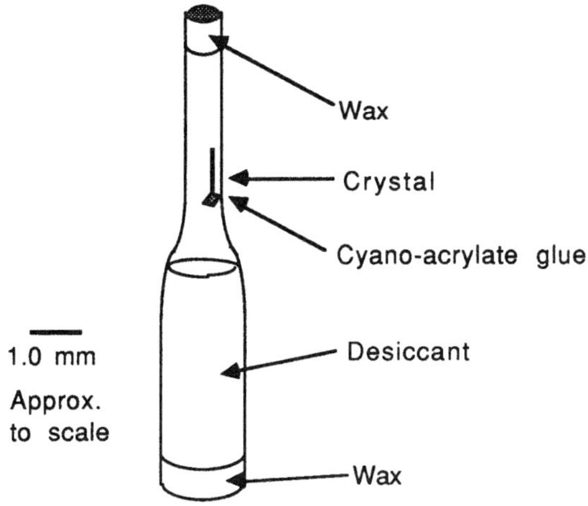

Figure 1. Experimental arrangement of a ribonuclease crystal in a glass capillary for controlled hydration and data collection.

The structure was solved by molecular replacement, using the package Merlot (Fitzgerald, 1988). Automated refinement was carried out using the program TNT (Tronrud et al., 1987). Manual modification of the structural model was accomplished using the program FRODO (Jones, 1978). Ten of 124 residues in the protein have not yet been placed in the electron density of the desiccated structure. Current refinement statistics for the desiccated structure are shown in Table 2.

The water content of crystalline ribonuclease was determined by thermal gravimetric analysis. Approximately 8 mg of crystalline material was incubated for at least 24 hr in the presence of a large excess of an appropriate saturated salt solution. Samples were heated from room temperature to 100C at 10C min^{-1} in a stream of dry nitrogen gas in a Perkin-Elmer TGA. Weight changes were reproducible when samples were recovered, rehydrated, and reanalyzed. Nevertheless, some thermal degradation of the protein at elevated temperature may have occurred. Bound water in less-hydrated samples (relative humidity < 11%) could not be determined because significant hydration from air occurred during the several seconds required to transfer samples from vial to apparatus. Measurement precision was ±10%.

Table 2. Refinement statistics.

	60% Methanol	Dried, $CaSO_4$
Refinement resolution range, Å	10 – 1.6	5 – 1.9
R-factor	0.150	0.204
RMS deviation from ideal geometry		
Bond length, Å	0.010	0.022
Bond angle, degrees	1.93	5.10
Planar groups, Å	0.010	0.035
Trigonal planarity, Å	0.016	0.027
Solvent molecules in current model		
Water	82	0
Sulfate	2	2
Methanol	4	0

RESULTS AND DISCUSSION

The x-ray structure of this crystal form of bovine pancreatic ribonuclease in 60% methanol has been determined to 1.6 Å resolution (1BEL). The structure is very similar to that of this protein when found in other, more frequently studied crystal forms (M. H. Dung and J. A. Bell, in preparation).

Incubation of the crystals at lower humidity is associated with a decrease in the volume of the unit cell (Figure 2). The amount of water bound varies by more than a factor of 4 over the range of conditions from 98% to 11% relative humidity (Figure 2). The change in the unit cell does cause the crystals to shrink in size and, sometimes, crack. A large increase in mosaic spread is also observed (Table 1).

When crystals are dried in the air, as described above, and then allowed to absorb water at 98% relative humidity, the x-ray crystal structure is very similar to the structure in aqueous methanol (results not shown). Thus, changes in the crystal structure of the desiccated protein are probably not just the result of the trauma of dehydration and removal of methanol.

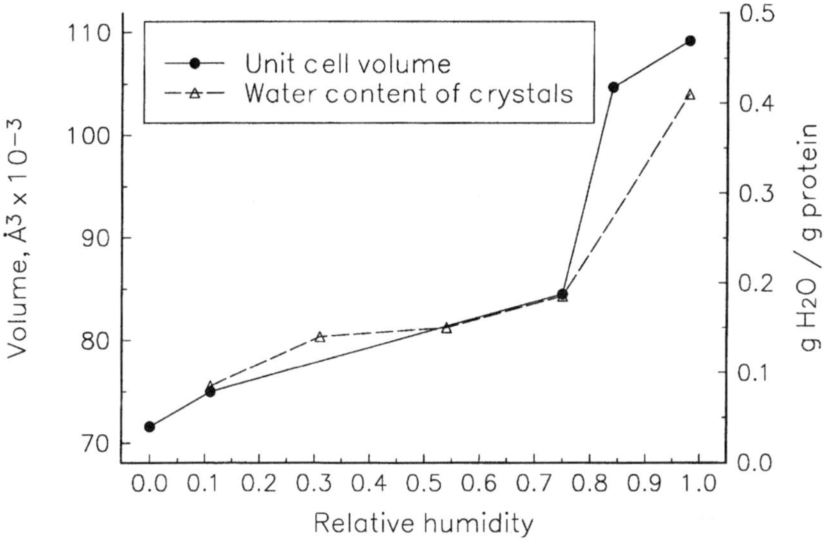

Figure 2. Unit cell volume and level of hydration of bovine pancreatic ribonuclease crystals after incubation at room temperature and at various levels of relative humidity. 0.5 g H_2O g^{-1} protein corresponds to approximately 3 water molecules per amino acid residue.

Figure 3 compares the structure of 1BEL with the structure of the desiccated protein for the main-chain atoms of the region that comprises the largest β-sheet in this protein. The second strand, consisting of residues 80 to 85, is clearly displaced in the hydrated versus the dry structures. The position of the third strand is displaced along the entire length by approximately 1Å. The same comparison has been applied between 1BEL and 7RSA, another fully solvated crystal form of bovine pancreatic ribonuclease (Wlodawer et al., 1988). The two structures would be essentially indistinguishable if both were shown in Figure 3.

Superposition of the main-chain atoms of this region of 1BEL and of the desiccated structure results in an RMS difference in position of 0.60 Å. By way of comparison, superposition of this region of 1BEL with 7RSA results in an RMS difference in main-chain atom positions of 0.20 Å.

Conformational changes of the same magnitude as those shown in Figure 3 were observed in other parts of the protein structure.

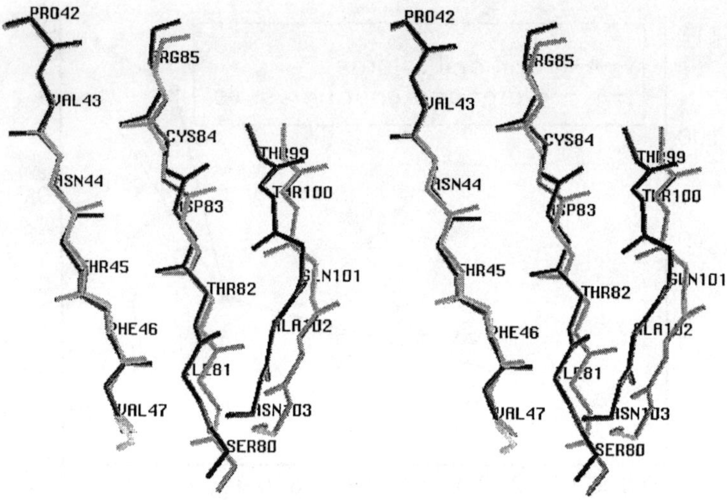

Figure 3. Structural differences observed, in the largest β-sheet of ribonuclease, between the structure in 60% methanol (light bonds) and the calcium-sulfate-desiccated structure (dark bonds), shown as a stereo pair. Main-chain atoms from residues 42 to 47 have been superimposed (Kabsch, 1978).

Temperature factors were high for the desiccated protein structure, especially in regions between secondary structure elements. While the regular secondary structure of the molecule was clear in the electron density map, loop regions between secondary structure elements were sometimes hard to trace in the electron density. For residues 66-71 and 86-94, the electron density appeared particularly tenuous at this stage of refinement.

A few unidentified electron density peaks in the dried structure may indicate residual, tightly bound water or counter ions. Two sulfate ions were identified in 1BEL, and several electron density peaks in the desiccated structure suggest bound ions by their height, by their distance from hydrogen-bonding groups, and by the charge of surrounding amino acid side chains.

ACKNOWLEDGMENTS

This work was supported by NIH grant GM50817. The use of the x-ray diffraction equipment of the Macromolecular Crystallographic Core

Facility of the Wadworth Center, New York State Department of Health, Albany, New York is gratefully acknowledged.

REFERENCES

Abola, EE, FC Bernstein, SH Bryant, TF Koetzle, and J Weng. **1987**. Protein data bank, pp. 107-132. In *Crystallographic Databases—Information Content, Software Systems, Scientific Applications*, F.H. Allen, G. Bergerhoff, and R. Sievers, eds. Data Commission of the International Union of Crystallography, Bonn/Cambridge/Chester.

Fitzgerald, PMD. **1988**. MERLOT, an integrated package of computer programs for the determination of crystal structures by molecular replacement. J. Appl. Crystallogr. 21:273-278.

Fitzpatrick, PA, D Ringe, and AM Klibanov. **1994**. X-ray crystal structure of cross-linked subtilisin Carlsberg in water vs. acetonitrile. Biochem. Biophys. Res. Comm. 198:675-681.

Jones, TA. **1978**. A graphics model building and refinement system for macromolecules. J. Appl. Crystallogr. 11:268-272.

Kabsch, W. **1978**. A discussion of the solution for the best rotation to relate two sets of vectors. Acta Crystallogr. A34:827-828.

Kachalova, GS, VN Morozov, TY Morozova, ET Myachin, AA Vagin, BV Strokopytov, and YV Nekrasov. **1991**. Comparison of structures of dry and wet hen egg-white lysozyme molecule at 1.8 Å resolution. FEBS Lett. 284:91-94.

King, MV, J Bello, EH Pignataro, and D Harker. **1962**. Crystalline forms of bovine pancreatic ribonuclease. Some new modifications. Acta Crystallogr. 15:144-147.

Klibanov, AM. **1989**. Enzymatic catalysis in anhydrous organic solvents. Trends Biochem. Sci. 14:141-144.

Kodandapani, R, CG Suresh, and M Vijayan. **1990**. Crystal structure of low humidity tetragonal lysozyme at 2.1-Å resolution. J. Biol. Chem. 265:16126-16131.

Rockland, LB. **1960**. Saturated salt solutions for static control of relative humidity between 5° and 40° C. Anal. Chem. 32:1375-1376.

Tronrud, DE, LF Ten Eyck, and BW Matthews. **1987**. An efficient general-purpose least-squares refinement program for macromolecular structures. Acta Crystallogr. A43:489-501.

Wlodawer, A, LA Svensson, L Sjölin, and GL Gilliland. **1988**. Structure of phosphate-free ribonuclease A refined at 1.26 Å. Biochemistry 27:2705-2717.

Yennawar, HP, NH Yennawar, and GK Farber. **1995**. A structural explanation for enzyme memory in nonaqueous solvents. J. Am. Chem. Soc. 117:577-585.

STABILIZATION OF VERATRYL ALCOHOL CATION RADICAL BY LIGNIN PEROXIDASE

A. Khindaria and S. D. Aust

Biotechnology Center, Utah State University, Logan, Utah

Key words: *Veratryl alcohol cation radical, lignin peroxidases,* Phanerochaete chrysosporium, *white rot fungi*

ABSTRACT

Lignin peroxidase (LiP) catalyzes the H_2O_2-dependent, two-electron oxidation of veratryl alcohol (VA) to veratryl aldehyde. The one-electron-oxidized intermediate, the veratryl alcohol cation radical ($VA^{\cdot+}$), is an important redox mediator for indirect oxidations by LiP. We have investigated the properties of $VA^{\cdot+}$, formed by oxidation with Ce(IV) and by LiP, by electron spin resonance spectroscopy (ESR). The $VA^{\cdot+}$ formed by oxidation with LiP exists as a complex with LiP. Only a fraction of the $VA^{\cdot+}$ (2.2 μM) formed by LiP (10 μM) could be detected by EPR under normal turnover conditions of the enzyme. However, when the same reaction mixture was acid-quenched, the $VA^{\cdot+}$ concentration increased to 8.9 μM. Additionally, the decay constant of the $VA^{\cdot+}$ was 1.2 X 10^3 s^{-1} when formed by oxidation with Ce(IV) and 1.82 s^{-1} when formed by oxidation with LiP. This supported the hypothesis that LiP not only forms a complex with $VA^{\cdot+}$ but also stabilizes it. This was further supported by the fact that when the LiP reaction was acid-quenched and the protein denatured, the decay constant of $VA^{\cdot+}$ increased to 9.9 X 10^2 s^{-1}, similar to one obtained with Ce(IV) as an oxidant. The interaction of $VA^{\cdot+}$ with LiP was further investigated by molecular modeling techniques. It was found that VA bound equally well near the δ-meso edge of the heme and in the site near the heme propionates. However, $VA^{\cdot+}$ bound in a proposed site by the heme propionates more stable (by 47 kcal/mol). Further, it was found by molecular dynamic studies that VA bound at the δ-meso edge of the heme, once oxidized to $VA^{\cdot+}$, would move at a diffusion-limited rate to the site by the heme propionates. Negative electrostatic potential, due to aspartates 88 and 183, glutamates 40 and 89, the C-terminus carboxyl of alanine 343, and the heme propionates, provides the electrostatic force for the radical to move to this site. Aspartate 183 should be particularly instrumental in this movement. Electrostatic potential calculations indicate that this negatively charged site creates a field around the $VA^{\cdot+}$; the calculated free energy change for the $VA^{\cdot+}$ binding to this site is -37 kcal/mol. This is more than sufficient to account for the stability of the $VA^{\cdot+}$, which our estimates suggest is stabilized by 14.3 kcalmol relative to the $VA^{\cdot+}$ in bulk solution.

INTRODUCTION

The white rot fungus *Phanerochaete chrysosporium*, is widely known for its ability to degrade lignin and a large number of environmental pollutants (Barr and Aust, 1994). This ability of the fungus is due in large part to its extracellular lignin-degrading system secreted upon nutrient starvation (Tien, 1987). This system consists of small molecules, VA and oxalate, enzymes that generate H_2O_2, and the heme-containing peroxidase enzymes. These peroxidases are divided into two categories, the lignin peroxidases (LiP) and the Mn-dependent peroxidases (Tien, 1987). Pertinent to this study are the lignin peroxidases.

Veratryl alcohol is a substrate for LiP and is oxidized by two electrons to veratryl aldehyde. The one electron-oxidized intermediate, the veratryl alcohol cation radical ($VA^{\cdot+}$) (Khindaria et al., 1995a), is an important redox mediator in LiP catalysis (Chung and Aust, 1995a; Goodwin et al., 1995). It acts as an indirect oxidant for molecules that cannot gain access to the heme in LiP. In this role, VA is oxidized to $VA^{\cdot+}$, which oxidizes secondary molecules and is itself reduced back to VA. It has been hypothesized that this role of $VA^{\cdot+}$ is central in degradation of lignin and various environmental pollutants (Chung and Aust, 1995b).

In this study we have compared the decay of $VA^{\cdot+}$ formed by oxidation with Ce(IV) versus LiP. Important differences between the two systems indicate that LiP stabilizes the $VA^{\cdot+}$. Furthermore, a negatively charged site near the heme propionates has been identified that has the negative electrostatic potential required to stabilize the cation radical.

RESULTS AND DISCUSSION

We have recently reported the formation of $VA^{\cdot+}$ during turnover of LiP with VA as a substrate (Khindaria et al., 1995a). The $VA^{\cdot+}$ was detected in a flow system using election spin resonance (ESR) spectroscopy. The identity of the radical was confirmed by generating it in an enzyme-free system with Ce(IV) as a peroxidase mimic (one electron oxidant) and by computer simulation of the ESR signal. The magnetic hyperfine splitting constants for the $VA^{\cdot+}$ are given in Table 1. Also tabulated (Table 1) are the spin densities on the $VA^{\cdot+}$. It would seem, then, that the $VA^{\cdot+}$ is a π orbital stabilized cation radical with asymmetric charge distribution, the α-carbon and vicinity being the most electron-deficient area. This asymmetric charge distribution would make the $VA^{\cdot+}$ unstable in bulk solution (Hammerich and Parker, 1984). The radical could decay by the attack of water on the positively charged area (nucleophilic attack), dimerization due to asymmetric charge distribution and deprotonation, besides other mechanisms. The unstable nature of $VA^{\cdot+}$ in bulk solution was proposed as the reason why it could not be detected by oxidation with Ce(IV) at pH above 2.5. (Rates of deprotonation and nucleophilic attack would be expected to increase with increase in pH.) (Khindaria et al., 1995a). However, the $VA^{\cdot+}$ could be detected in a flow system during LiP turnover even at pH 4.5. This was the first indication that $VA^{\cdot+}$ was being stabilized by LiP.

Table 1. Coupling constants for the protons and spin densities on the carbon bonded to the protons of $VA^{\cdot+}$

Position of Protons or Carbons	Coupling Constant *aH* (G)	Spin Density on the Carbon[b]
CH_2	6.14	0.166
$C_{2\ ring}$	2.85	0.029
OCH_3 (C_3)[a]	0.49	0.116
OCH_3 (C_4)	7.76	0.131
$C_{5\ ring}$	3.05	0.009
$C_{6\ ring}$	3.38	0.050

[a]The carbon number in parentheses denotes the position of the methoxy group on the ring.
[b]The higher the number, the greater the spin density at the site.

Later, it was found that during turnover of LiP (10 μM) with excess H_2O_2 (200 μM) only 2.2 μM of $VA^{\cdot+}$ could be detected at pH 4.5. This figure increased to 5.9 μM at pH 3.5. This may have been for various reasons, including the fact that the turnover rate of LiP at pH 3.5 is twice that at pH 4.5 (k_{cat} = 16 vs 8). It would then be expected that at pH 2.5 (k_{cat} = 78) the $VA^{\cdot+}$ concentration detected by ESR would be at least 22 μM; this was not observed (Khindaria et al., 1995b). Furthermore, it was found that at all pH the concentration of $VA^{\cdot+}$ that could be detected was less than the concentration of LiP, suggesting that the free $VA^{\cdot+}$ was not detected. In addition, it was observed that upon addition of the 10% HNO_3 the concentration of $VA^{\cdot+}$ detected by ESR at pH 4.5 increased to 8.9 μM (with 10 μM LiP). The concentration of $VA^{\cdot+}$ as a function of LiP concentration, under steady state at pH 4.5 and under acid-quenched conditions, is shown in Figure 1. Under acid-quenched conditions the concentration of $VA^{\cdot +}$ that could be detected was four times the concentration detected during steady state at all concentrations of LiP. However, it must be noted that the concentration of $VA^{\cdot +}$ that could be detected by ESR never exceeded the concentration of LiP. This was strong evidence to support the assumption that LiP-$VA^{\cdot +}$ complex was being detected. This suggested that the $VA^{\cdot +}$ formed by oxidation with LiP during turnover could exist in two forms, one ESR-active, which could be detected during turnover, and the other ESR-silent. The ESR-silent form (under normal turnover) could be detected when the reaction was acid-quenched.

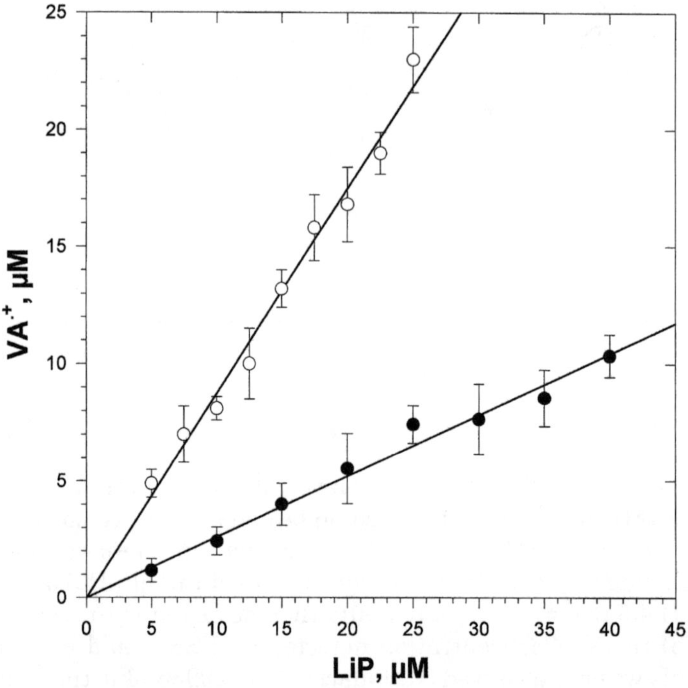

Figure 1. Steady-state concentration of $VA^{\cdot +}$ as a function of lignin peroxidase (LiP) concentration at pH 4.5. The veratryl alcohol ($VA^{\cdot +}$) concentrations were calculated from electron spin resonance spectra recorded during flow. Closed circles represent data from spectra during steady state; open circles represent data from the spectra obtained immediately after reaction was acid-quenched.

This observation could be explained by the fact that the ESR-silent form of $VA^{\cdot +}$ was either constrained, not allowing for physical tumbling of the radical, or was spin-coupled to the heme iron. In either case, the radical signal would be broadened beyond detection. At present it is difficult to say what caused the broadening of the radical signal. Upon acid-quenching the enzyme would be denatured, releasing the $VA^{\cdot +}$ (the ESR-silent form) in bulk solution, where it could be detected.

It was subsequently found that both the forms of the $VA^{\cdot +}$ formed by oxidation with LiP existed as a complex with lignin peroxidase compound II (LiPII-$VA^{\cdot +}$). One form of this complex (LiPII-$VA^{\cdot +}$; ESR-silent) could react with an additional molecule of VA, resulting in the formation of veratryl aldehyde. This was evident by kinetic data, both single-turnover and steady-state, by ESR data, and by spectral data. However, the exact mechanism for the formation of veratryl aldehyde by the reaction of LiPII-$VA^{\cdot +}$ (ESR-silent) with VA is not clearly understood. This could happen either by dismutation of two $VA^{\cdot +}$, or deprotonation of one and oxidation of the neutral radical by another $VA^{\cdot +}$, or by dismutation of two neutral radicals.

The ESR-silent $VA^{\cdot +}$ existed in equilibrium with the ESR-active $VA^{\cdot +}$, with the distribution constant being equal to 0.25 (Khindaria et al., 1995b). The decay of the ESR-active LiPII-$VA^{\cdot +}$ complex was studied by stopped-flow visible absorption spectroscopy. The complex decayed with a half-life of 0.54 s^{-1} (or k_d = 1.8 s^{-1}). This decay constant of LiPII-$VA^{\cdot +}$ was confirmed by stopped-flow ESR studies, and an identical decay constant to the one obtained by kinetic studies was calculated from the ESR data. However, the decay constant of the $VA^{\cdot +}$ under acid quenched studies was 9.9×10^2 s^{-1}, which was similar to the one obtained for the $VA^{\cdot +}$ formed by oxidation with Ce(IV) (1.2×10^3 s^{-1}).

These results indicated that the $VA^{\cdot +}$ formed by oxidation with Ce(IV) had a different property (lifetime) from that formed by oxidation with LiP. This could be explained if the $VA^{\cdot +}$ is stabilized by LiP. With this knowledge, we decided to explore the crystal structure of LiP (Poulos et al., 1993) with the intention of identifying a site in LiP capable of stabilizing the cation radical. The site would be negatively charged, would be sufficiently large to accommodate the cation radical, and the negative charges in the site would be symmetrically distributed so as not to constrain the physical tumbling of the cation radical.

A site near the heme propionates was identified that satisfied the above-mentioned criteria. However, this was not the site that was proposed as the binding site for VA (Poulos et al., 1993), which was thought to be near the δ-meso edge of the heme. We investigated both of these sites for their theoretical ability to bind VA and $VA^{\cdot+}$ separately. It was found by energy minimization calculations that VA could bind equally well at both the site near the heme propionates and near the δ-meso edge of the heme. However, the $VA^{\cdot+}$ would bind better in the site by the heme propionates (by 47 kcal/mol). We therefore investigated, by molecular dynamics, the possible movement of the $VA^{\cdot+}$ during catalysis. The VA was docked to the site by the δ-meso edge of the heme, minimized ($\Delta E < 1$ kcal/mol), then given a charge of + 1. This was done to mimic the binding of VA to LiP and its oxidation to $VA^{\cdot+}$, and served as the starting conformation for the molecular dynamics calculation.

It was observed that the $VA^{\cdot+}$ bound near the δ-meso edge moved during the course of the dynamics run to the site by the heme propionates; by 38 ps the $VA^{\cdot+}$ was found to be in the site by the heme propionates. This movement of the $VA^{\cdot+}$ was possibly due to the electrostatic pull of the negative charges by the heme propionates. It was observed that Asp 183 is a key amino acid that moves to pull the $VA^{\cdot+}$. The molecular dynamic was run for an additional 20 ps, during which the $VA^{\cdot+}$ did not leave the site. The calculated electrostatic potential of this site (by the heme propionates) was sufficient (-37 kcal/mol) to account for the stability of the $VA^{\cdot+}$, which our estimates suggest is stabilized by 14.3 kcal/mol in bulk solution.

In conclusion, we have shown that (1) $VA^{\cdot+}$ is stabilized by LiP; (2) there is a site, negatively charged, that can stabilize $VA^{\cdot+}$; (3) the VA bound near the δ-meso edge of the heme, once oxidized, moves to the site by the heme propionates, which stabilize it, and (4) the electrostatic potential at this site is sufficient to stabilize the $VA^{\cdot+}$.

ACKNOWLEDGMENTS

Aditya Khindaria is generously supported by a Utah State University Presidential Fellowship. This research was supported in part by NIEHS Superfund Research and Training grant number ES04922. The authors wish to thank Dr. I. Yamazaki for helpful discussions, Terri Maughan for excellent secretarial assistance and Carrie Liechty for help with the experiments.

REFERENCES

Barr, DP and SD Aust. **1994**. Conversion of lignin peroxidase compound III to active enzyme by cation radicals. Arch. Biochem. Biophys. 312:511-515.

Chung, N and SD Aust. **1995a**. Inactivation of lignin peroxidase by hydrogen peroxide during the oxidation of phenols. Arch. Biochem. Biophys. 316:851-855.

Chung, N and SD Aust. **1995b**. Veratryl alcohol mediated indirect oxidation of phenol by lignin peroxidase. Arch. Biochem. Biophys. 316:733-737.

Goodwin, DG, SD Aust, and TA Grover. **1995**. Evidence for veratryl alcohol as a redox mediator in lignin peroxidase-catalyzed oxidation. Biochemistry 34:5060-5065.

Hammerich, O and VD Parker. **1984**. Kinetics and mechanisms of reactions of organic cation radicals in solution, pp. 55-189. In: *Advances in Physical Organic Chemistry*, V. Gold, and D. Bethell, eds.), Academic Press, London.

Khindaria, A, TA Grover, and SD Aust. **1995a**. Evidence for formation of the veratryl alcohol cation radical by lignin peroxidase. Biochemistry 34:6020-6025.

Khindaria, A, I Yamazaki, and SD Aust. **1995b**. Veratryl alcohol oxidation by lignin peroxidase. Biochemistry 34:16860-16869.

Poulos, TL, SL Edwards, H Wariishi, and MH Gold. **1993**. Crystallographic refinement of lignin peroxidase at 2Å. J. Biol. Chem. 268:4429-4440.

Tien, M. **1987**. Properties of ligninase from *Phanerochaete chrysosporium* and their possible applications. CRC Crit. Rev. Microbiol. 15:141-168.

BIOREMEDIATION:
BIOTIC DIVERSITY AND BIOENGINEERING

MICROBIAL DIVERSITY IN SUBSURFACE ENVIRONMENTS AS REVEALED BY SMALL SUBUNIT (16S) RIBOSOMAL RNA GENE SEQUENCE ANALYSES

J. K. Fredrickson,[1] N. K. Fry,[2] S. Fishbain,[2] and D. A. Stahl[2]

[1]Pacific Northwest National Laboratory, Richland, Washington

[2]Northwestern University, Evanston, Illinois

Key Words: *Aquifer, anaerobic, alkaline, DNA, RNA, methanogenic, sulfate-reducing, metal-reducing*

ABSTRACT

The microbial community structure of two deep (1270 and 316 m), alkaline (pH 1 0.2 and 9.7), anaerobic (Eh = -400 and -305 mV) aquifers within the Columbia River Basalt Group in southeastern Washington State was characterized using rRNA-based analyses. Previous studies indicated that ground water from these aquifers (wells designated DC-06 and DB-11) was dominated by a sulfidogenic (DC-06) and methanogenic (DB-11) microbial metabolism, respectively. RNA was extracted from microorganisms collected from ground water by ultrafiltration, using hollow-fiber membranes, and hybridized to RNA-targeted, taxon-specific oligonucleotide probes. This revealed that each aquifer has a distinct and diverse community. Of the three domains (Bacteria, Eukarya, and Archaea), Bacteria dominated both communities, comprising 92.0% of the total rRNA from the sulfidogenic well and 64.4% from the methanogenic well. Eukarya comprised 5.7% and 14.4%, and Archaea, 1.8% and 2.5%, respectively. Gram-positive organisms were found in both wells: 11.71% in the sulfidogenic well and 7.6% in the methanogenic well. Two probes were used to target sulfate-reducing bacteria within the delta subdivision of the Proteobacteria, the *Desulfobacter* group and the *Desulfovibrio*/metal-reducing group. The *Desulfobacter* group was present (0.3%) only in the ground water with the high sulfate concentrations. However, comparable hybridization to a probe selective for the *Desulfovibrio* and some metal-reducing bacteria was found in both groundwater samples; 2.53% from the sulfidogenic ground water and 2.86% in the methanogenic ground water. These results indicate that these deep aquifer microbial communities are phylogenetically diverse and are composed of Gram-positive bacteria, sulfate- and metal-reducing bacteria, Archaea, and eukaryotes.

INTRODUCTION

The existence of active microbial populations in the deep subsurface (hundreds to thousands of meters below the surface) has been recognized for decades (Bastin et al., 1926; Kuznetsov et al., 1963). However, most of the early research focused on sulfur and oil deposits, generating little information on the microbiology of the deep subsurface environments that were relatively poor in potential electron donors (Ghiorse and Wilson, 1988). Only recently have more detailed studies of the microbiology of subsurface environments been conducted (Harvey et al., 1984; Harvey and George, 1987; Chapelle et al., 1988; Balkwill, 1989; Hirsch and Rades-Rohkohl, 1983). The general findings from these more recent studies confirmed that microorganisms are present in geologically diverse subsurface environments and identified how their metabolic activities impact regional groundwater geochemistry (Chapelle et al., 1988; Murphy et al., 1992; Pederson and Ekendahl, 1990; Stevens et al., 1993).

Active microbial communities have been described in several deep, alkaline, anaerobic aquifers within the Columbia River Basalt (CRB) group in southcentral Washington State (Stevens et al., 1993). Microorganisms from these aquifers that could grow on hydrogen and carbon dioxide greatly outnumbered those that could grown on organic compounds (Stevens and McKinley, 1995). Hydrogen concentrations in CRB ground waters as high as 60 μM were reported, and laboratory experiments demonstrated that hydrogen gas is produced by reactions between freshly exposed basalt surfaces and water. It has been proposed that the CRB harbor geochemical, energy-based, anaerobic microbial communities that are independent of photosynthetic reactions (Stevens and McKinley, 1995).

Despite considerable interest in the microbiology and biogeochemistry of deep subsurface environments, relatively little is known about the phylogeny of the indigenous microorganisms. The scant information that is available (Boone et al., 1995a; Fredrickson et al., 1995a; Reeves et al., 1995) is for pure cultures. It is well established that only a small fraction of the microbial communities in most environments can be cultured, and this also holds true for deep subsurface environments (Kieft et al., 1995). Methods involving the direct extraction of nucleic acids and/or lipids are beginning to provide insights into the phylogenetic composition of deep, subsurface, microbial communities (Ogram et al., 1995; Fredrickson et al., 1995b).

The aim of this study was to gain additional insights into the phylogeny of the active microbial populations in deep, alkaline aquifers using taxon-specific oligonucleotide probes targeted against extracted rRNA.

MATERIALS AND METHODS

Site Description

The study site was located in southeastern Washington State within the CRB. The CRB cover an area of approximately 200,000 km^2 in the northwestern United States. The CRB are Miocene Tholeiitic flood basalt flows up to 6 km of (20 to 150 m thick), interspersed with sedimentary interbeds (0.3 to 25 m thick), overlain by 60- to 250-m-thick alluvial and loess sediments. These basalt flows can act as confining layers which isolate individual aquifers. Temperature and pH within these confined aquifers increase with depth throughout the site, and ground water is anaerobic. The geochemistry of the CRB group has been studied extensively because of its previous consideration as a potential site for long-term storage of high-level radioactive waste (Hearn et al., 1990; U.S. DOE, 1988). Interest in the microbiology and geochemistry of these aquifers has also been generated by concern over potential contamination from the overlying unconfined aquifer on the Hanford Site, and from the finding that these aquifers harbor a novel, subsurface, microbial ecosystem that is based on geochemically produced H_2 (Stevens and McKinley, 1995). The two deep, confined aquifers that were selected for study are within the Grande Ronde (1270 m) and the Priest Rapids (316 m) formations. Previous studies of ground water from these two aquifers indicated that sulfidogenesis and methanogenesis were the principal respiratory processes in the Grande Ronde and Priest Rapids formations, respectively (Stevens et al., 1993).

Sampling

Groundwater samples were collected from two different artesian wells on the Hanford Site in southcentral Washington. One well, in the Grande Ronde formation, designated DC-06, was screened at a depth of 1270 m; the other well, DB-11, in the Priest Rapids formation, was screened at 316 m (Stevens et al., 1993). Before sampling, water was purged from each well for a minimum of one well volume (ca. 3471 L for DC-06 and 1134 L for DB-11) and until temperature, pH, and Eh (as measured in the field through an in-line flow cell) had stabilized. Additional physical and chemical characteristics were taken during sampling and/or from previous measurements. The general approach for nucleic acid analysis employed in this study is outlined in Figure 1.

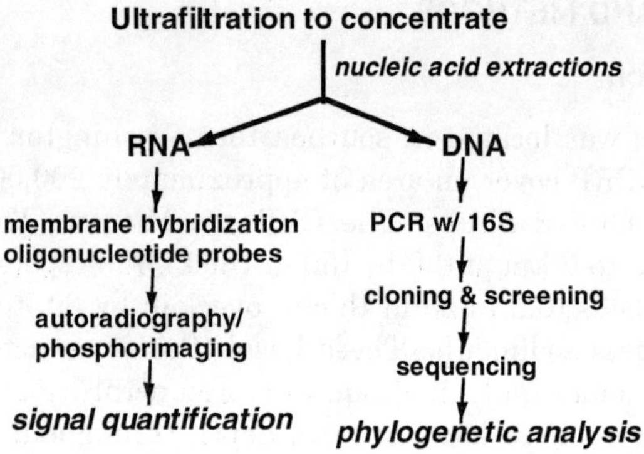

Figure 1. Approaches used for nucleic-acid-based analyses of microbial biomass collected from Columbia River basalt aquifer ground waters.

In order to obtain sufficient biomass for nucleic acid extractions, cells were concentrated to a volume of ca. 50 L by ultrafiltration (DC30P, Amicon), using 100,000 molecular weight-cutoff, hollow-fiber filters. A small-capacity, hollow-fiber ultrafiltration unit (DC10L, Amicon) was then used to further reduce the volume to 0.5 to 1.0 L. Aliquots of the retentate were immediately frozen on site, using dry ice, or filtered through sterile, 142-mm-diameter, 0.2-mm-pore-size filters (Durapore, Millipore) by pressure filtration. The filters were frozen immediately in the field on dry ice, transported to the laboratory and maintained at -80°C until processed. Cleaning and disinfection of both Amicon units was carried out using procedures recommended by the manufacturer (Amicon). Recirculation of 0.1% (w/v) sodium dodecyl sulfate (SDS), followed by rinsing with deionized water until all traces of foam were removed. Sodium hypochlorite (200 mg L^{-1}) was recirculated through the pump and filter for 10 min. This solution was then diluted to 100 mg L^{-1} and remained in contact with the filter during storage. The sodium hypochlorite was removed before the next collection by rinsing thoroughly with deionized water.

Extraction and Analysis of RNA

All glassware was baked at 200°C for at least 18 hr to inactivate RNases. New, unopened plasticware was assumed to be nuclease-free.

Solutions were made with double-distilled water taken directly from the still. A modification of the method of Stahl et al. (1988) was used (Lin and Tsen, 1995) to extract ribosomal RNA from the aquifer concentrates or filters, using mechanical disruption in conical screw-cap polypropylene vials (Sarstedt, Inc., Newton, NC) on a reciprocating shaker (Mini-Beadbeater; Biospec Products, Bartkesvile, OK) with zirconium/silica beads (0.1 mm diameter). Each vial contained approximately 0.5 g zirconium/silica beads, 50 ml of 20% (w/v) SDS, 500 ml of phenol equilibrated with a low-pH buffer (1 × low-pH buffer is 50 mM sodium acetate, 10 mM EDTA, pH 5.1), together with 500 ml of sample in a 1 × low-pH buffer. Concentrate biomass was processed in ca. 500-ml aliquots, and frozen filters were broken into small pieces (< 5 mm^2) using a baked glass rod. Samples were thawed in the vials containing the phenol/low-pH buffer/SDS mixture immediately before extraction.

Following reciprocation of the vials for 2 min at room temperature, samples were incubated for 10 min at 60°C, disrupted for an additional 2 min, then re-extracted with low-pH buffer (pH 5.1) equilibrated with phenol. Organic and aqueous phases were separated by centrifugation at 10,000 × g in a microcentrifuge. Samples were extracted an additional two times with an equal volume of phenol/chloroform (4:1, [v/v] equilibrated with the same buffer. Nucleic acid (primarily RNA) was precipitated by the addition of one-half volume of 7.5 M ammonium acetate plus two volumes of 100% ethanol, followed by incubation at -20°C (> 1 hr to overnight). Following 20 min centrifugation at 10,000 × g in a microcentrifuge (Eppendorf; Brinkman Instruments, Westbury, NY) at 4C, the supernatant was removed, and the pellet washed with 80% (v/v) ethanol and resuspended in double-distilled water. The integrity and approximate yield of the rRNA was examined by denaturing polyacrylamide gel electrophoresis. Copurification of unidentified substances from the sulfidogenic well (e.g., minerals, salts, humic acids) affected the electrophoresis of these samples. These samples were further purified by column centrifugation (Sepharose G-50).

RNA was denatured and immobilized on 0.45-mm nylon membranes (Magna Charge, MSI Inc.) using a Minifold II Slot Blot System (Schleicher and Schuell, Keene, NH). Membranes were cut into rectangles ca. 22.5 cm × 6.5 cm. Each membrane could accommodate 24 × 3 samples. A reference series of known concentrations of RNA (from 20 ng to 0.16 ng) was applied to each membrane, together with the groundwater RNA samples. Ribosomal RNA from *Escherichia coli, Bacillus subtilis,*

Desulfovibrio sapovorans, Desulfovibrio desulfuricans, Dictyostelium discoideum, and *Methanosarcina acetovorans* was used as reference RNA.

One volume RNA was denatured by the addition of three volumes of 2% glutaraldehyde in 50 mM phosphate buffer (pH 7.0). After incubation at room temperature for 10 min the samples were diluted with 1 mg/ml poly[A] and 0.04% bromophenol blue. Diluted samples were generally applied to the membranes in triplicate (50 ml volume) and filtered under a slight vacuum. Membranes were air-dried and baked at 80°C for 1 to 2 hr. Taxon-specific DNA oligonucleotide probes complementary to small subunit rRNA (see Table 1) were 5´-end-labeled with [g-32P]ATP and purified as described by Stahl et al. (1988). Probes used in this study comprised a universal probe and domain-specific probes, Bacteria, Archaea and Eukarya (Raskin et al., 1994; Amann et al., 1990a,b). More-specific probes were used to quantify Gram-positive bacteria (S-P-Grps1200aR-13) (Toze and Stahl, unpublished data) and sulfate-reducing bacteria (SRB) within the delta Proteobacteria, the *Desulfobacter* group and the *Desulfovibrio*/metal-reducing group. A 5- to 10-fold excess of the amount of probe necessary to bind the total RNA applied to the membrane was labeled and used in the hybridization reaction. Membranes were prewetted in hybridization buffer, 0.9 M NaCl, 50 mM phosphate buffer (pH 7.0), 5 mM EDTA, 10 × Denhardt's solution, 0.5 mg/ml poly[A], 0.5% SDS, then placed in screw-cap hybridization tubes (Robbins Scientific, Sunnyvale, CA) with approximately 10 ml hybridization buffer and incubated for 1 to 2 hr at the hybridization temperature (25°C for the Gram-positive probe, 40°C for all others) in a rotating incubator (Robbins Scientific). The hybridization buffer was removed, and 10 ml fresh hybridization buffer (at the hybridization temperature) containing labeled probe (in a volume of less than 1 ml) was added. Incubation was continued for 14 to 18 hr at the hybridization temperature. Membranes were then washed at the hybridization temperature for 2 hr with 3 × 100 ml wash solution, (1 × SSC [0.15 M NaCl, 0.015 M sodium citrate, pH 7.0], 1% [w/v] SDS). Membranes were then washed in 300 ml of 2 × wash solution at the T_d for 30 min. Optimal hybridization and wash temperatures were previously determined. Membranes were allowed to air-dry, then were wrapped in Saran® wrap. Bound probe was quantified by exposure of membranes to Storage Phosphor Screens (Molecular Dynamics, Sunnyvale, CA) and scanning exposed screens, using a Model 400 Series PhosphorImager (Molecular Dynamics). The least-squares method was used to calculate the slopes and intercepts of the reference RNA standard curves, using replicate samples. The abundance of each taxon was expressed as the fraction of total RNA in the sample. Total rRNA abundance was assumed to be the total activity obtained using a universal probe.

Table 1. Geochemical analyses of groundwater samples and physical characteristics of aquifers from which samples used in this study were collected.

Property	Well DB-11	Well DC-06
Depth (m)	316	1270
Flow rate (L/min)	50	5
Temperature at well (°C)	26	22
Temperature *in situ* (°C)	27	59
Eh (mV)	-305	-400
pH	8.7	10.2
Pressure (kPa)	552	1632
Dissolved inorganic C (mg/L)	34.0	12.9
Dissolved organic C (mg/L)	1.9	4.7
SO_4^{2}- (mg/L)	<0.09	139.1
NO_3- (mg/L)	<0.06	<0.06
Cl (mg/L)	4.9	172.7
Fe (mg/L)	0.74	<0.004
Na (mg/L)	29.9	257.9
Sulfide (μm)	bd[(a)]	22.9
CH_4	272 (μM)	1 (mole %)
H_2 (μM)	0.34	0.38

[(a)]Below detection.

RESULTS

Site Characteristics

The physical and chemical properties of the sulfidogenic (Grande Ronde, DC-06) and methanogenic (Priest Rapids, DB-11) aquifers are shown in Table 1. Total microorganisms in the ground waters were previously enumerated by acridine orange direct microscopic counts (Stevens et al., 1993). The concentration of microorganisms in well DC-06 was 7.6 (11.4) $\times$ 10^5 organisms/ml; in well DB-11, 3.6 (11.4) $\times$ 10^3 organisms/ml. Values in parentheses are standard deviations.

Nucleic Acid Extraction

Intact, high-molecular-weight rRNA (both large and small subunit) was visualized following polyacrylamide gel electrophoresis on samples recovered from both aquifers. The yield of RNA (in ng L^{-1}) was 9.5 and 40.7 for the samples from the Priest Rapids and Grande Ronde aquifers, respectively. Samples of concentrated RNA from the Grande Ronde ground water contained substances which affected electrophoresis. However, samples which were collected onto filters were not affected.

Nucleic Acid Hybridization

The amount of rRNA recovered, as estimated from hybridization of the universal probe, was approximately 40.7 and 9.5 ng/L from the Grande Ronde and Priest Rapids samples, respectively. The percentage of total rRNA corresponding to the various probe target groups in the RNA extracted from the Grande Ronde and Priest Rapids samples is shown in Table 2. Of the three domains (Bacteria, Archaea, and Eukarya) Bacteria were the most dominant in ground waters from both formations. Archaea were the least abundant, although a slightly greater proportion of the total RNA originated from this group in ground water from the Priest Rapids formation (methanogenic) than in ground water from the Grande Ronde. Eukarya, detected in both wells, comprised 5.7% and 14.4% of the total rRNA in the Grande Ronde and Priest Rapids formations, respectively. Most of the rRNA present was accounted for by probes of the three domains cited. The sum of the domain probes for the Grande Ronde and Priest Rapids ground waters was 99.4% and 81.3%, respectively.

Table 2. Domain probe hybridization analyses (%) of RNA extracted from biomass collected from groundwater samples.

Probe Name	Target Group	DB-11 (Priest Rapids)	DC-06 (Grand Ronde)
Univ 1391	Universal	100	100
Bact 338	Bacteria	64.4	92.0
Arc915	Archaea	2.5	1.8
Euk1379	5.7	14.4	5.7
Total	81.3	99.5	

Two probes were used to target sulfate-reducing bacteria in the delta Proteobacteria, S-F-Dsv687aR-16 and S-Dsb804R-18. The percentage rRNA due to the organisms in target group S-F-Dsv687aR-16 was 2.5% in the Grande Ronde and 2.9% in the Priest Rapids samples. The percentage of total rRNA of the target group for SRB804 was 0.3% in water from the Grande Ronde (Table 3). No signal was detected in RNA extracted from the Priest Rapids sample using this probe (0%, i.e., below the limit of detection). The results from the Gram-positive probe indicated that 11.7% of the signal detected using the universal probe in the Grande Ronde sample and 7.6% of that in the Priest Rapids was due to this group. As percentages of bacterial RNA, the Gram-positive populations represented 12.7% and 11.8% of the bacterial signal for the Grande Ronde and Priest Rapids samples, respectively.

Table 3. Hypridization analyses (%) using probes for Gram-positive bacteria *Desulfovibrio* and *Desulfobacter* group

Probe Name	Target Group	DB-11	DC-06
Grps120aR	Gram-positive bacteria	7.6[(a)]	11.7
Dsv687aR	*Desulfovibrio*	2.9	2.5
Dsb804aR	*Desulfobacter* group[(b)]	bd[(c)]	0.3

[(a)]Normalized to the Universal probe signal.
[(b)]Includes *Desulfobacter, Desulfobulbus, Desulfobacterium, Desulfoarculus, Desulfosarcina, Desulfobotulus*.
[(c)]Below detection.

DISCUSSION

Hybridization of extracted RNA from the CRB aquifers studied revealed differences in microbial community composition. Bacteria dominated both communities, but Eukarya and low levels of Archaea were also detected. Although the amount of Archaea RNA isolated from either aquifer was relatively low, a slightly greater abundance was detected in samples from the Priest Rapids (methanogenic) than in those from the Grande Ronde (sulfidogenic). Although geochemical analyses of DB-11 ground water indicated that methanogenesis is the dominant respiratory pathway in the Priest Rapids formation, it does not necessarily follow that methanogenic bacteria will be the numerically dominant or the most active microorganisms. In fact, Stevens and McKinley (1995) reported that culturable acetogenic bacteria were 100-fold greater in number in DB-11 ground water than were methanogens. Although acetogenesis yields less energy than methanogenesis in organic-rich anaerobic sediments, acetogens may be more successful at competing in the CRB aquifers due to the high concentration of H^2, believed to be the result of abiotic reactions (Stevens and McKinley, 1995). The results presented here are consistent with these findings, in that most known acetogens would hybridize with the Gram-positive and bacterial probes and that the Gram-positive signal was approximately 12% of the total bacterial signal in both groundwater samples.

The *Desulfobacter* group was detected only in ground water from the Grande Ronde, which also contained high sulfate concentrations (approximately 1.7 mM). However, a probe selective for the *Desulfovibrio*/metal-reducing bacteria hybridized at comparable levels to RNA extracted from both aquifer samples. The phylogeny of organisms contributing to the *Desulfovibrio*/metal-reducing group was determined by recovery of partial rDNA sequences using the delta and *Desulfovibrio*/metal-reducing primers (N. Fry, unpublished data). Sequences closely related to *Desulfovibrio* spp. were identified in both aquifers, hence sulfate-reducing bacteria (*Desulfovibrio* spp.) were present in both low- and high-sulfate environments. This suggests that these organisms are active even when sulfate concentrations are very low, possibly by dissimilatory reduction of Fe(III). Signal from the *Desulfobacter* group was detected only in Grande Ronde ground water which had a high concentration of sulfate. However, the sulfate/metal-reducing bacteria targeted by the probe S-F-Dsv687aR-16 were identified in both samples at similar levels. Culturable dissimilatory Fe(III)-reducing bacteria (DIRB) and SRB have been detected in ground water from both the Grande Ronde and Priest Rapids formations (Stevens and McKinley, 1995). Although the S-F-Dsv687aR probe targets *Geobacter*, *Pelobacter*, and *Desulfuromusa* as well as the *Desulfovibrionaceae*, no non-*Desulfovibrio* sequences were recovered using the PCR primers. The presence of SRB in low concentrations of sulfate has been reported elsewhere, and it is becoming clear that some sulfate-reducing bacteria, including *Desulfovibrio*, are capable of coupling the oxidation of hydrogen and organic matter to the reduction of Fe(III) (Coleman et al., 1993). It is hypothesized that the dissolution of ferrosilicate minerals such as olivine and pyroxene in the CRB contribute to H_2 production *in situ* and, in the process, precipitate magnetite (Fe_3O_4). It has recently been demonstrated that some DIRB can reduce Fe(III) in magnetite (Kostka and Nealson, 1995).

Proportions of RNA from SRB and/or DIRB in ground water from the Grande Ronde formation, although relatively low, were significant. These findings support earlier observations that SRB and DIRB are active *in situ* and have a significant impact on groundwater geochemistry within the CRB aquifers. Stevens and McKinley (1995) found that populations of fermentative and acetogenic bacteria were at least as high as SRB in ground water from the Grande Ronde. As with ground water from the Priest Rapids formation, acetogenic bacteria probably account for most of the signal from both the Gram-positive and bacterial probes.

It is interesting to note the detection of eukaryotic RNA in the CRB samples, particularly in samples from the Priest Rapid Formation (DB-11), where >14% of the signal with the universal probe was attributed to Eukarya. The source of eukaryotic sequences in the CRB samples is currently unknown. However, Sinclair and Ghiorse (1989) detected fungi and protozoa in samples collected from depths as great as 250 m beneath the U.S. southeastern coastal plain. Protozoa and fungi have also been detected in relatively shallow subsurface sediments in many different regions (Ghiorse and Wilson, 1988). Ground water in the Priest Rapids aquifer is considerably younger that in the Grande Ronde, hence recharge from the surface occurs more rapidly. Therefore, the eukaryotic signal may be due to allocthonous microeukaryotes associated with the recharge water. There is currently little information regarding the nature of microeukaryotes in deep subsurface environments, but their presence at several sites indicates further studies are warranted to determine what role(s) these microorganisms may have in deep subsurface microbial communities.

It is generally believed that subsurface solids must be sampled to obtain representative samples for assessing microbiological characteristics of the subsurface. Population densities and community composition are generally different in groundwater samples than those in subsurface sediments obtained from the same formation (Hazen et al., 1991; Hirsch and Rades-Rohkohl, 1988). However, sampling of the subsurface, particularly in basaltic aquifers where the much of the water flow is through fractures, vesicles, and interbed sediments, is technically challenging. Coring to obtain samples for microbiological and geochemical analyses in the CRB would probably not have yielded representative samples. The likelihood of coring fractures or vesiculated zones would be relatively low and, even if they could be cored, it would be difficult to obtain samples free of contamination from drilling fluids. Hence, the sampling strategy employed in this study involved the collection of ground water and subsequent concentration of biomass using ultrafiltration. Previous research has demonstrated that the microbiological properties of the CRB aquifers, as determined by analyses of groundwater samples, are congruent with the groundwater geochemistry (Stevens et al., 1993; Stevens and McKinley, 1995). This indicates that ground water from wells can provide representative samples for analyses of select microbiological properties of the deep subsurface. Although changes in the community composition may have taken place during ultrafiltration in our experiments due to exposure to air and light, this step was necessary to obtain sufficient biomass for nucleic acid extractions and probably did not impart major shifts in community composition.

ACKNOWLEDGMENT

This research was supported by the Deep Microbiology Subprogram of the Subsurface Science Program, Office of Health and Environmental Research, U.S. Department of Energy (DOE). Pacific Northwest Laboratory is operated for the DOE by Battelle Memorial Institute under Contract DE-AC06-76RLO 1830.

REFERENCES

Amann, RI, L Krumholz, and DA Stahl. **1990b**. Fluorescent-oligonucleotide probing of whole cells for determinative, phylogenetic, and environmental studies in microbiology. J. Bacteriol. 172:762-770.

Balkwill, DL. **1989**. Numbers, diversity and morphological characteristic of aerobic, chemoheterotrophic bacteria in deep subsurface sediments from a site in South Carolina. Geomicrobiol. J. 7:33-52.

Bastin, ES, FE Greer, CA Merritt, and G Moulton. **1926**. The presence of sulphate reducing bacteria in oil field waters. Science 63:21-24.

Boone, DR, Y Liu, Z-J Zhao, DL Balkwill, GR Drake, TO Stevens, and HC Aldrich. **1995**. *Bacillus infernus* sp. nov., an Fe(III)- and

Mn(IV)-reducing anaerobe from the deep terrestrial subsurface. Int. J. Syst. Bacteriol. 45:441-448.

Chapelle, FH, JT Morris, PB McMahon, and JL Zelibor. **1988**. Bacterial metabolism and the del-13C composition of groundwater, Floridian aquifer, South Carolina. Geology 16:117-121.

Coleman, ML, DB Hedrick, DR Lovley, DC White, and K Pye. **1993**. Reduction of Fe(III) in sediments by sulphate-reducing bacteria. Nature 361:436-438.

Fredrickson, JK, DL Balkwill, GR Drake, MF Romine, DB Ringelberg, and DC White. **1995a**. Aromatic-degrading *Sphingomonas* isolates from the deep subsurface. Appl. Environ. Microbiol. 61:1917-1922.

Fredrickson, JK, JP McKinley, SA Nierzwicki-Bauer, DC White, DB Ringelberg, SA Rawson, S-W Li, FJ Brockman, and BJ Bjornstad. **1995b**. Microbial community structure and biogeochemistry of Miocene subsurface sediment: implications for long-term microbial survival. Mol. Ecol. 4:619-626.

Ghiorse, WC and JT Wilson. **1988**. Microbial ecology of the terrestrial subsurface. Adv. Microbiol. 33:107-172.

Harvey, RW and LH George. **1987**. Growth determinations for unattached bacteria in a contaminated aquifer. Appl. Environ. Microbiol. 53:2992-2996.

Harvey, RW, RL Smith, and L George. **1984**. Effect of organic contamination upon microbial distributions and heterotrophic uptake in a Cape Cod, Mass., aquifer. Appl. Environ. Microbiol. 48:1197-1202.

Hazen, TC, L Jimenez, GL de Victoria, and CB Fliermans. **1991**. Comparison of bacteria from deep subsurface sediment and adjacent groundwater. Microb. Ecol. 22:293-304.

Hearn, PP, Jr., WC Steinkampf, LD White, and JR Evans. **1990**. Geochemistry of rock-water reactions in basalt aquifers of the Columbia River Plateau, pp. 63-68. In: Proceedings, U.S. Geological Survey Workshop on Environmental Geochemistry, USGS Circ. 1033. USGS, Denver, CO.

Hirsch, P and E Rades-Rohkohl. **1983**. Microbial diversity in a groundwater aquifer in Northern Germany. Dev. Ind. Microbiol. 25:183-200.

Hirsch, P and E Rades-Rohkohl. **1988**. Some special problems in the determination of viable counts of groundwater microorganisms. Microbial Ecol. 16:99-113.

Kieft, TL, JK Fredrickson, JP McKinley, BN Bjornstad, SA Rawson, TJ Phelps, FJ Brockman, and SM Pfiffner. **1995**. Microbiological comparisons within and across contiguous lacustrine, paleosol, and fluvial subsurface sediments. Appl. Environ. Microbiol. 61:749-757.

Kostka, JE and KH Nealson. **1995**. Dissolution and reduction of magnetite by bacteria. Environ. Sci. Technol. 29:2535-2540.

Kuznetsov, SI, MV Ivanov, and NN Lyalikova. **1963**. Introduction to Geological Microbiology. McGraw-Hill, New York.

Lin, CK and HY Tsen. **1995**. Development and evaluation of two novel oligonucleotide probes based on 16S rRNA sequence for the identification of *Salmonella* in foods. J. Appl. Bacteriol. 78:507-520.

Murphy, EM, JA Schramke, JK Fredrickson, HW Bledsoe, AJ Francis, DS Sklarew, and JC Linehan. **1992**. The influence of microbial activity and sedimentary organic carbon on the isotope geochemistry of the Middendorf Aquifer. Water Resour. Res. 28:723-740.

Ogram, A, W Sun, FJ Brockman, and JK Fredrickson. **1995**. Isolation and characterization of RNA from low-biomass deep subsurface sediments. Appl. Environ. Microbiol. 66:763-768.

Pederson, K and S Ekendahl. **1990**. Distribution and activity of bacteria in deep granitic groundwaters of southeastern Sweden. Microb. Ecol. 20:37-52.

Raskin, L, JM Stromley, BE Rittmann, and DA Stahl. **1994**. Group-specific 16S rRNA hybridization probes to describe natural communities of methanogens. Appl. Environ. Microbiol. 60:1232-1240.

Reeves, RH, JY Reeves, and DL Balkwill. **1995**. Strategies for phylogenetic characterization of subsurface bacteria. J. Microbiol. Meth. 21:235-251.

Sinclair, JL and WC Ghiorse. **1989**. Distribution of aerobic bacteria, protozoa, algae, and fungi in deep subsurface sediments. Geomicrobiol. J. 7:15-31.

Stahl, DA, B Flesher, HR Mansfield, and L Montgomery. **1988**. Use of phylogenetically based hybridization probes for studies of ruminal microbial ecology. Appl. Environ. Microbiol. 54:1079-1084.

Stevens, TO and JP McKinley. **1995**. Geochemically-produced hydrogen supports microbial ecosystems in deep basalt aquifers. Science 270:450-454.

Stevens, TO, JP McKinley, and JK Fredrickson. **1993**. Bacteria associated with deep, alkaline, anaerobic groundwaters in southeast Washington. Microb. Ecol. 25:35-50.

U.S. DOE. **1988**. Consultation Draft, Site Characterization Plan, Reference repository location, Hanford Site, Washington, DOE/RW-0164. U.S. Department of Energy, Richand, Washington.

BIODEGRADATION OF TOXIC AROMATIC COMPOUNDS: AN EVOLUTIONARY PERSPECTIVE

C. E. Danganan, M. R. Parsek, S. M. McFall, D. L. Daubaras, and A. M. Chakrabarty

Department of Microbiology and Immunology, College of Medicine, University of Illinois at Chicago, Illinois

Key Words: *Environmental pollution, gene recruitment, gene expression, protein-DNA interaction*

ABSTRACT

The persistence of recalcitrant compounds in the environment has become a serious environmental problem both in the United States and abroad. Various microorganisms have evolved intricate biodegradative pathways to utilize such compounds as sole sources of carbon and energy. Many of these bacteria have developed such abilities by evolving complex regulatory systems to control the expression of genes in the presence of the toxic compound. By attempting to understand the mechanisms involved in the evolution of these biodegradative genes, a more efficient means of eliminating hazardous compounds through these natural processes may be developed. Various examples of how certain microorganisms may have developed these biodegradative genes through natural and selective evolutionary processes are discussed.

INTRODUCTION

The persistence of toxic xenobiotic compounds in the environment is a matter of concern. Such compounds, particularly the highly chlorinated aromatics, have been used as herbicides/ pesticides, refrigerants, organic solvents, and as industrially useful compounds. Because these compounds are synthetic and are not usually found in the natural environment, microbes have a difficult time degrading them due to the lack of an appropriate degradative enzyme system. As a consequence, these compounds tend to accumulate in the environment, where their toxic effects are often exerted on the flora and fauna. Certain microorganisms have adapted to this unfavorable situation by evolving catabolic genes which, when expressed, allow the microbes to utilize these compounds as sole sources of carbon and energy. In order to develop appropriate pollution control technologies using microorganisms, it is essential that we understand the evolutionary processes involved in the recruitment of these unique degradative genes. We will discuss the various parameters which may be involved in the evolution of various biodegradative systems in naturally occurring microorganisms as well as those in a strain developed in the laboratory.

EVOLVING BIODEGRADATIVE PATHWAYS: GENE RECRUITMENT

Over the last several decades, large amounts of chlorinated aromatic compounds have been released to the environment, creating a major source of pollution. As a result, microbes are being continually isolated which are capable of utilizing some of these compounds. Chlorinated compounds, such as polychlorinated biphenyls, chlorobenzoates, chlorobenzenes, and 2,4-dichlorophenoxyacetic acid (2,4-D), have been shown to be degradable by microbes to varying extents. Highly chlorinated compounds such as γ-hexachlorocyclohexane (-HCH) (Nagata et al., 1994), pentachlorophenol (PCP) (Xun and Orser, 1991), 2,4,5-trichlorophenoxyacetic acid (Sangodkar et al., 1988; Haugland et al., 1990; Danganan et al., 1994; Daubaras et al., 1995), and some nitrotoluenes (Suen and Spain, 1993) have been reported to be degraded by certain microorganisms as well, and many of the genes encoding the enzymes of these pathways have been cloned and sequenced. The sequence information of three operons encoding enzymes responsible for the degradation of chlorocatechols has been derived. The evolutionary interrelationships have been determined from their genetic organization and the extent of their nucleotide sequence homology (Daubaras and Chakrabarty, 1992). The three operons constitute plasmid-borne genes from microorganisms isolated from three different continents (North America, Australia, and Europe). The genetic organization and sequence homology among the *clc*ABD, *tfd*CDEF, and *tcb*CDEF genes, all of which are involved in chlorocatechol (Clc) degradation, are strikingly similar. This suggests that certain degradative pathways evolve through a common mechanism, allowing recruitment from degradative genes performing an analogous function. These chlorocatechol degradative genes not only share significant homology with one another, but also with the ancestral catechol degradative *cat*BC operon. Thus, we postulate that bacteria acquire new sets of genes through cell-to-cell contact and plasmid-mediated gene transfer, so that microorganisms in other parts of the world may be able to use the same operon for degradation of similar chlorinated compounds. No definitive mechanism of evolution has been proposed for these highly homologous sets of genes. However, scientists have proposed that such a mode of gene transfer may be mediated through transposable elements. Fulthorpe and Wyndham (1992) have shown that when the chlorobenzoate-catabolic transposon (Tn*5271*) is introduced, as part of a conjugative plasmid (pBRC60) in the natural host (*Alcaligenes* sp. strain BR60) into lake water and sediment flow-through microcosms, the populations of Tn*5271*-carrying bacteria were significantly higher in microcosms exposed to chlorinated compounds. Thus, it appears that horizontal gene transfer among bacterial consortia is mediated not only by mobile genetic elements, but also through the selective effects of a particular chemical contaminant.

EVOLUTION OF BIODEGRADATIVE REGULATORY SYSTEMS

An essential feature among biodegradative pathways for toxic compounds is the presence of regulatory sequences and/or regulatory gene(s). Such regulatory genes contribute to the optimal expression of the biodegradative genes they control. All three chlorocatechol gene clusters, *clc*, *tcb*, and *tfd*, have evolved their own positive regulators (*clcR*, *tcbR*, and *tfdR*). All three of these regulatory genes encode polypeptides which belong to the LysR-type of positive regulators which bind a stretch of nucleotide residues with a common motif, T-N_{11}-A. The *cat*BC operon is also activated by the product of a positive regulatory gene, *cat*R, which binds two regions upstream of the *cat*BC operon, termed the repression binding site (RBS) and the activation binding site (ABS) (Parsek et al., 1994a). For expression of the *cat*BC operon, CatR requires the inducer compound, *cis,cis*-muconate. The RBS is bound by CatR in the absence of the inducer molecule, while both the RBS and ABS are bound in the presence of the inducer. Interestingly, the recognition sequence of CatR contains the motif G-N_{11}-A. When the G residue is changed to a T (making a perfect T-N_{11}-A motif), using site-directed mutagenesis, both the binding of CatR to the promoter region and activation of the *cat*BC operon were enhanced, suggesting that the *cat*BC operon is suboptimally expressed in the wild-type strain (Parsek et al., 1994a). In contrast, both the *clc*ABD operon and the *tcb*CDEF operlon harbor a perfect T-N_{11}-A motif. These data suggest that any deficiencies resulting in the suboptimal expression of degradative genes are corrected during the evolutionary process to allow efficient expression of the evolved genes.

Evolutionary adaptation of pre-existing catabolic pathways to the presence of new recalcitrant compounds may require different levels of change to assure expression of the new pathway genes; that is, the development of a novel regulator, and/or alterations in regulatory elements (i.e., positional or nucleotide changes in the DNA binding sequences). An example of optimization of a degradative pathway through regulatory sequence divergence is the phenol degradation pathway of *Pseudomonas putida* strain PaW85 (Kivisaar et al., 1990). A strain was isolated from a phenol-contaminated site from which the phenol degradative *phe*BA operon was cloned. *phe*A encodes phenol monooxygenase and *phe*B encodes catechol 1,2-dioxygenase. When expressed, these enzymes convert phenol to *cis,cis*-muconate. The *phe*BA operon is flanked by two open reading frames, designated ORF1 and ORF2, which show homologies to genes encoding transposases (Nurk et al., 1991), suggesting that the *phe*BA operon was picked up by a transposon. It has been demonstrated that a chromasomally encoded CatR is necessary for expression of the *phe*BA operon (Kasak et al., 1993). Furthermore, the transcriptional start site of this operon was determined, and the promoter region was found to be highly homologous to that of the *cat*BC operon (Kasak et al., 1993). Indeed, CatR interacts with the *phe*BA promoter in a manner similar to its interaction with the *cat*BC promoter, as shown by DNaseI footprinting and gel shift assays (Parsek et al., 1995). *In vitro* transcription assays demonstrated the presence of *phe*BA transcript in the presence of CatR, both in the presence and even in the absence of the inducer, *cis,cis*-muconate (Figure 1B). In contrast, no *cat*BC transcript is produced when the inducer is absent (Figure 1A). The cause of this phenomenon may be DNA bending at the promoter regions. LysR-type regulators have previously been shown to cause an inducer-responsive change in DNA bending at the promoter region; this property is thought to be important in the activation mechanism (Hyrniewicz and Kredich, 1994; Wang et al., 1992).

The DNA-bending characteristic was examined for CatR using the circular permutation gel shift assay (Parsek et al., 1995). Using this technique, it is possible to ascertain the amount of bending a DNA-binding protein exerts on a particular DNA fragment by noting the degree of the gel shift of the DNA/protein complex and comparing this to a known standard. It was found that CatR bent both the *cat*BC and *phe*BA promoter regions. When *cis,cis*-muconate was present in the reaction, CatR introduced almost identical bends in the two promoter regions (37 and 38, respectively; Figure 2). In the absence of inducer, CatR bent the *cat*BC promoter region differently from the way it bent

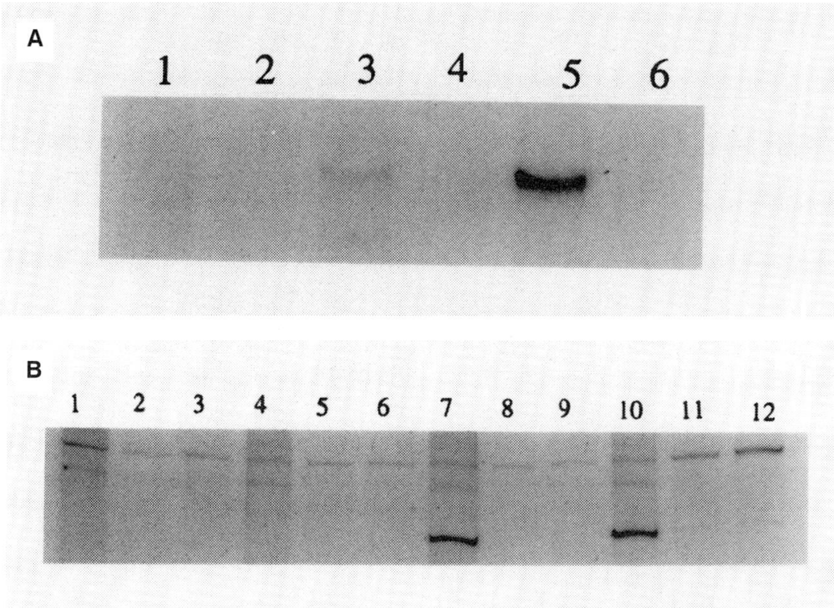

Figure 1. *In vitro* transcription assay demonstrating the importance of CatR, *cis,cis*-muconate, and α subunit RNA polymerase for expression of *cat*BC and *phe*BA transcripts. **A**, *In vitro* transcription of the *cat*BC promoter. Lanes 1,4,7 and 10 were *in vitro* transcription assays conducted in the presence of wild-type RNA polymerase. Lanes 2, 5, 8, 11 and 3, 6, 9, 12 were conducted in the presence of RNA polymerase bearing truncations at positions 235 and 256, respectively, at the C-terminal end of the α subunit. Lanes 1-3 were conducted in the absence of CatR and the inducer, *cis,cis*-muconate. Lanes 4-6 contained 55 ng of purified CatR. Lanes 7-9 contained 5.5 ng of CatR and 1-mM *cis,cis*-muconate. Lanes 10-12 contained 55 ng of purified CatR and 1-mM *cis,cis*-muconate. **B**, *In vitro* transcription of the *phe*BA promoter. Lanes 1,3 and 5 represent reactions conducted with wild-type RNA polymerase. Lanes 2,4 and 6 were reactions done with RNA polymerase bearing a truncation at position 235 of the C-terminal end of the α subunit. Lanes 1 and 2 contain RNA polymerase only. Lanes 3 and 4 contain 55 ng of purified CatR. Lanes 5 and 6 contain 55 ng of purified CatR and 1-mM *cis,cis*-muconate. The specific activities of the mutant and wild-type RNA polymerases were normalized using RNA1 transcript of the ColE1 replicon. Note the small amount of *phe*B transcript formed in absence of the inducer (lane 3, Figure 1B), while no *cat*B transcript was seen in absence of the inducer (lane 4, Figure 1A).

the *phe*BA promoter region (59° and 34°, respectively; Figure 2). Thus, CatR bent the *phe*BA promoter in the absence of inducer (34°) to approximately the same degree as it bent the *cat*BC promoter in the presence of inducer (37°). Although the binding of CatR to the *cat*BC and *phe*BA promoter regions is similar, the nature of these two protein/DNA interactions is quite distinct. As noted previously, changes in

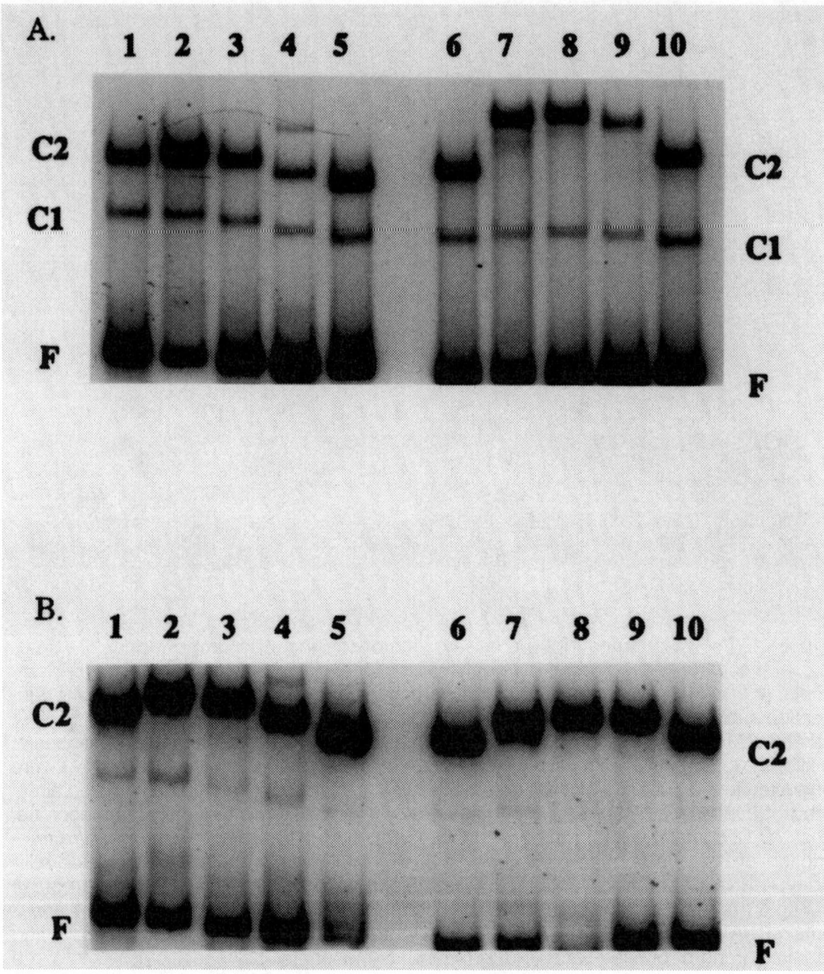

Figure 2. Gel shift assays depicting CatR bending of the *phe*BA and *cat*BC promoter regions in the presence and absence of inducer. **A**, Gel shift assay showing CatR bending of the *cat*BC and *phe*BA promoter regions in the absence of inducer. CatR forms two complexes with both promoter regions (designated C1 and C2). The free DNA is indicated by the letter F. CatR bends the *cat*BC promoter DNA (lanes 6-10) significantly more than the *phe*BA promoter DNA (lanes 1 to 5). **B**, Gel shift assay showing CatR bending of the *cat*BC and *phe*BA promoter regions in the presence of inducer. CatR forms a single complex (designated C2) with both promoter regions. Note the almost identical bending pattern of both *phe*BA and *cat*BC promoter regions by CatR in the presence of inducer.

the promoter regions of the evolved operons (*clcABD*, *tfdCDEF*, and *tcbCDEF*) may allow their optimal expression. Expression of the *phe*BA operon may have been optimized during evolution by the development of a perfect LysR family binding motif, T-N_{11}-A (Parsek et al, 1995). It appears that evolutionary divergence in the *phe*BA promoter has allowed optimal DNA bend by CatR, so that the *phe*BA genes can be partially expressed even in the absence of inducer. This is of physiological importance, since conversion of phenol to the inducer *cis,cis*-muconate by the basal levels of PheA and PheB must be rapid to induce the chromosomal *ortho* pathway of the host. It is important that the *phe*BA operon is expressed in absence of the inducer to rapidly generate the inducer molecule.

CROSS-TALK BETWEEN BIODEGRADATIVE REGULATORY PROTEINS

As mentioned above, CatR is able to recognize a heterologous promoter region and initiate transcription of the *phe*BA operon. It has also been demonstrated that heterologous regulators can functionally replace other regulators of the same family. This phenomenon has been observed in the NtrC-family of transcriptional regulators (Wanner, 1992; Dixon 1986). Dixon (1986) has shown that the *Escherichia coli* regulators NtrC and NifA can activate the upper pathway of the TOL plasmid, even though the *E. coli* and *P. putida* regulatory genes are known to be significantly divergent. The DmpR regulator of the phenol degrading plasmid pVI150 can cross-activate the upper pathway of the TOL plasmid (Fernandez et al., 1994). XylR, in turn, can activate the transcription of the *dmp* operon. High nucleotide homology was observed between the two operons in the regions upstream of their respective transcriptional start sites. When present in the same cell, both regulators exhibited their transcriptional activation properties upon the TOL plasmid upper pathway in the presence of the appropriate inducers (Fernandez et al., 1994). The ability to cross-regulate a heterologous operon has also been demonstrated with CatR and the *clcABD* operon. The *clcABD* operon from *P. putida* allows growth on 3-chlorocatechol and is under the regulatory control of the LysR-family member ClcR (Frantz and Chakrabarty, 1987; Coco et al., 1993). When introduced into a CatR$^-$ background containing a functional *cat*BC or *clcABD* operon, CatR was shown to allow growth of the host cells on benzoate or 3-chlorobenzoate (Parsek et al., 1994b). Although it is able to bind the *cat*BC promoter, ClcR was not able to complement the CatR$^-$ mutant for growth on benzoate. The reason for this may be the nature of contact between ClcR and the *cat*BC promoter. CatR binds the *cat*BC promoter region, making discrete contacts on the DNA in the form of RBS and ABS. By DNaseI footprinting analysis, it was shown that CatR fully contacted the G-N_{11}-A motif of the *cat*BC promoter region (Parsek et al., 1994b). As expected, CatR also contacted the T-N_{11}-A motif of the *clc*ABD promoter region. When incubated in the presence of the *clc*ABD promoter region, the CatR/DNA contacts are more extended. ClcR also binds the LysR-family binding motif, T-N_{11}-A, present in the promoter region of the *clc*ABD operon, and such contacts are very similar to that made by CatR (Parsek et al., 1994b). ClcR binds to the G-N_{11}-A motif of the *cat*BC promoter region, but the G residue is not contacted. The binding of ClcR to an ancestral promoter region may superficially seem to be similar to that of the CatR binding. However, a slight distinction in nucleotide contact exists between the two interactions which leads to the inability of ClcR to activate the *cat*BC operon. Thus, while CatR can activate both the *cat*BC and *clc*ABD operons, ClcR can activate only the *clc*ABD operon but not the *cat*BC operon. This difference is apparently enough to obviate a functional replacement of an essential regulatory protein by one from the same family and may contribute to the specificity of the evolved operon.

BIODEGRADATION OF 2,4,5-TRICHLOROPHENOXYACETIC ACID: AN EVOLUTIONARY PERSPECTIVE

The preceding discussion of biodegradative genes, their regulation, and development through evolution has been presented with regard to microorganisms isolated from the natural environment. In an effort to address the problem of toxic compound accumulation in the environment, our laboratory developed a technique termed "plasmid-assisted molecular breeding" (Kellogg et al., 1981) to isolate a microorganism which is able to completely mineralize the herbicide, 2,4,5-trichlorophenoxyacetic acid (2,4,5-T) (Kilbane et al., 1982). This strain, designated *Pseudomonas cepacia* strain AC1100, currently renamed *Burkholderia cepacia* AC1100, was isolated in a chemostat after several months of selection on 2,4,5-T.

In order to understand the evolution and gene organization of 2,4,5-T metabolism, our laboratory has been examining the genetics and biochemistry of the 2,4,5-T catabolic pathway in strain AC1100 (Figure 3).

To clone the genes necessary for 2,4,5-T degradation, a series of 2,4,5-T- mutants (strains unable to grow on 2,4,5-T) were made by simple growth on rich, nonselective media and by transposon mutagenesis (Tomasek et al., 1989; Haugland et al., 1990). By mobilizing a genomic library of wild-type AC1100 DNA into these mutants and screening for colonies able to grow on 2,4,5-T, appropriate 2,4,5-T degradative (*tft*) genes were cloned. A construct containing an 8.9-kb *Xba*I DNA fragment (pRHC89) complemented a 2,4,5-T- mutant, RHC22 (Haugland et al., 1990). Upon introduction into *P. aeruginosa* strain PAO1, this construct allowed the conversion of 2,4,5-T to 2,4,5-trichlorophenol (2,4,5-TCP), the first step of this metabolic pathway. This region was further subcloned as a 2.7-kb *Sst*I-*Xba*I DNA fragment, and part of this fragment was sequenced to reveal two open reading frames (ORF) with strong homology to bacterial biodegradative dioxygenases. The first ORF translational product shows a 37% amino acid sequence identity to the benzoate dioxygenase α-subunit, BenA, from *Acinetobacter calcoaceticus*, and 40% amino acid sequence identity to the α-subunit of the toluate dioxygenase from *P. putida*. The second ORF (*tftB*) gene product shows significant homology to the β-subunits of these dioxygenases, but this homology was less overall (Danganan, et al., 1994).

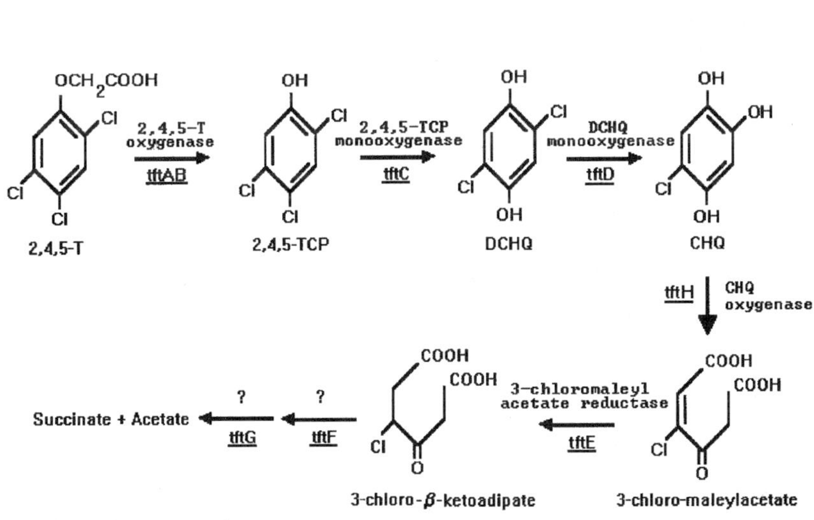

Figure 3. A tentative pathway of degradation of 2,4,5-trichlorophenoxyacetic acid (2,4,5-T) by *Burkholderia cepacia* strain AC1100. Although *tft*G and *tft*F are necessary for metabolism of 2,4,5-T, their functions in the pathway are not known.

These two open reading frames (now designated *tft*AB), encode polypeptides of approximately 49 kDa and 24 kDa, respectively, and these gene products quantitatively converted 2,4,5-T to 2,4,5-TCP and glyoxylate with the coconsumption of NADH and O2 (Xun and Wagnon, 1995). This enzyme also catalyzes the conversion of 2,4-dichlorophenoxyacetate (2,4-D) to 2,4-dichlorophenol (2,4-DCP) (Danganan et al., 1994; Xun and Wagnon, 1995). This reaction is similar to that of the 2,4-D α-ketoglutarate dioxygenase (TfdA) from *Alcaligenes eutrophus* strain JMP134, which also converts 2,4-D to 2,4-dichlorophenol (Fukumori and Hausinger 1993a,b). The 2,4,5-T oxygenase's rate of conversion of 2,4-D to 2,4-DCP is about 1000 times less than that of TfdA (Xun and Wagnon, 1995). This slow rate of conversion was also observed in whole-cell assays with AC1100 (Karns et al., 1983). The selection of a catalytically slower enzyme may have been important for growth and survival of AC1100 during chemostatic selection, since overproduction of 2,4,5-TCP is extremely toxic to AC1100. The development of an enzyme which slowly converts 2,4,5-T to 2,4,5-TCP can ensure that 2,4,5-TCP is not accumulated and is quickly converted to the next intermediate in the pathway. Although TftAB is catalytically slow compared to an analogous enzyme, its expression is constitutive (Karns et al.,1983).

Analysis of the -35/-10 region of *tft*AB shows a typical *E. coli*-type σ^{70} promoter. When a *lac*Z promoter-probe construct containing this region of DNA was introduced into both *E. coli* and wild-type AC1100 cells, these strains turned intensely blue on plates containing the chromogenic reagent X-gal (Danganan et al., 1995). Furthermore, the AC1100 strain containing this construct exhibited a high level of β-galactosidase activity in the presence or absence of 2,4,5-T throughout all phases of growth (Danganan et al., 1995). While AC1100 evolved an enzyme with low activity to safeguard against the overproduction of a potentially lethal but metabolically essential compound, it may have recruited a promoter region which ensured that the enzyme necessary for the catabolism of 2,4,5-T is expressed at constitutive levels.

Another 2,4,5-T$^-$ mutant, designated PT88, accumulated 5-chloro-2-hydroxy hydroquinone (CHQ) in the medium when grown in the presence of 2,4,5-T (Sangodkar et al., 1988). This mutant was complemented for growth on 2,4,5-T by an 8.0-kb DNA fragment. Through further subcloning experiments, this region was narrowed down to a 5.5-kb *EcoRI-BamHI* DNA fragment, which, when sequenced, revealed the presence of six ORF, designated ORF1 to ORF6 (Daubaras et al.,

1995). These ORF have recently been designated *tft*E, *tft*F, ORF3, ORF4, *tft*G, and *tft*H; four of these ORF were necessary for complementation of PT88. The deduced amino acid sequences of the genes, (*tft*E and *tft*H) show strong homology with the following proteins: maleylacetate reductases and catechol 1,2-dioxygenases, respectively (Daubaras et al., 1995). In gene clusters of *clc*, *tfd*, and *tcb* which have evolved in naturally occurring microorganisms, the chlorocatechol-1,2-dioxygenase genes (*clc*A, *tfd*C, and *tcb*C) are located at the 5′ end of the cluster, and the maleylacetate reductase genes *clc*F, *tcb*F, and, *tfd*F (Kasberg et al., 1995) are found at the 3′ end of the cluster. In contrast, the *tft* genes are organized in the opposite order; that is, *tft*H (ORF6) is located at the 3′ end of the cluster, and *tft*E (ORF1) is at the 5′ end (Figure 4). It is not understood why these genes are arranged in this manner. The purification and characterization of these gene products are in progress and should shed some light on this question.

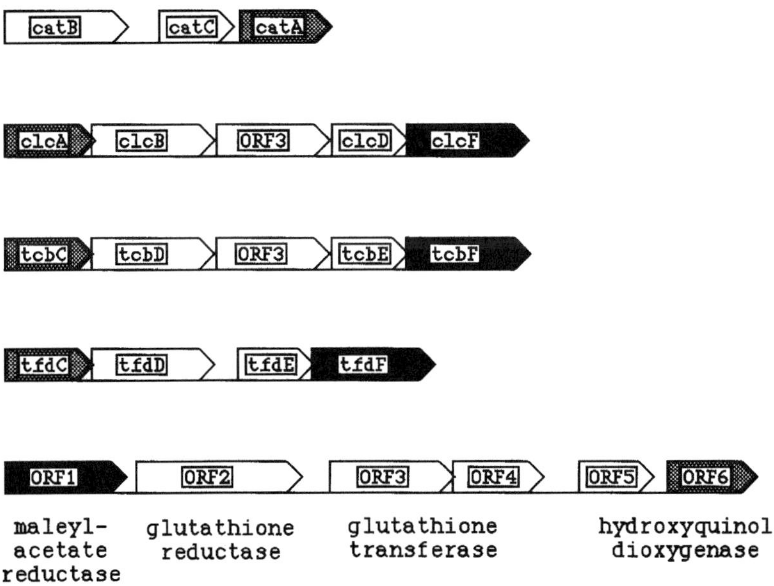

Figure 4. Organization of the structural genes of the catechol degradative *cat*BCA operon, 3-chlorocatechol degradative *clc*ABDF operon, trichlorocatechol degradative *tcb*CDEF operon, dichlorocatechol degradative *tfd*CDEF operon and the 2,4,5-T-derived CHQ degradative gene cluster ORF1(*tft*E)-ORF2 (*tft*F)-ORF3-ORF4-ORF5 (*tft*G)-ORF6 (*tft*H).

AC1100 has been shown to be extremely unstable with respect to its 2,4,5-T-degrading abilities; that is, this strain rapidly loses its ability to grow on 2,4,5-T when grown on rich media. The mutant used to clone the *tft*AB genes was isolated in this manner. It has since been demonstrated that this strain has insertion elements arranged throughout its genome (Tomasek et al., 1989). Furthermore, both the *tft* gene clusters *tft*AB and *tft*EFGH have copies of these elements in their upstream regions (Sangodkar et al., 1988; Haugland et al., 1991). These repeated sequences, designated IS931, appear to have properties of transposable elements. IS931 was able to transpose from the genome of AC1100 to the plasmid pKT240 and allowed the expression of a promoterless streptomycin phosphotransferase gene (Haugland et al., 1990), suggesting that IS931 has promoter-like activity. It is possible that these transposable elements may have been responsible for the recruitment of the *tft* genes. They may also contribute to the instability of the 2,4,5-T-degrading phenotype of AC1100 in the absence of selection pressure through recombination among various copies of IS931. Perhaps, when an alternative carbon source is available, strain AC1100 eliminates 2,4,5-T-metabolizing genes through IS-mediated recombinations or deletions.

CONCLUDING REMARKS

In the discussion above, we have attempted to address some of the important aspects of microbial evolution with respect to the biodegradation of synthetic chlorinated compounds. We have focused on the possible modes of evolution of both regulatory and structural genes of these naturally occurring and chemostat-derived microorganisms. Through this means we hope to gain a better understanding of how to enhance the microbial capabilities of mineralizing recalcitrant compounds. The data presented here comprise our initial efforts at understanding the underlying mechanisms behind this aspect of microbial evolution.

REFERENCES

Coco, WM, R Rothmel, S Henikoff, and AM Chakrabarty. **1993**. Nucleotide sequence and initial functional characterization of *clc*R, a LysR family activator of the *clc*ABD chlorocatechol operon in *Pseudomonas putida*. *J. Bacteriol*. 175:417-427.

Danganan, CE, RW Ye, DL Daubaras, L Xun, and AM Chakrabarty. **1994**. Nucleotide sequence and functional analysis of the genes encoding 2,4,5-trichlorophenoxyacetic acid oxygenase in *Pseudomonas cepacia* strain AC1100. *Appl. Environ. Microbiol.* 60:4100-4106.

Danganan, CE, S Shankar, RW Ye, and AM Chakrabarty. 1995. Substrate diversity and expression of the 2,4,5-trichlorophenoxyacetic acid oxygenase from *Burkholderia cepacia* strain AC1100. *Appl. Environ. Microbiol.* 61:4500-4504.

Daubaras, DL and AM Chakrabarty. **1992**. The environment, microbes and bioremediation: microbial activities modulated by the environment. *Biodegradation* 3: 125-135.

Daubaras, DL, CD Hershberger, K Kitano, and AM Chakrabarty. **1995**. Sequence analysis of a gene cluster involved in metabolism of 2,4,5-trichlorophenoxyacetic acid by *Burkholderia cepacia* AC1100. *Appl. Environ. Microbiol.* 61:1279-1289.

Dixon, R. **1986**. The xylABC promoter from the *Pseudomonas putida* TOL plasmid is activated by nitrogen regulatory genes in *Escherichia coli*. *Mol. Gen. Genet* 203:129-136.

Fernandez, S, V Shingler, and V de Lorenzo. **1994**. Cross-regulation by XylR and DmpR activators of *Pseudomonas putida* suggests that transcriptional control of biodegradative operons evolves independently of catabolic genes. *J. Bacteriol.* 176:5052-5058.

Frantz, B and AM Chakrabarty. **1987**. Organization and nucleotide sequence determination of a gene cluster involved in 3-chlorocatechol degradation. *Proc. Natl. Acad. Sci. USA*. 84:4460-4464.

Fukumori, F and R Hausinger. **1993a**. *Alcaligenes eutrophus* JMP134 "2,4-dichlorophenoxyacetate monooxygenase" is an a-ketoglutarate-dependent dioxygenase. *J. Bacteriol.* 175:2083-2086.

Fukumori, F and RP Hausinger. **1993b**. Purification and characterization of 2,4-dichlorophenoxyacetate α-ketoglutarate dioxygenase. *J. Biol. Chem.* 268:24311-24317.

Fulthorpe, RR and RC Wyndham. **1992**. Involvement of chlorobenzoate-catabolic transposon, Tn*5271*, in community adaptation to chlorobiphenyl, chloroaniline, and 2,4-dichlorophenoxyacetic acid in a freshwater ecosystem. *Appl. Environ. Microbiol.* 58:314-325.

Haugland, RA, UMX Sangodkar, and AM Chakrabarty. **1990**. Repeated sequences including RS1100 from *Pseudomonas cepacia* AC1100 function as IS elements. *Mol. Gen. Genet.* 220:222-228.

Haugland, RA, UMX Sangodkar, PR Sferra, and AM Chakrabarty. **1991**. Cloning and characterization of a chromosomal DNA region required for growth on 2,4,5-T by *Pseudomonas cepacia* AC1100. *Gene* 100:65-73.

Hyrniewicz, M and NM Kredich. **1994**. Stoichiometry of binding of CysB to the *cys*JIH, *cys*K, and *cys*P promoter regions of *Salmonella typhimurium*. *J. Bacteriol.* 176:3673-3682.

Karns, JS, S Duttagupta, and AM Chakrabarty. **1983**. Regulation of 2,4,5 trichlorophenoxyacetic acid and chlorophenol metabolism in *Pseudomonas cepacia* AC1100. *Appl. Environ. Microbiol.* 46:1182-1186.

Kasak, L, R Horak, A Nurk, K Talvik, and M Kivisaar. **1993**. Regulation of the catechol 1,2-dioxygenase and phenol monooxygenase-encoding *phe*BA operon in *Pseudomonas putida* strain PaW85. *J. Bacteriol.* 175:8038-8042.

Kasberg, T, DL Daubaras, AM Chakrabarty, D Kinzelt, and W Reineke. **1995**. Evidence that operons *tcb*, *tfd*, and *clc* encode maleylacetate reductase, the fourth enzyme of the modified *ortho* pathway. *J. Bacteriol.* 177:3885-3889.

Kellogg, ST, DK Chatterjee, and AM Chakrabarty. **1981**. Plasmid-assisted molecular breeding: new technique for enhanced biodegradation of persistent toxic chemicals. *Science* 214:1133-1135.

Kilbane, JJ, DK Chatterjee, JS Karns, ST Kellogg, and AM Chakrabarty. **1982**. Biodegradation of 2,4,5-trichlorophenoxyacetic acid by a pure culture of *Pseudomonas cepacia*. *Appl. Environ. Microbiol.* 44:72-78.

Kivisaar, M, R Horak, L Kasak, A Heinaru, and J Habicht. **1990**. Selection of independent plasmids determining phenol degradation in *Pseudomonas putida* and the cloning and expression of genes encoding phenol monooxygenase and catechol 1,2-dioxygenase. *Plasmid* 24:25-36.

Nagata, YR, R Ohtomo, K Miyauchi, M Fukuda, K Yano, and M Takagi. **1994**. Cloning and sequencing of a 2,5-dichloro-2,5-cyclohexadiene-1,4-diol dehydrogenase gene involved in the degradation of

-hexachlorocyclohexane in *Pseudomonas paucimobilis*. *J. Bacteriol.* 176:3117-3125.

Nurk, AL, L Kasak, and M Kivisaar. **1991**. Sequence of the gene (*phe*A) encoding phenol monooxygenase from *Pseudomonas sp.* strain EST1001: expression in *Escherichia coli* and *Pseudomonas putida*. *Gene* 102:13-18.

Parsek, MR, RW Ye, P Pun, and AM Chakrabarty. **1994a**. Critical nucleotides in the interaction of a LysR-type regulator with its target promoter region. *J. Biol. Chem.* 269:11279-11284.

Parsek, MR, SM McFall, DL Shinabarger, and AM Chakrabarty. **1994b**. Interaction of two LysR-type regulatory proteins CatR and ClcR with heterologous promoters: Functional and evolutionary implications. *Proc. Natl. Acad. Sci. USA* 91:12393-12397.

Parsek, MR, M Kivisaar, and AM Chakrabarty. **1995**. Differential DNA bending introduced by the *Pseudomonas putida* LysR-type regulator, CatR, at the plasmid-borne *phe*BA and chromosomal *cat*BC promoter. *Mol. Microbiol.* 15:819-828.

Sangodkar, UMX, PJ Chapman, and AM Chakrabarty. **1988**. Cloning, physical mapping and expression of chromosomal genes specifying degradation of the herbicide 2,4,5-T by *Pseudomonas cepacia* AC1100. *Gene* 71:267-277.

Suen, WC and JC Spain. **1993**. Cloning and characterization of *Pseudomonas sp.* strain DNT genes for 2,4-dinitrotoluene degradation. *J. Bacteriol.* 175:1831-1837.

Tomasek, PH, B Frantz, UMX Sangodkar, R Haugland, and AM Chakrabarty. **1989**. Characterization and nucleotide sequence determination of a repeat element isolated from a 2,4,5-T degrading strain of *Pseudomonas cepacia*. *Gene* 76:227-238.

Wang, L, JD Helmann, and SC Winans. **1992**. The *Agrobacter tumefaciens* transcriptional activator OccR causes a bend at a target promoter, which is partially relaxed by a plant tumor metabolite. *Cell* 69:659-667.

Wanner, BL. **1992**. Is cross regulation by phosphorylation of two-component response regulator proteins important in bacteria? *J. Bacteriol.* 174:2053-2058.

Xun, L and CS Orser. **1991**. Purification and properties of pentachlorophenol hydroxylase, a flavoprotein from *Flavobacterium* sp. strain ATCC 39723. *J. Bacteriol.* 173:4447-4453.

Xun, L and KB Wagnon. **1995**. Purification and properties of component B of 2,4,5-trichlorophenoxyacetate oxygenase from *Pseudomonas cepacia* AC1100. *Appl. Environ. Microbiol.* 61:3499-3502.

BIODEGRADATION OF XENOBIOTICS BY FLAVOMONOOXYGENASES

L. Xun

Microbiology Department, Washington State University at Tri-Cities, Richland, Washington

Key Words: *Chlorophenols, pentachlorophenol (PCP), nitrilotriacetate (NTA), and EDTA.*

ABSTRACT

Oxygenases play a major role in the biodegradation of xenobiotics. One group of oxygenases uses flavin as prosthetic groups. Four enzymes are compared: two flavin-adenine dinucleotide-containing enzymes that attack aromatic compounds, and two flavin-mononucleoxide-containing enzymes that oxidize substituted amines. Both pentachlorophenol 4-monooxygenase (PCPB) and chlorophenol 4-monooxygenase (TCP-Mo) oxidize chlorophenols. The PCPB is a single protein of 60 kDa, while TCP-Mo consists of two components of 58 and 22 kDa. Nitrilotriacetate monooxygenase (NTA-Mo) uses only NTA-metal complexes as substrate. On the other hand, EDTA monooxygenase (EDTA-Mo) degrades both NTA- and EDTA-metal complexes. The NTA-Mo consists of two polypeptides of 47 and 36 kDa, while EDTA-Mo has a 50-kDa component and an unidentified component. The similarities and differences of these enzymes imply the natural diversities of related enzymes.

INTRODUCTION

Many chemically synthesized compounds can be degraded by microorganisms (Chaudhry and Chapalamadugu, 1991). In most cases, oxygenases catalyze the initial attack to introduce hydroxyl groups. The hydroxylating oxygenases require prosthetic groups to activate molecular oxygen. P-450 monooxygenases use hemes; aromatic-ring-hydroxylating dioxygenases contain iron-sulfur centers; and flavooxygenases require flavin. They all play important roles in degrading different but sometimes overlapping spectra of chemicals. We have worked on four flavooxygenases over the past few years. Pentachlorophenol (PCP) 4-monooxygenase (10,14,15,16) and chlorophenol 4-monooxygenase contain flavin-adenine dinucleotide (FAD) and catalyze similar reactions. Nitrilotriacetate (NTA) monooxygenase uses only NTA-metal complexes as its substrates (Ueta et al., 1992). On the other hand,

EDTA-monooxygenase degrades both NTA- and EDTA-metal complexes (Xun, unpublished data). Both enzymes require flavin mononucleotide (FMN) as a prosthetic group (Table 1). In this communication, I will focus on chlorophenol 4-monooxygenase that is newly identified, purified, and characterized.

Table 1. Comparison of four flavooxygenases.

Enzymes	Subunits Flavin	(kDa)
PCP 4-monoox	genase	FAD[a]y63
Chlorophenol 4-monooxygenase	FAD	58 & 22
NTA monooxygenase	FMN[b]	50 & 35
EDTA monooxygenase	FMN	50 & ?

[a]Flavin-adenine dinucleotide.
[b]Flavin mononucleotide.

MATERIALS AND METHODS

Materials and Bacterium

All chlorinated phenols and hydroquinones were purchased from Aldrich. 6-Chlorohydroxyquinol (CHQ) was a gift from Dr. Ronald L. Crawford of the University of Idaho. *Pseudomonas cepacia* AC1100 was kindly provided by Dr. A. M. Chakrabarty of University of Illinois at Chicago and was cultured with sodium glutamate (2 g/L) and 2,4,5-T (0.15 g/L) in a mineral medium (Kilbane et al., 1982).

Enzyme Assay

The chlorophenol 4-monooxygenase activity was assayed by measuring the production of tetrachloro-*p*-hydroquinone (TeCH) from 2,3,5,6-tetrachlorophenol (TeCP) by high-performance liquid chromatography (HPLC) analysis (Xun and Orser, 1991). The reaction was initiated by adding NADH to the reaction mixture.

Purification of Chlorophenol 4-Monooxygenase

The enzyme was extracted from *Ps. cepacia* Ac1100 by French press, precipitated by ammonium sulfate, and purified by chromatographic columns.

Analytical Methods

A Waters HPLC with a Nova-Pak C18 column (3.9 by 150 mm) was used to analyze chlorinated phenols and hydroquinones (Xun and Orser, 1991). The retention time and absorption spectrum of each compound or reaction product were compared with those of authentic standards for identification, and peak area was used for quantification.

Sodium dodecyl sulfate-polyacrylamide gel electrophoresis (SDS-PAGE) was done by the method of Laemmli (Laemmli, 1970). Gels were stained for proteins with Coomassie brilliant blue R-250. Protein concentrations were determined with a protein dye reagent (Bradford, 1976), with bovine serum albumin as the standard.

RESULTS

Identification of Chlorophenol 4-Monooxygenase Activities in Cell Extracts

The cell extract degraded 2,4,5-trichlorophenol (2,4,5-TCP) in the presence of NADH and O_2; however, the proposed product, 2,5-dichloro-*p*-hydroquinone (2,5-DiCH) (Sangodkar et al., 1990), was not detectable. This may have been due to the further degradation of 2,5-DiCH by subsequently used enzymes. Since it is easier to follow the formation of a product than to check the decrease of a substrate for activity assay, other chlorophenols were tested for the accumulation of a product. When TeCP was used as a substrate, TeCH was produced by the cell extract in the presence of NADH and O_2. Therefore, TeCP was used as a substrate to follow the enzyme activity during protein purification. Flavin adenine dinucleotide but not FMN was found to stimulate the reaction. NADPH could replace NADH, but the enzyme activity was significantly reduced.

Enzyme Purification

Chlorophenol 4-monooxygenase was purified from *Ps. cepacia* Ac1100 by monitoring TeCH production in reaction mixtures following each purification step. The enzyme activity was separated into two components by MonoQ. Neither of the components degraded TeCP alone, but the combination of the two components converted TeCP to TeCH. First, cB was further purified to homogeneity by using cA from the MonoQ column to reconstitute the enzyme activity. Then, cA was further purified by using purified cB to reconstitute the enzyme activity.

Enzyme Properties

The cA was apparently purified to homogeneity as indicated by SDS-PAGE analysis, which revealed a single 22-kDa band. The native cA was estimated to be a monomer by gel filtration. The purified cA was slightly yellowish, and FAD (but not FMN) stimulated its cytochrome-C reduction, indicating that cA uses FAD as a prosthetic group. A solution containing 68.2 μM cA (1.5 mg of protein ml^{-1}; molecular weight, 22,000) gave an A_{450} of 0.15. The flavin content was calculated to be 13.2 μM by using a molar extinction coefficient of 11,300 for FAD (Strickland and Massey, 1973). The molar ratio of flavin to protein was 0.19, indicating that only about 19% of purified cA contains FAD.

The SDS-PAGE analysis of purified cB revealed a protein of 58 kDa. The enzyme preparation was colorless. Gel filtration indicated that the purified cB is a monomer.

Enzyme Activity

When using the purified enzyme to hydroxylate 2,4,5-TCP, 2,5-DiCH accumulated only slightly during 2,4,5-TCP degradation. Subsequently, I found out that the enzyme used 2,5-DiCH as a substrate. Both 2,4,5-TCP hydroxylation and 2,5-DiCH consumption required cA and cB as well as NADH and oxygen; FAD stimulated both reactions. The final product from either 2,4,5-TCP or 2,5-DiCH was identified as CHQ by HPLC analysis. The final product and authentic CHQ had the same retention time of 6.2 min and the same absorption maximum of 288 nm. When using TeCP as the substrate, TeCH was produced and accumulated. The enzyme also quantitatively converted 2,4,6-TCP to 2,6-DiCH which was not further degraded. Although TeCH was slightly decreased, it was not conclusive whether TeCH was converted to 2,5,6-trichlorohydroxyquinol by the enzyme.

Kinetic Analysis

Enzymatic rates at low substrate concentrations are dependent on the specificity constant k_{cat}/K_M (Fersht, 1985). The bigger the constant, the higher the reaction rate. The kinetic parameters were determined for 2,4,5-TCP and 2,5-DiCH. Since the k_{cat}/K_M value for 2,5-DiCH was more than 10 times higher than that for 2,4,5-TCP, it explains why 2,5-DiCH was barely detectable during 2,4,5-TCP degradation.

DISCUSSION

This is the first time that chlorophenol 4-monooxygenase has been identified, purified and characterized. Since the enzyme oxidized TeCP to TeCH, 2,4,6-TCP to 2,6-DiCH, and 2,4,5-TCP to 2,5-DiCH, and since the hydroxylating position for these chlorophenols are all at the 4-position, the enzyme should be named chlorophenol 4-monooxygenase. In addition to chlorophenols, the enzyme oxidized 2,5-DiCH to CHQ, which also required NADH and O_2. The enzyme did not use 2,6-DiCH as a substrate, but it might use TeCH as a substrate at a very slow rate. Apparently, the single enzyme catalyzed the sequential hydroxylation of 2,4,5-TCP to 2,5-DiCH and then to CHQ (Figure 1).

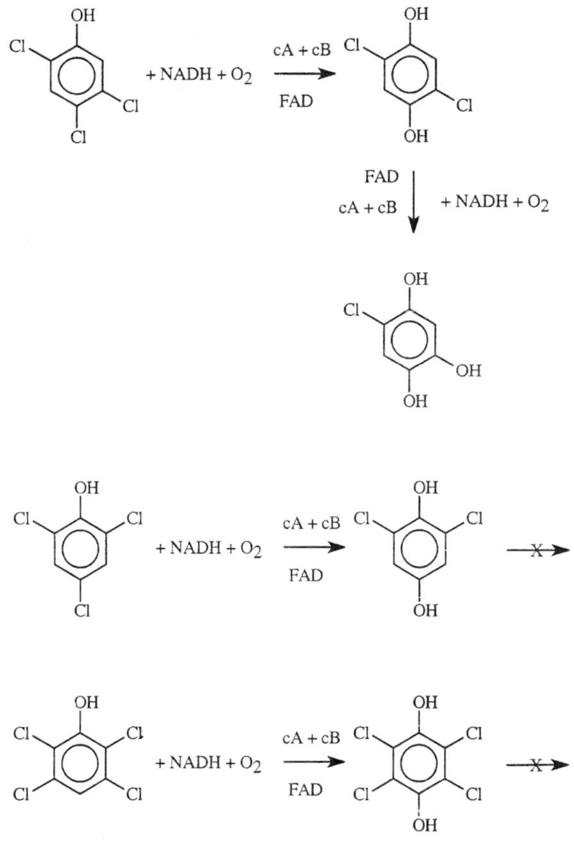

Figure 1. Proposed reactions catalyzed by chlorophenol 4-monooxygenase.

The substrates tested so far for chlorophenol 4-monooxygenase are similar but different from those of PCP 4-monooxygenase (Xun et al., 1992b). Both enzymes use PCP, TeCH, and 2,4,6-TCP, but PCP 4-monooxygenase does not use 2,4,5-TCP or any chlorinated hydroquinones (Xun et al., 1992a). PCP 4-monooxygenase uses many other substituted phenols as substrates (Xun et al., 1992a). Detailed substrate analysis of chlorophenol 4-monooxygenase will reveal important information about the enzyme. The apparent ability of the enzyme to oxidize hydroquinones in addition to chlorophenols is rather unusual.

For mono- or dichlorinated phenols, breakdown is usually through a catechol intermediate before ring cleavage (Haggblom, 1992). On the other hand, all polychlorinated phenols have been shown to be degraded through CHQ before ring cleavage (Apajalahti and Salkinoja-Salonen, 1987; Daubaras et al., 1995; and Zaborina et al., 1995). Chlorophenol 4-monooxygenase is the first enzyme reported that produces CHQ. The enzyme is probably not involved in PCP degradation because PCP was a poor substrate (unpublished data), and 2,6-DiCH, a PCP degradation intermediate, was not a substrate for chlorophenol 4-monooxygenase. The enzyme can effectively oxidize 2,4,6-TCP to 2,6-DiCH but not 2,6-DiCH to CHQ. Therefore, a different enzyme must be responsible for converting 2,6-DiCH to CHQ for PCP and 2,4,6-TCP degradation.

The diverse substrate range of these flavomonooxygenases is fascinating. Since the enzymes can catalyze some substrates more effectively than others, improving specific catalytic activity toward certain compounds through genetic engineering may be relatively easier than introducing a completely new activity. Natural evolution may constantly modify these enzymes to meet the need of some microorganisms to survive in the environment.

ACKNOWLEDGMENTS

This research was supported by NSF grant MCB-9218783. I thank Kenneth Wagnon and Caroline Auker for technical help.

REFERENCES

Apajalahti, JHA and MS Salkinoja-Salonen. **1987**. Complete dechlorination of tetrachlorohydroquinone by cell extracts of pentachlorophenol-induced *Rhodococcus chlorophenolicus*. J. Bacteriol. 169:5125-5130.

Bradford, MM. **1976**. A rapid and sensitive method for the quantitation of microgram quantities of protein using the principle of protein-dye binding. Anal. Biochem. 72:248-254.

Chaudhry, GO and S Chapalamadugu. **1991**. Biodegradation of halogenated organic compounds. Microbiol. Rev. 55:59-79.

Daubaras, DL, CD Hershberger, K Kitano, and AM Chakrabarty. **1995**. Sequence analysis of a gene cluster involved in metabolism of 2,4,5-trichlorophenoxyacetic acid by *Burkholderia cepacia* AC1100. Appl. Environ. Microbiol. 61:1279-1289.

Fersht, A. **1985**. *Enzyme Structure and Mechanism*, p. 347, Freeman, New York.

Haggblom, MM. **1992**. Microbial breakdown of halogenated aromatic pesticides and related compounds. FEMS Microbiol. Rev. 103:29-72.

Kilbane, JJ, DK Chatterjee, JS Karns, ST Kellogg, and AM Chakrabarty. **1982**. Biodegradation of 2,4,5-trichlorophenoxyacetic acid by a pure culture of *Pseudomonas cepacia*. Appl. Environ. Microbiol. 44:72-78.

Laemmli, UK. **1970**. Cleavage of structural proteins during the assembly of the head of bacteriophage T4. Nature (London). 227:680-685.

Orser, CS, CC Lange, L Xun, TC Zahrt, and BJ Schneider. **1993**. Cloning, sequence analysis, and expression of the *Flavobacterium* pentachlorophenol 4-monooxygenase gene in *E. coli*. J. Bacteriol. 175:411-416.

Sangodkar, UMX, PJ Chapman, and AM Chakrabarty. **1990**. Cloning, physical mapping and expression of chromosomal genes specifying degradation of the herbicide 2,4,5-T by *Pseudomonas cepacia* AC1100. Gene (Amsterdam). 71:267-277.

Strickland, S and V Massey. **1973**. The purification and properties of the flavoprotein melilotate hydroxylase. J. Biol. Chem. 248:2944-2952.

Ueta, T, R Schneider, M Snozzi, and T Egli. **1992**. Purification and characterization of a two-component monooxygenase that hydroxylates nitrilotriacetate from "*Chelatobacter*" strain ATCC 29600. J. Bacteriol. 174:1179-1188.

Xun, L and CS Orser. **1991**. Purification and properties of pentachlorophenol hydroxylase, a flavoprotein from *Flavobacterium* sp. strain ATCC 39723. J. Bacteriol. 173:4447-4453.

Xun, L, E Topp, and CS Orser. **1992a**. Diverse substrate range of a *Flavobacterium* pentachlorophenol hydroxylase and the reaction stoichiometries. J. Bacteriol. 174:2898-2902.

Xun, L, E Topp, and CS Orser. **1992b**. Confirmation of oxidative dehalogenation of pentachlorophenol by a *Flavobacterium* pentachlorophenol hydroxylase. J. Bacteriol. 174:5745-5747.

Zaborina, O, M Latus, J Eberspacher, LA Golovleva, and F Lingens. **1995**. Purification and characterization of 6-chlorohydoxyquinol 1,2-dioxygenase from *Streptomyces rochei* 303: comparison with an analogous enzyme from *Azotobacter* sp. strain GP1. J. Bacteriol. 177:229-234.

OBTAINING ENZYMES FOR BIODEGRADATION OF POLLUTANTS: SEARCH, SELECT, ASSEMBLE AND SYSTEMATIZE

L. P. Wackett,[1] C. C. Lange[1,2] and R. L. Ornstein[2]

[1]Department of Biochemistry and Biological Process Technology Institute, University of Minnesota, St. Paul, Minnesota

[2]Environmental Molecular Sciences Laboratory, Pacific Northwest National Laboratory, Richland, Washington

Key Words: *Bacteria, enzyme, atrazine, cytochrome P450, toluene dioxygenase, biodegradation, database*

ABSTRACT

The biodegradation of hazardous organic compounds, in nature and industrial practice, is based on microbial enzyme-catalyzed reactions that detoxify the substrate. There is great interest in using enzymes that metabolize the most toxic and persistent pollutants. Environmental persistence of a compound suggests that few enzymes exist to metabolize that substrate. In that case, what is the fastest route for obtaining a suitable enzyme or metabolic pathway: (1) search nature, (2) impose selective pressure on a microorganism to evolve enzymes in the laboratory, (3) assemble enzymes into a new metabolic pathway by genetic engineering, or (4) use computational tools in a knowledge-based approach to improve biodegradation? All four approaches will be discussed. For the first approach, a new microorganism was isolated for its ability to metabolize the herbicide atrazine. The key enzyme catalyzes a hydrolytic dechlorination of atrazine to yield a detoxified product with greatly reduced soil mobility. Approaches (2) and (3) are being conducted with cytochrome $P450_{CAM}$ and toluene oxygenases. In one example, the genes encoding cytochrome $P450_{CAM}$ and toluene dioxygenase were combined in a *Pseudomonas* strain to construct a metabolic pathway for polyhalogenated solvents. Cytochrome $P450_{CAM}$ catalyzes reductive dehalogenation reactions, yielding haloalkenes that are substrates for toluene dioxygenase. Lastly, the University of Minnesota Biocatalysis/Biodegradation Database has been under development since February, 1995, and is currently available on the World Wide Web (http://dragon.labmed.umn.edu/~lynda/index.html). It can serve as a tool for predicting how to optimize microbial metabolism for the biodegradation of hazardous compounds.

SEARCHING NATURE FOR NOVEL BIODEGRADATIVE ENZYMES

Atrazine [2-chloro-4-(ethylamino)-6-isopropylamino-1,3,5-triazine] is the most heavily used herbicide in the United States; over eighty million pounds are knowingly applied on soils annually. For 20 years, researchers have sought to better understand the biodegradation of atrazine. Despite significant efforts, pure bacterial cultures, and enzymes and genes that metabolize atrazine, were not identified until recently. We contributed to the recent successes by isolating *Pseudomonas* sp. ADP from atrazine-contaminated soil by selective enrichment culture (Figure 1). The isolated bacterium grows on atrazine, which is provided as its sole nitrogen source (Mandelbaum et al., 1995). *Pseudomonas* sp. ADP metabolizes atrazine to ammonia, carbon dioxide, and chloride. The first intermediate in the atrazine metabolic pathway is now firmly established as hydroxyatrazine (Mandelbaum et al., 1993; DeSouza et al., 1995). Thus, previous assertions that hydroxyatrazine was environmentally derived exclusively from abiotic reactions must be reinterpreted. The gene from *Pseudomonas* sp. ADP responsible for the conversion of atrazine to hydroxyatrazine has been cloned and expressed in *Escherichia coli*. Most recently, the enzyme has been purified from recombinant *E. coli* clones, and its specificity and reaction mechanism are being studied (DeSouza, Sadowsky and Wackett, unpublished data). The enzymatic conversion of atrazine to hydroxyatrazine constitutes a net detoxification of the herbicide, since the latter compound is not considered problematic by government regulatory agencies. Furthermore, hydroxyatrazine is less mobile in soil than atrazine, and it readily undergoes further biodegradation. Current research is devoted to applying enzyme or microbial cell technology for atrazine hydrolysis in soil and water.

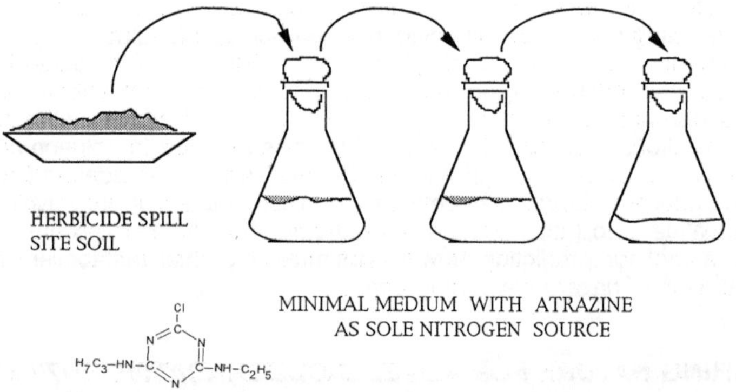

Figure 1. Method of isolating *Pseudomonas* sp. ADP from atrazine-containing soil by passage into consecutive flasks with growth medium containing atrazine as the sole source of nitrogen. Structure of atrazine shown at lower left.

SELECTION FOR OBTAINING IMPROVED BIODEGRADATION CATALYSTS

Genetic selection is nature's way of improving biocatalysis and it can be a powerful method for directed evolution in the laboratory. Before the atrazine-metabolizing organism described above could be obtained, the genes and enzymes had to evolve via natural selection. The set of known enzymes and genes is increasing exponentially, providing a greatly expanding set of building blocks for carrying out directed evolution experiments. The keys to success in improving enzymes by directed evolution are: (1) devising a strong genetic selection screening protocol and (2) producing large numbers of diverse types of mutations in the gene(s) of interest. The former method varies with the system under study. The second key, mutagenesis methods, has undergone a near revolution due to the introduction of powerful methods for randomly generating mutant genes for selection. The best methods are combinatorial and generate a mixture of point mutations, sequence deletions, and sequence insertions for exploring the widest possible range of sequence space.

In one example, a technique called DNA shuffling uses an error-prone polymerase chain reaction (PCR), DNA recombination, and genetic selection to generate and identify desirable new enzyme variants (Stemmer, 1994). The technique was applied to a *beta*-lactamase gene which encoded an enzyme with poor hydrolytic activity against the antibiotic cefatoxime. In fact, *beta*-lactamases with high activity against this antibiotic are not known to have evolved naturally. This provided a testing grounds for determining if laboratory evolution could achieve new enzymatic capabilities. After several rounds of DNA shuffling, a *beta*-lactamase containing multiple sites of mutation was obtained that showed a 64,000-fold increase in specific activity with cefatoxime. DNA shuffling is now being applied to enzymes involved in the detoxification of environmental pollutants.

ASSEMBLY OF GENES AND ENZYMES INTO NOVEL PATHWAYS

Polyhalogenated organic compounds are among the pollutants that are most resistant to environmental biodegradation. This is because metabolism of polyhalogenated compounds to intermediary metabolites requires the simultaneous evolution of multiple enzymatic dehalogenation reactions in the same bacterial cell. Generally, multiply chlorinated carbon atoms are most susceptible to an anaerobic reductive mechanism for cleavage of the carbon-to-halogen bond (Holliger, 1992). After one or two halogen substituents are removed, the most facile dehalogenation reactions then occur via hydrolytic or oxygenative mechanisms. The pace of natural evolution is likely slower when disparate enzymatic mechanisms must arise simultaneously, which is probably what underlies the observed environmental persistence of polyhalogenated compounds.

It is conceivable that an anaerobic reductive enzyme and a hydrolytic or oxygenative enzyme could be combined in the same cell by recombinant DNA techniques. This requires a knowledge of the enzymes that catalyze the respective dehalogenation reactions. There are currently few known examples of such enzymes. Only a few have been well-studied mechanistically or have had their tertiary structure elucidated at this date.

To begin to fill this gap, studies have been conducted with a bacterial heme-containing monooxygenase, cytochrome $P450_{CAM}$, for which significant mechanistic detail and an x-ray structure are currently known (Gunsalus and Wagner, 1978; Poulos et al., 1985). Cytochrome $P450_{CAM}$ has been studied mostly for its oxygenating reaction, but it has been known to fortuitously catalyze the reductive dehalogenation of suitably sized halogenated aliphatic compounds (Castro et al., 1985). In that study, it was proposed that the site of camphor oxygenation and the site of halogenated compound reduction were not the same. This was investigated in a subsequent study (Li and Wackett, 1993). Chlorinated substrates were shown to cause a change in the spin state of the active-site heme-iron atom similar to that induced by camphor and structurally related compounds. X-ray crystallographic and spectroscopic studies indicate that the substrate-induced, low-to-high spin conversion results from substrates binding at the top axial face of the heme-iron, leading to expulsion of a weak axial water ligand and the other water molecules that populate the active site in the native enzyme (Gunsalus and Wagner, 1978: Poulos and Raag, 1992). Thus, the observed spin state change with both types of substrates argued for a common binding site. Additional evidence was obtained by steady-state kinetic studies, conducted under anaerobic conditions in which camphor acted as an inhibitor in the reductive dechlorination of hexachloroethane. The inhibition pattern obtained in the treatment of the kinetic data was consistent with a competitive mode of inhibition. Competitive inhibition is typically observed in those cases where the substrate and inhibitor compete for binding to the same site on the enzyme. Most recently, molecular dynamics simulations have been performed with chlorinated substrates such as hexachloroethane docked in the camphor binding. The observed mobility of the substrates is consistent with the kinetic and spin state data described above (Paulsen and Ornstein, 1994).

The products of cytochrome P450-catalyzed reductive dechlorination of chlorinated ethanes that contain chlorine substituents on both carbon atoms are ethenes containing one less chlorine substituent on each carbon atom (Figure 2). The reaction is an overall 2-electron reduction, the vicinal carbon-to-chlorine bond being cleaved via an elimination reaction that occurs concomitantly with reduction. The heme-iron center undergoes one-electron redox change, so reduction of the carbon-to-chlorine bond likely proceeds through a one-electron reduced intermediate, although such a species has not yet been demonstrated. The overall reduction reactions are not a great deal faster than the corresponding reactions of reduced heme and chlorinated ethanes free in solution (Li and Wackett, 1993). However, one must consider that these reactions are a fortuitous function of heme chemistry, and the enzyme has likely not been acted on by selective pressure to evolve active-site functional groups that promote carbon-to-halogen bond cleavage. The selective pressure has optimized this enzyme for rapid regio- and enantiospecific hydroxylation of camphor and related terpenoid substrates. As an example, pentachloroethane is reduced at a steady-state rate of 1 per second and the same enzyme preparations oxidize camphor at a turnover rate of 20 per second.

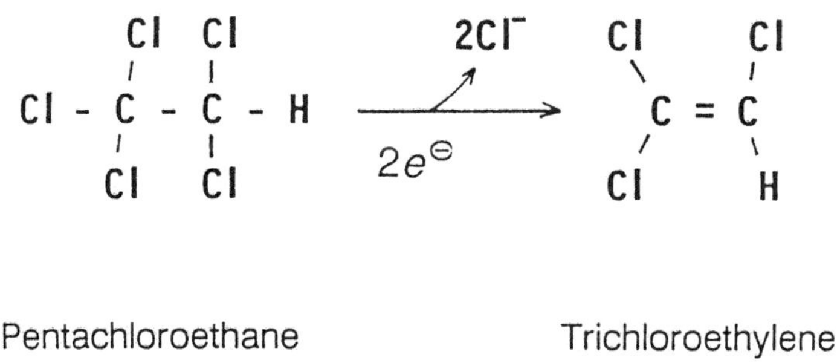

Figure 2. Representative 2-electron reduction reaction of a vicinally chlorinated ethane yielding an ethene product.

Pentachloroethane is stoichiometrically reduced by cytochrome $P450_{CAM}$ to trichloroethylene (TCE), and the latter compound is known to be oxidized by a growing number of nonspecific bacterial catabolic oxygenases. Purified toluene dioxygenase oxidizes TCE to formate and glyoxylate in a 2:1 ratio (Li and Wackett, 1992).

These observations provided the impetus for genetically assembling a *Pseudomonas* strain that could metabolize pentachloroethane, and related chlorinated alkanes, via sequential reductive and oxidative dechlorination reactions yielding nonchlorinated products. Experimentally, this required that the *Pseudomonas* strain had to contain and express seven genes necessary for functional cytochrome $P450_{CAM}$ and toluene dioxygenase enzyme systems. *Pseudomonas putida* G786 (pHG-2) contains two different plasmids: the CAM plasmid, containing the P450 enzyme system genes, and plasmid pHG-2, containing toluene dioxygenase genes, under control of the *tac* promoter (Wackett et al., 1994). The organism was grown under conditions in which both enzyme systems were expressed and then shown to stoichiometrically reduce pentachloroethane to TCE under anaerobic conditions. Then, introduction of air into the reaction mixtures resulted in the disappearance of TCE. In separate experiments, the same strain was shown to oxidize TCE to carbon dioxide and glyoxylate in a 2:1 ratio. Based on the data with purified toluene dioxygenase, it is likely that formate, initially produced by toluene dioxygenase, was further oxidized *in vivo* to carbon dioxide.

It was perhaps more surprising that perhalogenated ethanes were also substrates for sequential reductive and oxygenative dehalogenation by the recombinant organism (Wackett et al., 1994; Hur et al., 1994). For example, 1,1,1-trichloro-2,2,2-trifluoroethane is reduced by cytochrome P450 to 1,1-dichloro-2,2-difluoroethene, and the latter compound is oxidized stoichiometrically to oxalate by toluene dioxygenase. These studies highlight the potential for assembling novel collections of genes to construct metabolic pathways that may not have been constructed by natural selection in the environment. Furthermore, the approach illustrated here was based on an understanding of the respective enzyme reaction mechanisms that led to recruiting fortuitous reactions for a new biochemical purpose.

SYSTEMATIZING BIODEGRADATION REACTIONS VIA A WORLD WIDE WEB COMPUTER DATABASE

The range of microbial catabolism is enormous, making it an impossible task to assimilate all the details, even in a lifetime of study. The rate of discovery of new enzymes and gene sequences is now outstripping the ability for humans to become familiar with them. For example, GenBank currently contains 370,000 sequences, with over 1,000,000 sequences anticipated in only 1.5 years. There are currently on the order of 8,000,000 known organic compounds, and many of them are metabolizable by microorganisms. New organic compounds are being discovered or synthesized faster than their respective microbial catabolism can be investigated. Thus, it becomes imperative that the aggregate information on microbial catabolism be systematized to allow for generalizable knowledge that can be applied predictively to new compounds.

The University of Minnesota Biocatalysis/Biodegradation Database (UM-BBD) was initiated in February, 1995, to meet the need for a consolidated information source on microbial metabolism for: (1) the application of existing information for bioremediation, (2) predicting the metabolic fate of new compounds, and (3) providing a knowledge-base for creative metabolic engineering of new catabolic pathways (Ellis and Wackett, 1995). The UM-BBD contains information on metabolic pathways, enzymes, genes, microorganisms, and chemical compounds in a highly accessible format and is currently available on the World Wide Web. The Web address, or URL, is: http://dragon.labmed.umn.edu/~lynda/index.html

A representative pathway map in the UM-BBD is shown for the starting compound 1,2-dichloroethane (Figure 3). Clicking with the mouse on the Hypertext link 1,2-dichloroethane transports the user to physicochemical and toxicological information on that chemical compound. Clicking on the name of the first enzyme in the biodegradative pathway, haloalkane dehalogenase, brings the user to an enzyme reaction page. From there, one can obtain: (1) a GIF file showing the chemical structures of reactants and products, (2) Hypertext links to Kyoto Ligand (Enzyme) Database, Genobase, and GenBank; and (3) Medline references. Via the latter, one can readily display the x-ray structure of haloalkane dehalogenase in various projections.

The endpoint of the 1,2-dichloroethane pathway is glycolic acid. The latter compound is an intermediary metabolite, and the UM-BBD links to databases of intermediary metabolism to allow the users to follow its potential fate via degradative or biosynthetic pathways. The encoding of this information in compatible computer databases will allow the development of software that can search for desired metabolic routes for disposition of any given starting compound. For example, the linkage of biodegradative and biosynthetic pathways can theoretically transform 1,2-dichloroethane into any cellular metabolite, some of which have commercial value. This exercise in computer science can have utility for developing novel approaches to transform hazardous wastes into salable chemicals. This outcome makes both environmental and economic sense, and it puts microorganisms to work doing what they do best, recycling the earth's carbon to allow the perpetuation of life.

The following is a text-format 1,2-dichloroethane degradation pathway map. Follow the links for more information on compounds or reactions.

```
1,2-Dichloroethane
          |
          |
          |  haloalkane dehalogenase
          |
          |
          v
2-Chloroethanol
          |
          |
          |  pyrrolo-quinoline quinone-containing
          |  alcohol dehydrogenase
          |
          v
Chloroacetaldehyde
          |
          |
          |  NAD-dependent aldehyde dehydrogenase
          |
          |
          v
Chloroacetic acid
          |
          |
          |  2-haloacid dehalodenase
          |
          |
          v
   Glycolate
```

[Reactions] [Compounds] [BBD Main Menu]

Figure 3. A Web page from the University of Minnesota Biocatalysis/Biodegradation Database (BBD), showing the pathway for the bacterial metabolism of 1,2-dichloroethane. Underlined terms are Hypertext links to additional information. Pathway contributed by R. Li, University of Minnesota.

ACKNOWLEDGMENTS

This research was supported by National Institutes of Health grant GM41235, U.S. Department of Agriculture grant 94-34339-112, Binational Agricultural Research and Development Fund grant US-2364-93, and a research award from Ciba-Geigy to LPW. CCL was supported as an Associated Western Universities Postdoctoral Fellow by grant KP0402 (to RLO) from the Health Effects and Life Science Division's Office of Health and Environmental Research, Office of Energy Research, U.S. Department of Energy.

REFERENCES

Castro, CE, RS Wade, and NO Belser. **1985**. Biodehalogenation: Reactions of cytochrome P450 with polyhalomethanes. Biochemistry 24:204-210.

DeSouza, ML, LP Wackett, KL Boundy-Mills, RT Mandelbaum, and MJ Sadowsky. **1995**. Cloning, characterization, and expression of a gene region from *Pseudomonas* sp. strain ADP involved in the dechlorination of atrazine. Appl. Environ. Microbiol. 61:3373-3378.

Ellis, LBM and LP Wackett. **1995**. A microbial biocatalysis database. Soc. Ind. Microbiol. News 45:167-173.

Gunsalus, IC and GC Wagner. **1978**. Bacterial $P450_{CAM}$ methylene monooxygenase components: Cytochrome *m*, putidaredoxin and putidaredoxin reductase, pp. 166-188. S. Fleischer and L. Packet, eds. In: *Methods in Enzymology*, vol. LII. Academic Press, New York.

Holliger, C. **1992**. Reductive dehalogenation by anaerobic bacteria, Ph.D. thesis. Wageningen University, The Netherlands.

Hur, H-G, MJ Sadowsky, and LP Wackett. **1994**. Metabolism of chlorofluorocarbons and polybrominated compounds by *Pseudomonas putida* G-786 (pHG-2) via an engineered metabolic pathway. Appl. Environ. Microbiol. 60:4148-4154.

Li, S and LP Wackett. **1992**. Trichloroethylene oxidation by toluene dioxygenase. Biochem. Biophys. Res. Commun. 185:443-451.

Li, S and L Wackett. **1993**. Reductive dehalogenation by cytochrome $P450_{CAM}$: Substrate binding and catalysis. Biochemistry 32:9355-9361.

Mandelbaum, RT, DL Allan, and LP Wackett. **1993**. Rapid hydrolysis of atrazine to hydroxyatrazine by soil bacteria. Environ. Sci. Technol. 27:1943-1946.

Mandelbaum, RT, DL Allan, and LP Wackett. **1995**. Isolation and characterization of a *Pseudomonas* sp. that mineralizes the *s*-triazine herbicide atrazine. Appl. Environ. Microbiol. 61:1451-1457.

Paulsen, MD and RL Ornstein. **1994**. Active site mobility inhibits reductive dehalogenation of 1,1,1-trichloroethane by cytochrome $P450_{CAM}$. J. Comput. Aided Mol. Des. 8:389-404.

Poulos, TL and R Raag. **1992**. Cytochrome $P450_{CAM}$: Crystallography, oxygen activation, and electron transfer. FASEB J. 6:674-679.

Poulos, TL, BC Finzel, IC Gunsalus, GC Wagner, and J Kraut. **1985**. The 2.6 A crystal structure of *Pseudomonas putida* cytochrome P450. J. Biol. Chem. 260:16122-16130.

Stemmer, WPC. **1994**. Rapid evolution of a protein *in vitro* by DNA shuffling. Nature 370:389-391.

Wackett, LP, MJ Sadowsky, LM Newman, H-G Hur, and S Li. **1994**. Metabolism of polyhalogenated compounds by a genetically engineered bacterium. Nature 368:627-629.

MECHANISMS FOR THE METABOLISM OF ENVIRONMENTAL POLLUTANTS BY WHITE ROT FUNGI

S. D. Aust

Biotechnology Center, Utah State University, Logan, Utah

Key Words: *white rot fungi, veratryl alcohol, lignin peroxidases, Mn-dependent peroxidases, oxygen radicals*

ABSTRACT

Lignin, the three-dimensional structural polymer of wood, can be mineralized by white rot fungi. The lignin biodegradation system of these fungi is unique in many respects. We have proposed that the system may be applicable for the biodegradation of a great many environmental pollutants. The mineralization by white fungi of a wide variety of structurally unrelated, toxic and insoluble chemicals that are significant environmental pollutants has been demonstrated. The biodegradation system of white rot fungi is quite different from that of bacterial systems; it is based largely on free-radical reactions catalyzed by peroxidases, and is extracellular and quite nonspecific. The fungi are able to catalyze other reactions important for the free-radical nature of the biodegradation system. Thus, very few enzymes and other extracellular activities are able to degrade such a large number of structurally unrelated chemicals. Both oxidative and reductive reactions can be catalyzed; reactions can be direct or indirect and can even be catalyzed by a diffusible oxidant. The system favors complete oxidation rather than synthetic or polymerization reactions. This a natural yet biochemically unique system may be applicable for the bioremediation of a large number of environmental pollutants.

Lignin, the three-dimensional structural polymer in woody plants, is synthesized by the free-radical-mediated polymerization of phenylpropanoid units (Sarkanen and Ludwig, 1971). Various carbon-carbon and carbon-oxygen bonds with nonspecific stereochemistry exist in lignin. However, lignin can be completely mineralized by white rot fungi, but they obtain no energy from its biodegradation (Kirk et al., 1976). These fungi use some cellulose to obtain energy to degrade lignin in wood so that more cellulose is exposed.

The lignin degrading system is active when the fungi are starved for some nutrient (Tien and Kirk, 1984; Gold et al., 1984). In the laboratory, nutrient nitrogen starvation is usually employed for expression of the lignin degrading system (Tuisel et al., 1990). The degradation system is quite complex, consisting of a family of peroxidases, hydrogen peroxide and some additional small molecules, and some other activities related to the free-radical nature of the biodegradation system, all of which can function extracellularly (Barr and Aust, 1994a). The biodegradation of wood must occur, or at least start, extracellularly. Since the fungi get no energy from the lignin there is no reason to absorb and metabolize some depolymerization product for energy.

The fact that the lignin degradation system is extracellular may be significant for the biodegradation of many environmental pollutants (Barr and Aust, 1994a,b; Higson, 1991). Many of these chemicals, like lignin, are very insoluble; however, because of the extracellular biodegradation system, intracellular absorption is not a requirement for biodegradation. Additionally, an extracellular biodegradation system might also degrade adsorbed chemicals. Detoxification can also occur extracellularly.

Another advantage for the lignin degrading system in its application to environmental pollutants also distinguishes it from most microbial biodegradation systems. The lignin biodegradation system that evolved to degrade a nonstereoselective heteropolymer is free-radical in nature; that is, free radicals are used to catalyze a variety of reactions that result in the mineralization of lignin. These include both oxidative and reductive reactions, both of which must occur aerobically. Redox cycling and free-radical-based polymerization reactions must be avoided. Unique oxidizing and reducing equivalents must be used because they cannot be regenerated, as the reactions occur extracellularly. Once released into the environment it would be difficult to regenerate NADPH, for example.

Mineralization is accomplished by a very unique biodegradation system, based primarily on a family of peroxidases, divided into two types (Tien, 1987), but recent studies have shown that this division is complicated (Khindaria et al., 1995a; Sutherland et al., 1995). The lignin peroxidases oxidize a variety of organic molecules. Another group, called manganese-dependent peroxidases, oxidize manganese (Wariishi et al., 1989; Glenn et al., 1986); the manganese may serve as a diffusible oxidant. The lignin peroxidases seem to prefer veratryl alcohol as a substrate, and veratryl alcohol is produced by white rot fungi even when grown on glucose and ammonia (Barr et al., 1992). This substrate for the lignin peroxidases may be produced by the fungus for what we termed mediation of free-radical reactions (Goodwin et al., 1995). That is, the veratryl alcohol cation radical, generated by the lignin peroxidases, serves to oxidize other chemicals which are not oxidized directly by the enzymes. We call this indirect oxidation, depicted as follows, where Fe is the iron of protoporphyrin IX, represented as P, and VA and $VA^{\cdot+}$ are veratryl alcohol and its cation radical, respectively. The veratryl alcohol cation radical then mediates the indirect oxidation of other chemicals, designated here as R:

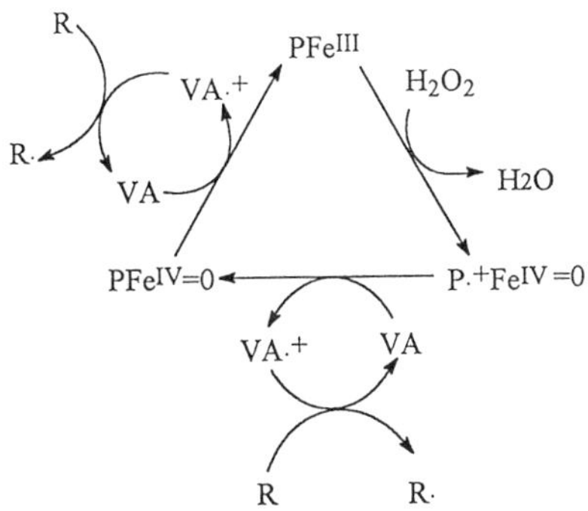

It should be noted that upon analysis for veratryl alcohol oxidation to veratryl aldehyde, which is used to assay lignin peroxidases, one might conclude that chemicals represented by R in this scheme are inhibiting the enzyme. This is not correct; the enzyme is turning over, but it is simply using veratryl alcohol as a mediator to oxidize other chemicals.

Veratryl alcohol can have other important functions. Peroxidases are susceptible to their own oxidation when the concentration of hydrogen peroxide is greatly in excess of the concentration of chemical to be oxidized (i.e., reductants). Thus, the production of veratryl alcohol by the fungi may protect the enzymes from self-destruction (Barr and Aust, 1994c). For this reason veratryl alcohol is usually added to cultures of the fungi used for enzyme production (Tuisel et al., 1990). Another use for veratryl alcohol is to complete the enzymatic catalytic cycle. This is the case for anisyl alcohol oxidation, for example (Koduri and Tien, 1994), which cannot be oxidized by compound II. The addition of veratryl alcohol provides a reductant for compound II and completes the catalytic cycle. Compound I could oxidize anisyl alcohol, or the alcohol may be oxidized indirectly by the veratryl alcohol cation radical, depending on the relative concentrations of each. The lignin peroxidases can also oxidize manganese; however, only by compound I

(Sutherland et al., 1995); LiPH2 is an exception (Khindaria et al., 1995a). Again, the addition of veratryl alcohol again completes the catalytic cycle, sustaining manganese oxidation at a rate equivalent to the oxidation of veratryl alcohol by compound II (Sutherland et al., 1995).

Another use of veratryl alcohol may be to "rescue" the enzyme if, in fact, it is inactivated by excess hydrogen peroxide. The reaction of compound II with hydrogen peroxide first results in the production of a complex of the enzyme with superoxide. A second reaction, resulting in destruction of the heme, is irreversible. During turnover a fraction of the enzyme can be converted to the first inactive form, called compound III. This form of the enzyme can be "rescued" by reduction with the veratryl alcohol cation radical, as follows (Barr and Aust, 1994c):

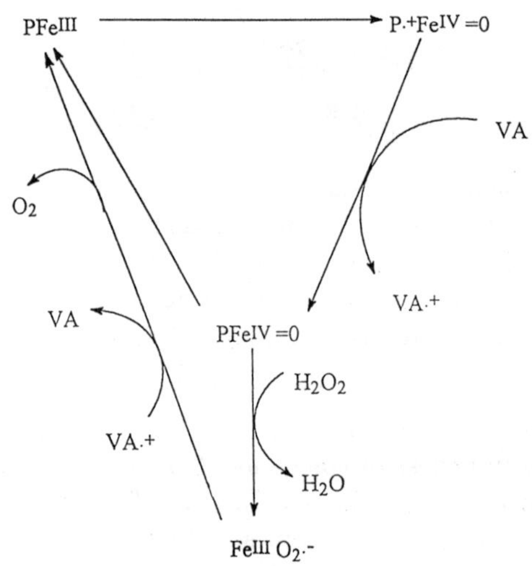

The manganese-dependent peroxidases oxidize manganese at a very significant rate (k_{cat} = 200-300 s^{-1}). Compound I can oxidize free manganese, but chelated manganese is required for compound II (Khindaria et al., 1995a). Since Mn^{III} is such a strong oxidant it must be chelated so that it can separate from the enzymes to oxidize other chemicals, rather than the enzyme. It would appear that the chelator used by the fungi is oxalate (Khindaria et al., 1995a). Oxalate-chelated

Mn^{III} would be an excellent diffusible oxidant produced by the manganese-dependent peroxidases in the following manner:

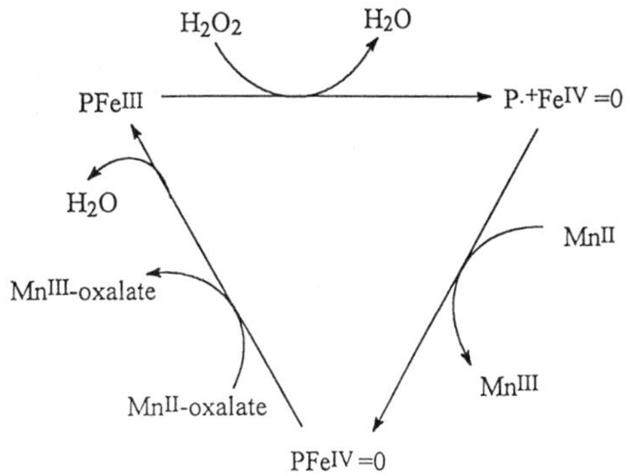

Oxalate is also physiologically significant otherwise. It is very susceptible to oxidation (i.e., it is an excellent reductant because it is actually a di—keto acid), and the oxidation product quickly decomposes to provide the carboxylate anion radical, (an excellent reductant), and CO_2, which entropically "pulls" the reaction. Oxalate can be oxidized by either Mn^{III} or the veratryl alcohol cation radical:

$$\text{oxalate} + Mn^{III} \rightarrow CO_2 + CO^{\cdot-} + Mn^{II}$$

$$\text{oxalate} + VA^{\cdot+} \rightarrow CO_2 + CO_2^{\cdot-} + VA.$$

Thus, both types of peroxidases can use oxalate to produce an excellent reducing radical (Khindaria et al., 1994; Shah et al., 1993). The carboxylate anion radical can be used to reduce other chemicals or to reduce molecular oxygen to superoxide. Fenton reagent can be generated if the carboxylate anion radical is used to reduce molecular oxygen to superoxide. The dismutation of superoxide can regenerate hydrogen peroxide for the peroxidases. With judicious choice of concentrations one can add catalytic amounts of H_2O_2 for continuous iron reduction (Khindaria et al., 1994).

Other reductive reactions, such as reductive dechlorinations, can be catalyzed by the carboxylate anion radical. Reductive dechlorination of haloalkanes, such as carbon tetrachloride (Shah et al., 1993) and trichloroethylene, has been demonstrated. We also demonstrated the aerobic mineralization of trichloroethylene by *Phanerochaete chrysosporium* (Khindaria et al., 1995b).

Hydroquinones are readily oxidized to semiquinones by manganese. Quinones are intermediates of lignin degradation (Valli and Gold, 1991; Schoemaker, 1990), and white rot fungi can make quinones from veratryl alcohol (Haemmerli et al., 1987; Schmidt et al., 1989). White rot fungi also have both intracellular (Brock et al., 1995) and extracellular quinone reductases (Westermark and Eriksson, 1974). The cell surface membrane redox potential of *P. chrysosporium* can also reduce quinones (Stahl et al., 1995). The oxidation of hydroquinones by manganese peroxidases results in the production of semiquinones that can be used for reductions (Rasmussen et al., 1995).

The cell surface redox potential can be used for metabolizing some environmental pollutants that require reduction before they can be oxidized by the peroxidases. The best example is TNT, which is first reduced (Stahl and Aust, 1993a), and consequently detoxified (Stahl and Aust, 1993b), by the cell surface membrane potential. The resultant amines are then oxidized by the peroxidases (Stahl and Aust, 1995). Thus TNT is a candidate for bioremediation by white rot fungi (Stahl and Aust, 1994).

White rot fungi seem to have other cell surface activities that are important for bioremediation; an example is methylation reactions. Methylation of phenols makes them more amenable to oxidation to cation radicals rather than to phenoxyl radicals, which would be simply reduced back to phenols, resulting in a futile cycle. Such is the case for pentachlorophenol, which is rapidly methylated by *P. chrysosporium* to pentachloroanisole, which is then dechlorinated to form tetrachlorobenzoquinone and, ultimately, mineralized (Chung and Aust, 1995). Phenols can be polymerized by pure lignin peroxidases but the object of degradation by the fungi is mineralization. Thus, phenols are first methylated then oxidized. Likewise, quinones can be reduced, methylated, then oxidized.

The oxidation of veratryl alcohol to veratryl aldehyde, which is used as an assay to quantitate the activity of lignin peroxidase, could theoretically occur by two mechanisms. The first reaction is the same in both cases: Veratryl alcohol is oxidized to the cation radical by compound I. In one case, the veratryl alcohol cation radical would also be oxidized by the enzyme. In the second, the cation radical could rapidly deprotonate and be oxidized by molecular oxygen. In the first case the stoichiometry would be one veratryl aldehyde produced per H_2O_2 consumed. In the latter, the stoichiometry would be something greater than two. The reduction of molecular oxygen to superoxide would provide more H_2O_2 by its dismutation. However, the stoichiometry for veratryl alcohol oxidation by the enzyme is just slightly above 1.0. However, the decay of the cation radical in solution is extremely fast, $1.3 \times 10^3\ s^{-1}$ (Khindaria et al., 1995c). A question therefore arises: Why doesn't the cation radical react with molecular oxygen during enzyme catalysis, which is actually quite slow ($1.8\ s^{-1}$)? Additionally, the pKa of the cation radical in bulk solution was estimated to be about -1.0 (Khindaria et al., 1995d). This also suggests rapid deprotonation and reaction with molecular oxygen, which does not seem to occur to any great extent. Obviously, the cation radical must be stabilized by the enzyme (Khindaria et al., 1995d). It can therefore be readily reduced by organic chemicals and thus is a good mediator (Goodwin et al., 1995). Alternatively, the oxidation of the cation radical by the enzyme would lead to veratryl aldehyde. We are currently investigating the mechanism of this reaction.

We propose that the stabilization of the cation radical of veratryl alcohol by the enzyme results from the fact that the active site in the enzyme is very acidic. We propose a different site than that suggested by Poulos et al. (1993). The site we propose is surrounded by the two heme proprionates, aspartates 88 and 183, glutamates 40 and 89, and the C terminal alanine 343. Electrostatic potential calculations show that this site is very negative and that the charge distribution is symmetrical, completely surrounding the cation radical (Figure 1). Also note that the pH optimum for this enzyme is about 2.5 (Tuisel et al., 1990); thus, the cation radical is a good mediator of other oxidations but does not serve as a diffusible oxidant, as does manganese.

In summary, white rot fungi use some very unusual mechanisms to degrade lignin. The unusual nature of this system is applicable for the biodegradation of many environmental pollutants. The use of white rot fungi for bioremediation of environmental pollution sites is being investigated. We have established a company, Intech One-Eighty Corporation, to assist in the transfer of this technology to the remediation industry.

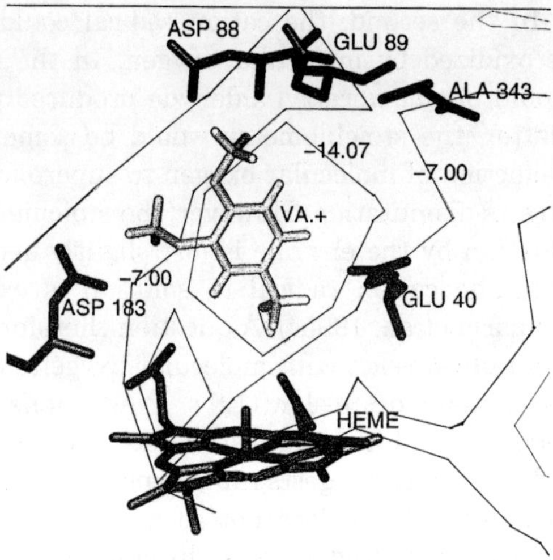

Figure 1. Calculated electrostatic potential due to lignin peroxidase surrounding a proposed bonding site for (veratryl alcohol cation radical (VA·+). In light grey is VA+; in dark grey are the acidic amino acid residues around the active site, including the proprionates of the heme. Calculated electrostatic potentials due to the enzyme are shown as contour plots with the numbers denoting the potential in eV. Electrostatic potential was calculated using DELPHI.

ACKNOWLEDGMENTS

The author wishes to thank the many graduate students, postdoctoral fellows, and technicians who actually conducted much of the research reviewed in this manuscript. I also thank Terri Maughan for excellent secretarial assistance and acknowledge support from NIEHS Superfund Basic Research and Training Grant ES04922 and Intech One-Eighty Corporation.

REFERENCES

Barr, DP and SD Aust. **1994a**. Mechanisms white rot fungi use to degrade pollutants. Environ. Sci. Technol. 28:78A-87A.

Barr, DP and SD Aust. **1994b**. Pollutant degradation by white rot fungi. pp. 49-72. In: *Reviews of Environmental Contamination and Toxicology*, G Ware, ed., vol. 138. Springer-Verlag, New York.

Barr, DP and SD Aust. **1994c**. Conversion of lignin peroxidase compound III to active enzyme by cation radicals. Arch. Biochem. Biophys. 312:511-515.

Barr, DP, MM Shah, TA Grover, and SD Aust. **1992**. Production of hydroxyl radical by lignin peroxidase from *Phanerochaete chrysosporium*. Arch. Biochem. Biophys. 298:480-485.

Brock, BJ, S Rieble, and MH Gold. **1995**. Purification and characterization of a 1,4-benzoquinone reductase from the basidiomycete *Phaernochaete chrysosporium*. Appl. Environ. Microbiol. 61:3076-3081.

Chung, N and SD Aust. **1995**. Degradation of pentachlorophenol in soil by *Phanerochaete chrysosporium*. J. Hazardous Materials 41:177-183.

Glenn, JK, L Akileswaran, and MH Gold. **1986**. Mn(II) oxidation is the principal function of the extracellular Mn-peroxidase from *Phanerochaete chrysosporium*. Arch. Biochem. Biophys. 251:688-696.

Gold, MH, M Kuwahara, AA Chiu, and JK Glenn. 1984. Purification and characterization of an extracellular H_2O_2-requiring diarylpropane oxygenase from the lignin-degrading basidiomycete, *Phanerochaete chrysosporium*. Arch. Biochem. Biophys. 234:353-362.

Goodwin, DC, SD Aust, and TA Grover. **1995**. Evidence for veratryl alcohol as a redox mediator in lignin peroxidase-catalyzed oxidation. Biochemistry 34:5060-5065.

Haemmerli, SD, HE Schoemaker, HWH Schmidt, and MSA Leisola. **1987**. Oxidation of veratryl alcohol by the lignin peroxidase of *Phanerochaete chrysosporium*, involvement of activated oxygen. FEBS Lett. 220, 149-154.

Higson, FK. **1991**. Degradation of xenobiotics by white rot fungi. Rev. Environ. Contam. Toxicol. 122:111-141.

Khindaria, A, TA Grover, and SD Aust. **1994**. Oxalate dependent reductive activity of manganese peroxidase from *Phanerochaete chrysosporium*. Arch. Biochem. Biophys. 314:301-306.

Khindaria, A, DP Barr, and SD Aust. **1995a**. Lignin peroxidases can also oxidize manganese. Biochemistry 34:7773-7779.

Khindaria, A, TA Grover, and SD Aust. **1995b**. Reductive dehalogenation of aliphatic halocarbons by lignin peroxidase of *Phanerochaete chrysosporium*. Environ. Sci. Technol. 29:719-725.

Khindaria, A, I Yamazaki, and SD Aust. **1995c**. Veratryl alcohol oxidation by lignin peroxidase. Biochemistry 34:16860-16869.

Khindaria, A, TA Grover and SD Aust. **1995d**. Evidence for formation of the veratryl alcohol cation radical by lignin peroxidase. Biochemistry 34:6020-6025.

Kirk, TK, WJ Connors, and JG Zeikus. **1976**. Requirement for a growth substrate during lignin decomposition by two wood-rotting fungi. Appl. Environ. Microbiol. 32:192-194.

Koduri, RS and M Tien. **1994**. Kinetic analysis of lignin peroxidase: Explanation for the mediation phenomenon by veratryl alcohol. Biochemistry 33:4225-4230.

Poulos, TL, SL Edwards, H Wariishi, and MH Gold. **1993**. Crystallographic refinement of lignin peroxidase at 2Å. J. Biol. Chem. 268:4429-4440.

Rasmussen, SJ, N Chung, A Khindaria, TA Grover, and SD Aust. **1995**. Reductions catalyzed by a quinone and peroxidases from *Phanerochaete chrysosporium*. Arch. Biochem. Biophys. 320:243-249.

Sarkanen, KV and CH Ludwig. **1971**. *Lignins: Occurrence, Formation and Structure*. Wiley-Interscience, New York.

Schmidt, HW, SD Haemmerli, HE Schoemaker, and MSA Leisola. **1989**. Oxidative degradation of 3,4-dimethoxylbenzyl alcohol and its methyl ether by the lignin peroxidase of *Phanerochaete chrysosporium*. Biochemistry 28:1776-1783.

Schoemaker, HE. **1990**. On the chemistry of lignin biodegradation. Rech. Trav. Chim. Pays-Bas. 109:255-272.

Shah, MM, TA Grover, and SD Aust. **1993**. Reduction of CCl_4 to the trichloromethyl radical by lignin peroxidase H2 from *Phanerochaete chrysosporium*. Biochem. Biophys. Res. Commun. 191:887-892.

Stahl, JD and SD Aust. **1993a**. Plasma membrane dependent reduction of 2,4,6-trinitrotoluene by *Phanerochaete chrysosporium*. Biochem. Biophys. Res. Commun. 192:471-476.

Stahl, JD and SD Aust. **1993b**. Metabolism and detoxification of TNT by *Phanerochaete chrysosporium*. Biochem. Biophys. Res. Commun. 192:477-482.

Stahl, JD and SD Aust. **1994**. Concepts for the bioremediation of trinitrotoluene contaminated wastes by the white rot fungus *Phanerochaete* chrysosporium, pp. 172-185. In: *Proceedings of the 9th Annual Conference on Hazardous Waste Remediation*. Kansas State University, Manhattan, KS.

Stahl, JD and SD Aust. **1995**. Biodegradation of 2,4,6-trinitrotoluene by the white rot fungus Phanerochaete chrysosporium, pp. 117-133. In: *Biodegradation of Nitroaromatic Compounds*, J Spain, ed. Plenum Publishing Corp., New York.

Stahl, JD, SJ Rasmussen, and SD Aust. **1995**. Reduction of quinones and radicals by a plasma membrane redox system of *Phanerochaete chrysosporium*. Arch. Biochem. Biophys. 322:221-227.

Sutherland, GRJ, A Khindaria, N Chung, and SD Aust. **1995**. The effect of manganese on the oxidation of chemicals by lignin peroxidase. Biochemistry 34:12624-12629.

Tien, M. **1987**. Properties of ligninase from *Phanerochaete chrysosporium* and their possible applications. CRC Crit. Rev. Microbiol. 15:141-168.

Tien, M and TK Kirk. **1984**. Lignin-degrading enzyme from *Phanerochaete chrysosporium*: Purification, characterization, and catalytic properties of a unique H_2O_2-requiring oxygenase. Proc. Natl. Acad. Sci. USA 81:2280-2284.

Tuisel, H, R Sinclair, JA Bumpus, W Ashbaugh, BJ Brock, and SD Aust. **1990**. Lignin peroxidase H2 from *Phanerochaete chrysosporium*: Purification, characterization and stability to temperature and pH. Arch. Biochem. Biophys. 279:158-166.

Valli, K and MH Gold. **1991**. Degradation of 2,4-dichlorophenol by the lignin-degrading fungus *Phanerochaete chrysosporium*. J. Bacteriol. 173:345-352.

Wariishi, H, HB Dunford, ID MacDonald, and MH Gold. **1989**. Manganese peroxidase from the lignin-degrading basidiomycete *Phanerochaete chrysosporium*: transient-state kinetics and reaction mechanism. J. Biol. Chem. 264:3335-3340.

Westermark, U and K-E Eriksson. **1974**. Carbohydrate-dependent enzymic quinone reduction during lignin degradation. Acta Chem. Scand. B28:204-208.

TRANSIENT KINETIC ANALYSIS OF THE CATALYTIC MECHANISM OF HALOALKANE DEHALOGENASE: IDENTIFICATION OF THE RATE-DETERMINING STEP

J. P. Schanstra, J. Kingma, and D. B. Janssen

Department of Biochemistry, Groningen Biomolecular Sciences and Biotechnology Institute, University of Groningen, The Netherlands

Key Words: *Haloalkane dehalogenase, pre-steady-state kinetics, conformational change, product release, halide*

ABSTRACT

We have studied the rate-determining step(s) in the conversion (k_{cat}) of 1,2-dibromoethane by haloalkane dehalogenase, a bacterial enzyme involved in degradation of halogenated aliphatic compounds. Analysis of solvent kinetic isotope effects and stopped-flow fluorescence experiments with enzyme under single-turnover conditions indicated that release of the charged bromide ion out of the buried active site cavity could be the slow step in the reaction sequence. Stopped-flow fluorescence analysis of bromide binding and release confirmed this and showed that a slow enzyme-isomerization step preceding the actual bromide release was the slow step that limited the k_{cat}. We propose that this isomerization is a conformational change of part of the cap domain of the dehalogenase that is necessary to allow water to enter the active site and solvate the charged halide ion.

INTRODUCTION

Haloalkane dehalogenase from *Xanthobacter autotrophicus* GJ10 is a bacterial enzyme capable of cleaving carbon-halogen bonds in xenobiotic compounds. The enzyme converts a broad range of chlorinated (C_2-C_6) and brominated (C_2-C_8) alkanes to the corresponding alcohols and halides (Keuning et al., 1985; Pries et al., 1994a).

The x-ray structure of the dehalogenase was solved at 1.9 Å (Verschueren et al., 1993a). The enzyme is a globular protein and consists of two domains, a main domain and a cap domain. The main domain has a α/β-hydrolase fold (Ollis et al., 1992), consisting of a central β-sheet connected by helices. The other domain lies like a cap on top of the main domain and is largely α-helical. Between these two domains lies a hydrophobic buried active site cavity, 37 Å^3 in size. The catalytic residues Asp124, His289 and Asp260 are the only charged residues in this active site cavity.

The reaction mechanism for 1,2-dichloroethane has been solved using x-ray crystallography and site-directed mutagenesis studies (Verschueren et al., 1993b, Pries et al., 1994b, 1995; Figure 1). The reaction is initiated by binding of the substrate into the Michaelis complex (E.RX) followed by nucleophilic attack of Asp124 on the carbon atom to which the halogen is bound, leading to formation of an alkyl-enzyme intermediate (E-R.X). This covalent intermediate is subsequently hydrolyzed by activated water, with His289 acting as a general base catalyst, leading to free enzyme and products (E+ROH+X).

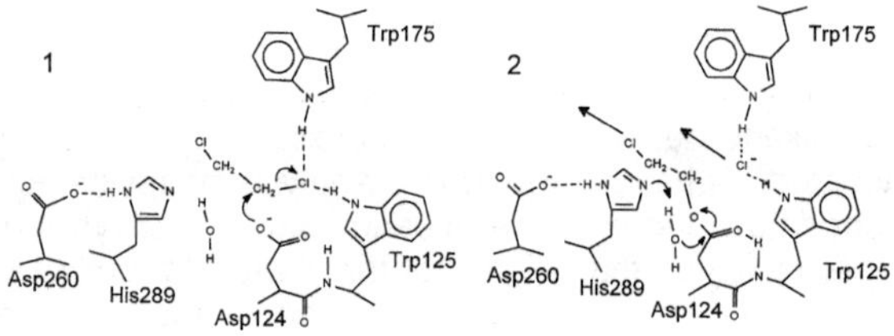

Figure 1. Reaction mechanism of haloalkane dehalogenase.

Applying initial rate conditions and observing that the dehalogenase reaction is essentially irreversible yield the following reaction scheme (Scheme I): The fluorescence of the two binding site tryptophans is quenched upon binding of halides (Verschueren et al., 1993c) and substrate (Pries et al., 1994b, 1995). Titration of the enzyme with halide results in a typical ligand binding curve from which the dissociation constant (K_d) of this ligand can be calculated. The fraction of accessible tryptophans (f_a) for the ligand can also be determined in this way. For the wild-type dehalogenase, this is one-third, in agreement with the total number of six tryptophans in the enzyme (Verschueren et al., 1993c, Kennes et al., 1995).

The aim of the work presented in this paper is to identify the rate-determining step(s) of 1,2-dibromoethane conversion. For this, we used stopped-flow measurements of the quenching of the tryptophan fluorescence to study substrate conversion and halide binding, and rapid-quench-flow experiments to follow alcohol production.

MATERIALS AND METHODS

Haloalkane dehalogenase was expressed and purified as described by Schanstra et al. (1993). Steady-state kinetic experiments were performed as described by Keuning et al. (1985). Solvent kinetic isotope effects were determined with substrate dissolved in ^2H_2O. Assays were performed as described above, using increasing concentrations of ^2H_2O. Stopped-flow fluorescence experiments were performed on an Applied Photophysics model !SX17MV, fitted with a xenon arc lamp with excitation at 290 nm. Fluorescence emission from Trp residues was observed through a 320-nm cut-off fliter. All reported concentrations were in the reaction chamber. Rapid-quench-flow experiments were performed on a Kintek RQF-63 instrument. All reported concentrations were in the reaction line of the rapid-quench-flow instrument.

RESULTS AND DISCUSSION

Which Step(s) Can Be Overall Rate-Determining?

In order to identify the step(s) that can be rate-limiting in the dehalogenase reaction we determined the steady-state kinetic parameters in the presence of ^2H_2O as the solvent. In increasing concentrations of ^2H_2O, the k_{cat} decreased (Figure 2A) for 1,2-dibromoethane conversion. A ^2H_2O solvent kinetic isotope effect on k_{cat} often indicates the cleavage of H-X (X = C, N, O) bonds in the rate-determining step. Cleavage of a H-X bond in the dehalogenase reaction occurs during hydrolytic cleavage of the alkyl-enzyme intermediate, which is facilitated by proton-abstraction from water by histidine 289 (Verschueren et al., 1993b; Pries et al., 1995). For a three-step mechanism

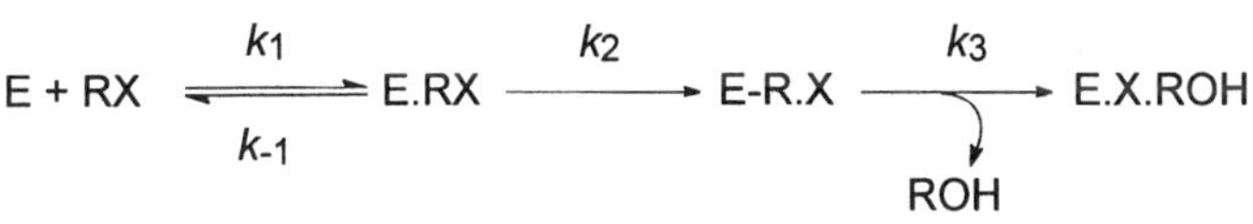

(Scheme I)

$K_m = K_s k_3/(k_2 + k_3)$, $k_{cat} = k_2 k_3/(k_2+k_3)$ and $k_{cat}/K_m = k_2/K_s$. Thus, the absence of a solvent kinetic isotope effect on the k_{cat}/K_m may indicate that only k_3 is affected, and that step 3 is mainly rate-limiting.

Hydrolysis of the alkyl-enzyme intermediate to free enzyme and products was also found to be rate-limiting in a stopped-flow fluorescence experiment with enzyme in excess over substrate (Figure 2B). Fitting the fluorescence transient with numerical simulation (Frieden, 1994) of Scheme I yielded a slow k_3 (5 s^{-1}).

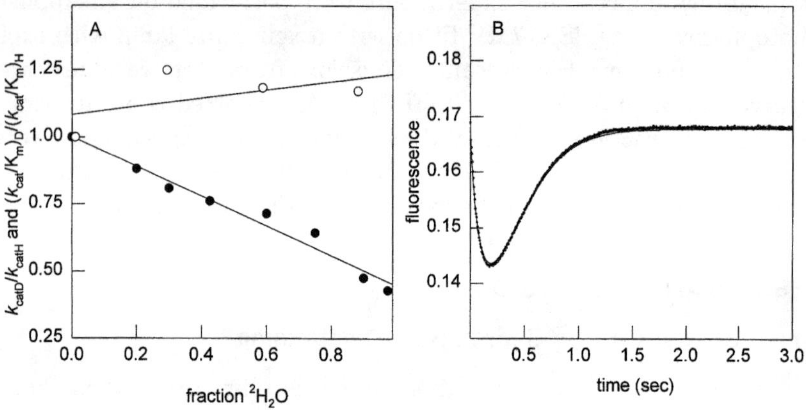

Figure 2. A, Effect of increasing ^2H_2O concentrations on the k_{cat}, k_D/k_H (-•-) and k_{cat}/K_m, $(k_{cat}/K_m)_D/(k_{cat}/K_m)_H$ (-O-) of 1,2-dibromoethane conversion. **B**, Fluorescence transient obtained after rapid mixing of haloalkane dehalogenase (15 μM) and 1,2-dibromoethane (12.5 μM). Solid line represents fit of data to Scheme I as described in the text.

The slow step in conversion of 1,2-dibromoethane is thus most likely k_3, which can be split into (1) hydrolysis of the alkyl-enzyme intermediate, (2) alcohol release, and (3) halide release.

In a rapid-quench-flow experiment with 1,2-dibromoethane in excess over enzyme, a transient burst of alcohol formation was observed (Figure 3). Thus, hydrolysis of the alkyl-enzyme is not rate-limiting. Alcohol release is unlikely to be the rate-limiting step since the alcohol was never found back in the x-ray structure (Verschueren et al., 1993b). Furthermore, the enzyme has a very low affinity for the alcohol produced from 1,2-dibromoethane (JP Schanstra and DB Janssen, unpublished results). Halide release is a good candidate for the rate-determining step since halide ions were observed in the x-ray structure after completion of the chemical part of the reaction. Therefore, this step was studied separately.

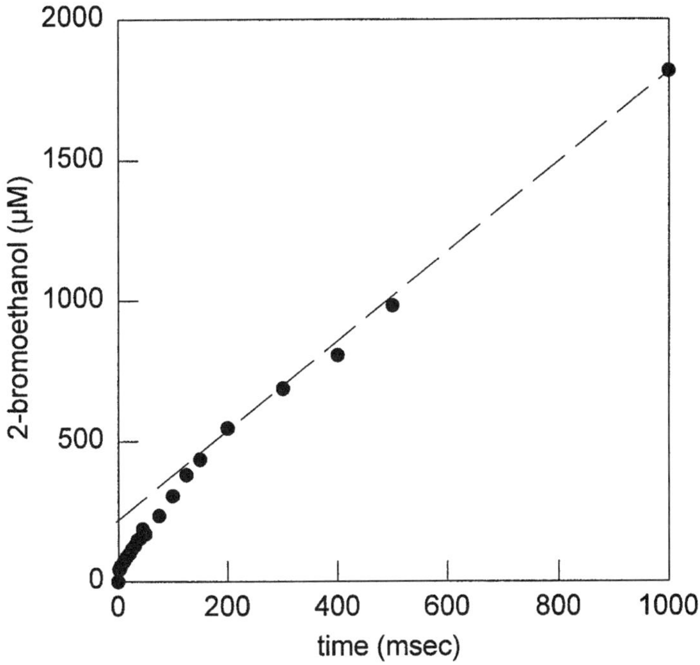

Figure 3. Burst experiment with 1,2-dibromoethane (10 mM) in excess over enzyme (550 μM).-•-, 2-bromoethanol production. Dashed line represents steady-state alcohol production (identical to k_{cat}).

Kinetics of Halide Binding and Release

Stopped-flow fluorescence experiments were used to study the kinetics of bromide binding under pseudo first-order conditions (1 to 1000 mM Br^-) to the dehalogenase (5 μM). The typical fluorescence transients obtained after rapid mixing of enzyme with various concentrations of bromide (1 to 50 mM) are shown in Figure 4A. The dissociation constant calculated from the steady-state fluorescence reached at the end of each transient resulted in a K_d (10 ± 1 mM), which is identical to the one obtained by steady-state fluorescence quenching measurement of bromide binding (K_d = 9.5 ± 0.5 mM), indicating that the conditions of the stopped-flow and steady-state experiments were similar. All fluorescence transients could be fitted to a single exponential. Analysis of the observed rates (k_{obs}) over the complete concentration range

(1 to 1000 mM KBr) showed a complicated dependence on the halide concentration (Figure 4B).The k_{obs} decreased from 8 s^{-1} at 1 mM KBr to 5 s^{-1} between 15 and 25 mM KBr. There was an increase in the k_{obs} with increasing bromide concentration above 25 mM KBr. At high bromide concentrations (above 600 mM) the k_{obs} leveled off again. Rate and equilibrium constants for this complex dependence of the k_{obs} on the bromide concentration could be fitted with numerical simulation of Scheme II, which was chosen to explain the observed dip since single exponentials were found in all cases and no burst was observed upon dilution of enzyme premixed with halide. The rate constants derived are shown below in Scheme II.

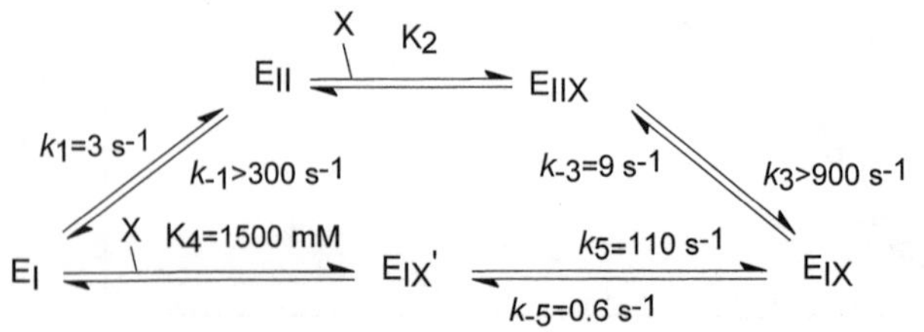

(Scheme II)

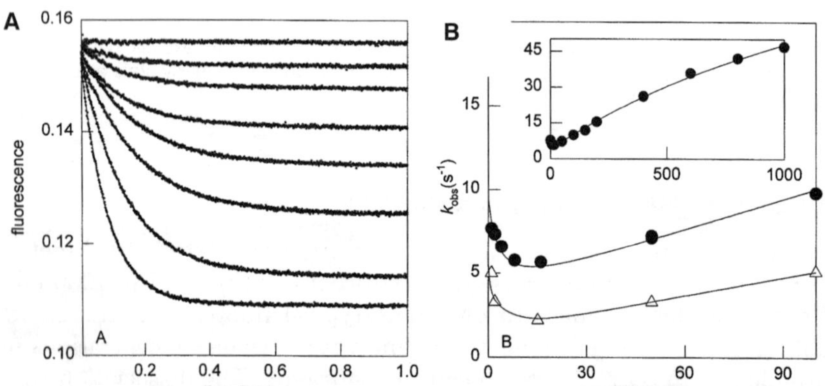

Figure 4. Kinetics of bromide binding (1 to 1000 mM) to haloalkane dehalogenase (5 μM). **A**, Fluorescence transients obtained upon binding of bromide. Haloalkane dehalogenase (5 μM) was mixed with KBr to give final concentrations of 0, 1, 2, 4, 8, 16, 50 and 150 mM KBr (from top to bottom). Solid lines represent best fit to a single exponential. **B**, Concentration dependence of k_{obs} for bromide binding (1 to 100 mM, -•-). Inset displays complete bromide concentration range used in experiment (1 to 1000 mM, -•-). Solid line is best fit of data in Scheme II. -Δ-, k_{obs} for bromide binding (1, 2, 15, 50 and 100 mM) in ^2H_2O (94.5%, v/v). Solid line is best fit of the data to Scheme II.

Steps 2 (K_2) and 4 (K_4) in Scheme II were in rapid equilibrium. The slow steps in bromide binding and release were the enzyme isomerizations preceding the actual halide binding and release. Bromide release mainly followed the upper route of scheme II. This implies a slow bromide off-rate (9 s^{-1}), which was close to the k_{cat} (3 s^{-1}) for 1,2-dibromoethane conversion and therefore can limit the k_{cat}.

We also studied the kinetics of halide binding in ^2H_2O (Figure 4B). All fluorescence transients could be fitted by single exponentials, and a clear effect of ^2H_2O on the k_{obs} was found. Extraction of the rate and equilibrium constants in ^2H_2O showed that both slow isomerization steps in Scheme II, (k_1) and (k_{-3}), were affected by use of ^2H_2O. The isomerization step mainly determined that the bromide off-rate (k_{-3}) was reduced in ^2H_2O from 9 to 5 s^{-1}. The k_{cat} in ^2H_2O for 1,2-dibromoethane conversion reduced in a similar way (Figure 2A). Viscosity controls with glycerol indicated that the higher viscosity of ^2H_2O could not account for the observed effect on the rate of halide binding.

Possible Routes for Halide Export

The active site of the dehalogenase is buried in the protein, with no obvious route for substrate entrance or product release. However, binding of the uncharged substrate (JP Schanstra and DB Janssen, unpublished results) and probably also release of the uncharged alcohol product are fast. Only release of the charged halide ion out of the buried active site was found to be slow. The route for substrate entrance and alcohol release is apparently not suitable for halide release. A probable explanation for this is that the uncharged alkyl halide and alcohol can diffuse out through openings formed by fluctuations or small motions, whereas the halide ion can only leave the buried active site after solvation by water. Halide ions can not freely diffuse through proteins, which is illustrated by the observation that halides are not aspecific collisional quenchers of buried tryptophan residues, unlike the polar nonionic compound acrylamide, which quenches all tryptophans in proteins (Eftink and Chiron, 1976).

Transport of halide from and to the surrounding water from the active site in haloalkane dehalogenase may need a specific route, or water may be required to enter the active site to solvate the halide ion during binding and release. For this, two routes may be envisaged. One route could involve the entrance of water *via* a tunnel extending from the solvent close to the active-site cavity, as proposed by x-ray crystallography

(Verschueren et al., 1993a). For entrance, only Leu262, which appears to block this tunnel, would have to move away. A second route may be formed by movement of parts of the cap domain. This domain possesses considerable flexibility, since it allows binding and conversion of large substrates such as 1-bromooctane (Pries et al., 1994a), which do not fit in the active site cavity as it is determined by x-ray crystallography (Verschueren et al., 1993a). Large substrate-induced structural changes were observed in mobile surface loops of several lipases which, like the dehalogenase, belong to the α/β-hydrolase fold enzymes (Ollis et al., 1992). Whether these two possible routes correlate with one of the kinetically detected routes is not yet known. It is unlikely that movement of part of the cap domain involves positioning of charged residues such as arginine close to the halide ion to assist export out of the active site cavity, a mechanism for halide transport in halorhodopsin (Rüdiger et al., 1995). No arginine residue seems to be close enough to interact with the halide ion.

We therefore propose that the slow enzyme isomerization predicted by the kinetic analysis of halide binding is a conformational change of a part of the cap domain, allowing water to enter and solvate the halide. The halide release sequence would then be as follows: opening of a segment of the cap domain (slow); entrance of water and solvation of halide (rapid); followed by reversal of the conformational change in the cap domain (fast). This proposal is supported by the effect of ^2H_2O on the rate of the enzyme isomerization. When ^2H_2O was used as the solvent, the rate of the enzyme isomerization preceding halide release and binding was slowed. This is in agreement with the increasing strength of hydrogen bonds in ^2H_2O, and it is well known that ^2H_2O protects several proteins against thermal denaturation. This overall stabilizing effect of ^2H_2O and its effect on the kinetics of halide release support a conformational change as the rate-limiting step.

REFERENCES

Eftink, MR and CA Chiron. **1976**. Exposure of tryptophanyl residues in proteins. Quantitative determination by fluorescence quenching studies. Biochemistry 15:672-680.

Frieden, C. **1994**. Analysis of kinetic data: practical applications of computer simulation and fitting programs. Meth. Enzymol. 240:311-322.

Keuning, S, DB Janssen, and B Witholt. **1985**. Purification and characterization of hydrolytic haloalkane dehalogenase from *Xanthobacter autotrophicus* GJ10. J. Bacteriol. 163:635-639.

Kennes, C, F Pries, GH Krooshof, E Bokma, J Kingma, and DB Janssen. **1995**. Replacement of tryptophan residues in haloalkane dehalogenase reduces halide binding and catalytic activity. Eur. J. Biochem. 228:403-407.

Ollis, DL, E Cheah, M Cygler, BW Dijkstra, F Frolow, SM Franken, M Haral, SJ Remington, I Silman, J Schrag, JL Sussman, KHG Verschueren, and A Goldman. **1992**. The /-hydrolase fold. Protein Eng. 5:197-211.

Pries, F, AJ van den Wijngaard, R Bos, M Pentenga, and DB Janssen. **1994a**. The role of spontaneous cap domain mutations in haloalkane dehalogenase specificity and evolution. J. Biol. Chem. 269:17490-17494.

Pries, F, J Kingma, M Pentenga, G van Pouderoyen, CM Jeronimus-Stratingh, AP Bruins, and DB Janssen. **1994b**. Site-directed mutagenesis and oxygen isotope incorporation studies of the nucleophilic aspartate of haloalkane dehalogenase. Biochemistry 33:1242-1247.

Pries, F, J Kingma, GH Krooshof, CM Jeronimus-Stratingh, AP Bruins, and DB Janssen. **1995**. Histidine 289 is essential for hydrolysis of the alkyl-enzyme intermediate of haloalkane dehalogenase. J. Biol. Chem. 270:10405-10411.

Rüdiger, M, U Haupts, K Gerwert, and D Oesterhelt. **1995**. Chemical reconstitution of a chloride pump inactivated by a single point mutation. EMBO J. 14:1599-1606.

Schanstra, JP, R Rink, F Pries, and DB Janssen. **1993**. Construction of an expression and site-directed mutagenesis system of haloalkane dehalogenase in *Escherichia coli*. Protein Expr. Purif. 4:479-489.

Verschueren, KHG., SM Franken, JH Rozeboom, KH Kalk, and BW Dijkstra. **1993a**. Refined crystal structures of haloalkane dehalogenase at pH 6.2 and pH 8.2 and implications for the reaction mechanism. J. Mol. Biol. 232:856-872.

Verschueren, KHG, F Seljée, HJ Rozeboom, KH Kalk, and BW Dijkstra. **1993b**. The catalytic mechanism of haloalkane dehalogenase:

crystal structures of the enzyme with substrate, covalently bound intermediate and product. Nature 363:693-698.

Verschueren, KHG, J Kingma, HJ Rozeboom, KH Kalk, DB Janssen, and BW Dijkstra. **1993c**. Crystallographic and fluorescence studies on the interaction of haloalkane dehalogenase with halide ions. Biochemistry 32:9031-9037.

APPENDIX A

ABSTRACTS OF PLATFORM PRESENTATIONS

(not submitted as full papers)

STRUCTURE-BASED REDESIGN OF CARBONIC ANHYDRASE IN THE DEVELOPMENT OF A ZINC BIOSENSOR

D.W. Christianson

Department of Chemistry, University of Pennsylvania, Philadelphia, Pennsylvania

ABSTRACT

Human carbonic anhydrase II (CAII) can be used to sense and quantify nanomolar concentrations of zinc. However, the properties of the wild-type enzyme (e.g., zinc affinity and protein-zinc equilibration kinetics) are not ideal for a real-time biosensor capable of continuously monitoring metal ion concentrations in an aqueous environment. In collaboration with Prof. Carol Fierke (Duke University Medical Center), we have used X-ray crystallographic methods to dissect the structural features of the CAII active site which contribute to zinc affinity in the nanomolar to femtomolar range. We have discovered variants that exhibit vastly improved protein-zinc equilibration kinetics for biosensor development. Additionally, in collaboration with Prof. Richard Thompson (University of Maryland School of Medicine), we have determined the binding modes of novel chromophores designed to transduce a signal reflecting the concentration of bound zinc. The structure-based design optimization of both CAII (the sensor) and the chromophoric ligand (the analyte transducer) leads toward the development of a robust, CAII-based zinc biosensor.

ENGINEERING SUPRAMOLECULAR SELF-ASSEMBLY REACTIONS: METAL-DEPENDENT FORMATION OF PROTEIN TUBES, CABLES AND SHEETS

M.J. Dabrowski, P.M. Schurke, and W.M. Atkins

Department of Medicinal Chemistry, University of Washington, Seattle, Washington

ABSTRACT

Escherichia coli glutamine synthetase (GS) is a dodecamer of identical subunits arranged as two face-to-face hexameric rings. In the presence of various divalent transition metal ions, the dodecamers "stack" to form tubes, which subsequently associate laterally to generate cables. The X-ray crystal structure of the nearly identical *S. typhimurlum* GS indicates that the N-terminal helices of each subunit project away from the 6-fold axis of symmetry which runs through the center of the dodecameric ring structure and form the "spokes of a wheel" on each exposed hexameric face. The residues His-4, Met-8, and His-12 from N-terminal helices in "stacked" dodecamers provide potential ligands for interdodecameric metal binding sites. Site-directed mutants in which His-4, Met-8, and His-12 individually have been putative metal-binding sites. Light scattering and electron microscopy experiments indicate that these mutants, which retain their dodecameric ring structure, do not stack in the presence of divalent metal ions. Furthermore, the structural model of a dodecamer-dodecamer complex suggests that the intermolecular metal binding site is binuclear. In order to further define the metal binding sites, model peptides based on the sequence of the N-terminal helices of the wild type and M8C mutant have been used. Electrospray ionization mass spectrometry with the model peptides indicates that Cu^{2+} or Zn^{2+} ions bind to amino acids at positions 4, 8, and 12, with a stoichiometry of 2 metal ions/2 peptides. In addition, the mutant protein H12D, which does not form tubes, undergoes metal-dependent lateral self-assembly to form hexagonally packed protein sheets. These results demonstrate that protein-based supramolecular assemblies with useful shapes, sizes, and symmetry may be engineered from a single wild-type protein.

PROGRESS REPORT OF THE BATTELLE UHF NMR PROJECT

P.D. Ellis

Environmental Molecular Science Laboratory, Pacific Northwest National Laboratory, Richland, Washington

ABSTRACT

The ultrahigh field (UHF) NMR magnet project is about a year away from delivering a high-resolution NMR magnet with a field of at least 21.15 T (900 MHz for H1 resonance frequency). The presentation will summarize progress to date for the project while discussing several of the planned operational differences between the UHF magnet and conventional magnets, i.e., 17.6 T and below. If time permits, I will outline how the magnet can be accessed by the general research community and our own plans for the resulting spectrometer.

THE CALCIUM INDUCED STRUCTURAL CHANGE THAT TRIGGERS MUSCLE CONTRACTION

S. M. Gagné, S. Tsuda, M.X. Li, C.M. Slupsky, L.B. Smillie, and B.D. Sykes

Department of Biochemistry and MRC Group in Protein Structure and Function, University of Alberta, Edmonton, Alberta, Canada

ABSTRACT

In the thin filament of skeletal muscle, the binding of calcium to the calcium-binding protein troponin-C (TnC) induces a conformational change in the protein which alters its interaction with the inhibitory protein troponin-I (TnI) and ultimately leads to muscle contraction. We have focused on several aspects of this calcium-triggering, including the structures of the calcium-binding domains of TnC, the nature of the conformational change that occurs in TnC upon calcium binding, and the structure and location of TnI binding to TnC. To accomplish this we have studied peptide fragments and domains of TnC and TnI, along with the intact TnC, and binary and ternary complexes of these peptides and proteins. Many of these have been cloned and expressed in *Escherichia coli,* so they have been enriched with ^{13}C and ^{15}N for use in multinuclear, multidimensional NMR spectroscopy and structure determination.

The structures of apo Tr_1C fragment (residues 12-87) and apo N-domain residues (1-90) of TnC and of the calcium-saturated N-domain have been determined by triple resonance 3D NMR experiments. The comparison of the structures of the regulatory domain of TnC in the apo and calcium-saturated states allows us to define the primary structural change responsible for triggering muscle contraction, and to propose a mechanism for the direct coupling of calcium binding to this conformational change. We have determined the solution structure of intact troponin-C in the calcium-saturated form in the presence of trifluoroethanol to keep it as a monomer in solution. The solution structure of the N-domain portion of the TnC is the same as that of NTnC, but the linker between the two domains is flexible in solution. We have also studied the structure and calcium-binding properties of the E41A mutant of NtnC, which allows us to understand the mechanism of coupling between calcium binding and the resultant conformational changes induced in the protein.

HUMAN CELL REPAIR OF LETHAL AND MUTAGENIC LESIONS BY GLYCOSYLASES

B. Hange, A. Chenna, and B. Singer

Donner Laboratory, Lawrence Berkeley National Laboratory, University of California, Berkeley, California

ABSTRACT

Cyclic adducts are formed by a variety of naturally occurring carcinogens. This laboratory has focused on the vinyl chloride and benzene metabolites, chloroacetaldehyde (CAA) and p-benzoquinone (p-BQ), which react with dA and dC (as well as with dG), forming additional new exocyclic ring(s). These modifications lead to the cleavage of the glycosyl bond in defined oligonucleotides or DNA when acted upon by human cell extracts.

In the case of 1, N^6-etheno dA (εA) the enzyme has been identified as an m^3A DNA-glycosylase. Separate and different human glycosylase activities are being studied in defined 25-mers containing a single nucleoside modified by naturally occurring processes such as deamination, oxidation, etc. Modification by the agents we are studying does not lead to removal of such bases by any of the previously characterized glycosylases, e.g., uracil, hypoxanthine, m^3A, etc. or, detectably, through the excision repair pathway.

We have separated a 3, N^4-etheno dC (εC) glycosylase from human cells which differs completely in substrate specificity compared to the εA glycosylase. In the case of 3,N^4-benzetheno dC (pBQ-C) from p-BQ, the enzyme responsible for this glycosylase activity has now been purified to homogeneity. These new enzymes do not overlap in specificity with each other or with m^3A DNA-glycosylase, which excises 1,N^6-εA. No activity toward 1,N^6-benzetheneo dA (pBQ-A) is found in the pBQ-C glycosylase, suggesting a separate enzyme. This is supported by rapid inactivation of pBQ-A glycosylase nicking activity when heated.

At this point we do not know whether other, yet untested, cyclic derivatives are recognized by the new glycosylases. Neither do we know why such chemically closely related derivatives require separate glycosylases. These questions may be answered when the 3-D structure of such enzymes are known, as well as the active site of each enzyme and how it interacts with the substrate. There are likely to be additional glycoslase(s) to be added to this enzyme class when additional cyclic substrates are synthesized and tested in the same manner.

KINETIC CHARACTERIZATION OF A TETRACHLOROETHYLENE UTILIZING MICROBIAL CONSORTIUM

B.S. Hooker, R.S. Skeen, J. Gao, and M.M. Shah

Bioprocessing Group, Environmental Technology Division, Pacific Northwest National Laboratory, Richland, Washington

ABSTRACT

The kinetics of successive biological dechlorination of tetrachloroethylene (PCE) by a microbial consortium that derives energy from reductive dehalogenation have been characterized. A set of batch experiments have been completed, under a variety of conditions, to determine the effect of variable PCE levels as well as the presence of the chlorinated intermediates trichloroethylene, (TCE), dichloroethylene (DCE), and vinyl chloride (VC) on dechlorination rate. Experiments were completed using a VOC-tight batch reactor to ensure proper mass balance closure on volatile chlorocarbons. The experimental design, along with results, for PCE, chlorinated intermediates, ethylene, methanol (substrate), biomass, organic acids, and hydrogen levels will be reported. In addition, the development and use of a preliminary kinetic model will be discussed.

The treatability of PCE biodegradation was tested under anaerobic culture conditions with three contaminated-site soil samples and one river sediment. Various substrates, for example methanol, formate, sucrose, acetate, and lactate, were added to all these soil/sediment cultures with and without sulfate addition. The substrate consumption of substrate PCE, TCE, and other volatile chlorinated compounds, as well as acetate and methane formation were monitored during the cultures.

FOUR-STRANDED DNA STRUCTURES

C. H. Kang[1] and A. Rich[2]

[1]Department of Biochemistry and Biophysics, Washington State University, Pullman, Washington

[2]Department of Biology, Massachusetts Institute of Technology, Cambridge, Massachusetts

ABSTRACT

The telomeres found at the end of chromosomes contain hundreds of repeats of a simple motif. The motif has a segment containing clusters of guanine residues on one strand and clusters of cytosine residues on the other strand. We have crystallized telomeric sequences containing guanine or cytosine residues. These structures have been solved by X-ray diffraction methods.

Both of these strands individually form four-stranded structures. However, the four-stranded structures containing guanine are quite different in their organization from those containing cytosine. In the guanine quartets the four guanine residues are organized in one plane, using a system of cyclic hydrogen bonding. The cytosine-containing quartets are made of two duplexes pointing in opposite directions that are held together by $C \bullet C^{+}$ hydrogen bonds. The two sequences intercalate so that a base pair from one duplex is surrounded by base pairs from the other duplex. The structure of several of these four-stranded complexes will be described.

ENZYMATIC CATALYSIS IN ORGANIC SOLVENTS

A.M. Klibanov

Department of Chemistry, Massachusetts Institute of Technology, Cambridge, Massachusetts

ABSTRACT

It is now well established that many enzymes can function as catalysts in anhydrous organic solvents and catalyze new chemical reactions in this unnatural milieu. Among other benefits of this nonaqueous enzymology, it has become possible to address experimentally the question, "Can enzyme selectivity be substantially altered by switching from one organic solvent to another as the reaction medium?" Recent studies by us and others have answered this question affirmatively. In particular, we have discovered that stereoselectivity, regioselectivity, chemoselectivity, and substrate selectivity of enzymes can all be profoundly affected, and sometimes even reversed, simply by changing the solvent. In this presentation, selected examples, the mechanistic rationale, and synthetic implications of this phenomenon will be discussed.

ELECTRONIC STRUCTURE CALCULATIONS ON THE REDOX SITE OF RUBREDOXIN

J.B. Koerner and T. Ichiye

Department of Biochemistry and biophysics, Washington State University, Pullman, Washington

ABSTRACT

Ab initio quantum mechanical calculations on the $Fe(SCH_3)_4^{n-}$ complex (where n=1 or 2), a model for the redox site for the electron transfer protein rubredoxin, were performed. The effects of the electron properties of the system due to the internal structure of the redox site, and of water molecules close to this site, were determined. Specifically, different conformations of the methylate ligands were examined to determine the effect of rotations about these dihedral angles on the electron densities in the models. Optimizations of the different conformational geometries will be done at the both HF and MP2 levels. The effects of basis set size on the HF optimized geometries were also examined. From the electronic wave function, the Mulliken and ESP charges were calculated. In addition, since water molecules have been observed in classical molecular dynamics simulations to closely approach the redox site, various orientations of an added water molecule, with respect to the $Fe(SCH_3)_4^{n-}$ subunit, were examined. The effect of the water on the electronic properties of the system was determined.

PERTURBATIONS OF DNA STRUCTURE BY RADIATION-INDUCED LESIONS

J.H. Miller and K. Miaskiewicz

Environmental Molecular Science Laboratory, Pacific Northwest National Laboratory, Richland, Washington

ABSTRACT

Quantum-chemical calculations were carried out with GAUSSIAN 90 to determine the electronic energy and optimum geometry of all possible stereoisomers of selected adenine and thymine base lesions. Residual electronic charges and other parameters of the AMBER all-atom forced field for the modified residues were deduced from these molecular-orbital calculations. Isomers with the lowest energy (including hydration) were introduced into the dodecamer d(CGC-GAATTCGCG)[2] at either A5 or T7. Molecular dynamics simulations were carried out using the AMBER 4.0 molecular-modeling package with explicit water and counterions included under periodic boundary conditions. Trajectories, typically 100-200 ps in duration, were analyzed by the CURVES algorithm and its implementation in the DIALS and WINDOWS program. Lesions investigated by these methods include the intrastrand and cross-link 8,5'-cyclodeoxyadenine, the cis,syn-cyclobutane thymine dimer, thymine glycol, dihydrothymine, 5-hydroxy-5,6-dihydrothymine, thymine hydrate, urea and formamido. Whenever possible, results are compared with experimental data from solution-phase, high-resolution NMR spectroscopy. In all cases, calculated perturbations to DNA structure were localized to the immediate neighborhood of the damage base.

The most common lesion-induced changes were in helical parameters that affect base stacking and base-pairing. Significant perturbations to backbone torsion angles and sugar puckering occurred only in the case of 8,5'-cyclodeoxyadenine due to the stereospecific requirements to form the new C5' to C8 covalent bond. The magnitude of distortions introduced by ring-saturated thymine lesions was determined primarily by the orientation of the methyl group at C5. Addition of hydroxyl at this position displaces the methyl group from its native in-plane configuration to an orientation that is nearly perpendicular to the pyrimidine ring. Hence, thymine glycol and 5-hydroxy-5,6-dihydrothymine induced significantly greater distortions than thymine hydrate and dihydrothymine. This is consistent with observations by Wallace and coworkers that thymine glycol is a substrate for nucleotide excision repair while dihydrothymine is not. The absence of significant kinking or bending of the DNA backbone as a result of the cis,syn-cyclobutane thymine dimer is in agreement with the recent NMR study by Taylor and coworkers on the sequence CGTATTATGC containing a thymine dimer.

Results from our simulations indicate that H1' to H6 distances in pyrimidines or H1' to H8 distances in purines significantly exceed 5 A only at the A6/T7 step. Taylor observed H1'-H6 or H1'-H8 NOE cross peaks at every base step except A4/T5 which, like the A6/T7 step of our dodecamer, contains the 5' thymine of the dimer. These findings suggest that the primary distortions induced by the cis, syn-cyclobutane thymine dimer involve reorientation of the 5' thymine relative to the helical axis and a corresponding weakening of hydrogen bonds to the complementary adenine base.

DUPLICATION OF GENES INVOLVED IN DEGRADATION OF AROMATICS

M.F. Romine

Pacific Northwest National Laboratory, Richland, Washington

ABSTRACT

Bacterial degradation of aromatic compounds can proceed by an initial hydroxylation of the aromatic ring. Known dioxygenase enzymes that carry out this reaction are made up of either three or four different protein subcomponents, including a ferredoxin, ferredoxin reductase, and a terminal oxygenase. The terminal oxygenase, which is involved in substrate binding, contains either four proteins that are identical or two each of two different proteins (a large and a small component). The corresponding plasmid-encoded genes for the terminal oxygenase have been sequenced from a bacterium isolated from sediments 410 m below the subsurface at the U.S. Department of Energy-managed Savannah River Site. A plasmid (>100 kb) from this isolate, designated *Sphingomonas* sp. F199, encodes the information necessary for degradation of naphthalene, toluene, all isomers of xylene, dibenzothiophene, and fluorene. Unlike other enzymes belonging to this family, the corresponding genes in F199 are distributed on multiple operons, suggesting a unique regulatory mechanism.

Additionally, two different genes with sequence homology to other large components of the terminal oxygenase were found in close proximity. This finding suggests that expression of the duplicated genes could result in formation of multiple different dioxygenases, each containing different combinations of these components, thereby resulting in the capability to degrade a much wider array of aromatic compounds than other typical aromatic degraders. The weak homology of these genes to known genes further suggests that additional substrates can be targeted by F199.

TIME-AVERAGED CRYSTALLOGRAPHIC REFINEMENT OF BPTI

C.A. Schiffer, W.F. van Gunsteren, and A.A. Kossiakoff

Genetech Incorporation, San Francisco, California

ABSTRACT

Time-averaged crystallographic refinement was performed on the crystal structure of bovine pancreatic trypsin inhibitor (BPTI). This method fits an ensemble of structures, rather than a single structure, against the crystallographic data. This allows, given sufficient resolution, regions in the crystal to be more thoroughly characterized in the refinement. BPTI is a small, 58 residue, highly studied protein. Although crystals of this protein diffract to high resolution, many of the side chains remain disordered in the electron density, thus providing a good system to test this technique. In this study we examine the conformational diversity of side chains and the order of water molecules within the crystal. The occupancy, B-factor, and anisotropy can be determined for each crystallographic water site from the dynamics of the water during the refinement. Time-averaged crystallographic refinement thus provides a way of describing an ensemble of motions which fit within the envelope of the electron density in crystals of macromolecules.

THE ROLE OF CHROMATIN IN DNA DAMAGE AND REPAIR

D.L. Springer, B.D. Thrall, A.O. Murad, D.B. Mann, V.L. Burnett

Biology and Chemistry Department, Pacific Northwest National Laboratory, Richland, Washington

ABSTRACT

Studies of the binding of chemical carcinogens to DNA reconstituted with histone core particles were conducted to determine the influence that nucleosome structure has on covalent adduct formation. Reconstitution of the 5S rRna gene from *Xenopus borealis* produced a characteristic nucleosome as determined by hydroxyl radical footprinting, micrococcal nuclease digestion and gel retardation assays. Analysis of benzo[a]pyrene diol epoxide (BPDE) adduction at the sequence level by primer extension indicated that the amount of adduction within the 5S nuclesome was suppressed by 50% relative to naked DNA.

The amount of adduction at specific sites could not be accounted for by the rotational setting of guanines, since guanines in close proximity to the histone core were equally susceptible to BPDE as guanines on the outer surface of the nucleosome. DNA near the 5S nucleosome dyad, where modulations in minor groove width are known to occur, were the least susceptible to covalent adduction. These results demonstrate that structural features of nucleosomal DNA contribute to the susceptibility of DNA to modifications by BPDE.

MICROBIALLY MEDIATED URANIUM REDUCTION AND PRECIPITATION

M.J. Truex, B.M. Peyton, Y.A. Gorby, and N.B. Valentine

Pacific Northwest National Laboratory, Richland, Washington

ABSTRACT

Many microorganisms are able to reduce metals. Among these, dissimilatory metal reducers can efficiently remove uranium from aqueous streams. These microorganisms use uranium as an electron acceptor and reduce soluble uranium (VI) to insoluble uranium (IV). Data are presented that describe uranium reduction rates under nongrowth conditions, using lactate as the energy source. In addition, a continuous-flow system for uranium bioprecipitation is described.

THE EFFECT OF SOLVENT ON PROTEIN STRUCTURE AND DYNAMICS IN NONAQUEOUS ENVIRONMENTS

Y.J. Zheng and R.L. Ornstein

Environmental Molecular Sciences Laboratory, Pacific Northwest National Laboratory, Richland, Washington

ABSTRACT

Many potential benefits may be realized by a better understanding of nonaqueous enzymology, including improving by rational redesign the remediation capacity of enzymes for direct use in concentrated waste plumes in deep subsurface environments. Crystal structures have recently appeared for the enzyme subtilisin Carlsberg in anhydrous acetonitrile and in water. To gain a mechanistic understanding of how the solvent environment affects protein structure, dynamics, and function, we have performed molecular dynamics simulations, using the all-atom AMBER forcefield, with either a large cap of solvent or under periodic boundary conditions, on subtilisin Carlsberg in water, in acetonitrile, in carbon tetrachloride, and in DMSO. Our preliminary results and their implications will be presented.

APPENDIX B

ABSTRACTS OF POSTER PRESENTATIONS

(not submitted as full papers)

QUATERNION RIBBONS: A TOOL FOR THE VISUALIZATION OF NONLOCAL CONTACTS IN PROTEINS

K. Albrecht,[1] J. Hart,[1] A. Shaw,[2] and A.K. Dunker[2]

[1]Department of Electrical Engineering and Computer Sciences, Washington State University, Pullman, Washington

[2]Department of Biochemistry/Biophysics, Washington State University, Pullman, Washington

ABSTRACT

Protein folding involves nonlocal contacts, e.g., interactions between amino acids that are well-separated along the sequence. Understanding such nonlocal contacts is important for knowledge-based studies of protein structure and folding. If, indeed, a "protein folding code" exists, such nonlocal contacts must be at the heart of that code, yet there has been little systematic study of such contacts. Here we describe a procedure for inserting ribbons or surfaces that bisect the distances between main-chain atoms of nonlocally interacting segments of secondary structure, followed by quaternion-based deformations of the ribbons or surfaces to conveniently display the zones of contact between the side chains. We have developed the algorithms so far to display helix/helix quaternion contact ribbons. For helix/helix contacts, we first identify the axes of the contacting helices, using Kahn's algorithm. Next, a ribbon is constructed midway between the helical axes, using quaternions to describe the orientation of each ribbon segment. Quaternions have been used recently for computer animation because of their advantages for describing rotations. Taking advantage of the properties of quaternions, the contact ribbon is untwisted while largely preserving the various side-chain/ribbon spatial relationships. Applying this tool to a bZIP protein contact demonstrates the tool's utility and reveals new insights regarding the nonlocal protein contacts within a typical "leucine zipper" (e.g., the GCN4 contact).

STRUCTURE AND THERMODYNAMICS IN POLYELECTROLYTE SOLUTIONS: INTEGRAL EQUATION AND SIMULATION RESULTS

C.S. Babu and T. Ichiye

Department of Biochemistry/Biophysics and of Chemistry, Washington State University, Pullman, Washington

ABSTRACT

The structure of a model polyelectrolyte solution was investigated, using a new integral equation theory. The merit of the new theory over existing theories of fluids is that it can treat the consequences of long as well as short-range repulsive and attractive forces in a fluid. This is important in biological systems such as a protein solution, where these interactions are primarily responsible for protein stability and folding, and their function. Our model consists of spherical macroions and small counterions in a dielectric continuum that interact with soft-repulsive and Coulomb forces. The results are also compared with Monte Carlo simulation results.

CORRELATION OF THE ELECTROSTATIC ENVIRONMENT OF FERREDOXIN REDOX SITE CLUSTERS WITH THEIR REDOX POTENTIALS: A MOLECULAR DYNAMICS STUDY

B.W. Beck and T. Ichiye

Department of Biochemistry/Biophysics, Washington State University, Pullman, Washington

ABSTRACT

The redox potential of electron transfer proteins is extremely important in characterizing their electron-transfer reactions. The redox potential for a given type of redox center often varies significantly, within families of related proteins as well as within a single protein that contains multiple sites. In addition, recent experimental evidence also indicates that proteins with multiple redox clusters often have a preference for the order in which the clusters are reduced. The results of a molecular dynamics study of ferredoxins containing two similar redox clusters are presented here. The ferredoxins were simulated under aqueous conditions at 300 K in each of four different oxidation state couples: oxidized(-2)/ oxidized (-2), oxidized(-2)/reduced(-3), reduced(-3)/oxidized(-2), and reduced(-3)/

reduced(-3). The electrostatic interactions of the protein and solvent with each redox cluster were calculated, and alterations in protein and solvent structure as a function of oxidation state were determined. The role of differential solvation of the clusters was determined, and correlations of the electrostatic environment of these ferredoxin redox site clusters with their differences in redox potentials were found.

SELECTIVE INHIBITION OF GLYCOSOMAL PHOSPHOGLYCERATE KINASE FROM TRYPANOSOMA

B.E. Bernstein, C.L.M.J. Verlinde, and W.G.J. Hol

Departments of Biochemistry and Biological Structure, and Howard Hughes Medical Institute, University of Washington, Seattle, Washington

ABSTRACT

Trypanosoma brucei, the causative agent of sleeping sickness, is completely reliant on glycolysis as its energy source while infecting the human host. Indeed, this protozoan utilizes a unique organelle to contain the enzymes which carry out this process. In the search for a better treatment for this disease we have focused on potential means for selectively inhibiting glycolysis in the trypanosome. Phosphoglycerate kinase (PGK) from T. brucei has been overexpressed and purified in our laboratory. Crystallization trials are underway, with variable success. Once determined, the structure of T. brucei PGK will be compared with a model of the analogous human enzyme, and structural differences will be exploited for the purpose of drug design.

In parallel with crystallization, kinetics studies have been performed. This work has successfully identified a micromolar inhibitor of PGK which is specific to the T. brucei version. The inhibitor is the azo-dye, 2-(p-sulfonophenylazo)-1, 8-dihydroxy-3, 6-naphthalene disulfonic acid (SPADNS). On the basis of kinetic and modeling experiments we propose that SPADNS interacts with the nucleotide binding site of PGK. Sequence and charge differences in this region between human and trypanosomal PGK may explain the preference SPADNS has for the trypanosomal enzyme.

HETERONUCLEAR NMR STUDIES OF FRACTIONATION FACTORS FROM HISTIDINE-CONTAINING PROTEINS *ESCHERICHIA COLI* (ECHPR) AND *BACILLUS SUBTILIS* (BSHPR)

P.M. Bowers and R.E. Klevit

Biomolecular Structure Center and Department of Biochemistry, University of Washington, Seattle, Washington

ABSTRACT

An equilibrium constant of isotope exchange is termed a fractionation factor. Within a protein, hydrogens, such as those associated with backbone and sidechain amides and threonine and serine hydroxyls, exchange with the solvent:

OH (protein) + D (Solvent) <===> OD (protein) + H (solvent).

A site with a fractionation factor less than one has a preference for protium relative to its content in the solvent. A large body of evidence, predominantly from the studies of small molecules, suggests that strong hydrogen bonds accumulate protium over deuterium and tritium relative to the solvent content. The strength of hydrogen bonds and their importance in protein stability has been hotly debated for decades. It remains unresolved due to the experimental difficulties in assessing the relative strength of protein-solvent versus protein-protein hydrogen bonds. A recent study of staphylococcal nuclease demonstrated that protium accumulates at specific sites within the protein, suggesting that these protein-protein hydrogen bonds are significantly stronger than protein-solvent hydrogen bonds.

We have used 1D and 2D heteronuclear NMR as well as MS to study fractionation factors in EcHPr and BsHPr and in an EcHPr mutant, S31A. Histidine-containing proteins are small (90 residue) heat-stable proteins which share a highly similar structure while sharing only ~50% sequence identity, making them ideal for these studies. Our results suggest that protium enrichment occurs at specific sites with the protein and that these sites of enrichment are conserved across species lines. The strong hydrogen bonds identified in these studies are likely to play an important role in protein stability and function.

EFFECTS OF CRYSTAL PACKING FORCES ON NUCLEIC ACID STRUCTURE AS DETERMINED BY MOLECULAR DYNAMICS

K. Brameld and G.A. Goddard III

Materials and Process Simulation Center, Beckman Institute Division of Chemistry and Chemical Engineering, California Institute of Technology, Pasadena, California

ABSTRACT

X-ray crystallography remains the key analysis tool for gaining detailed structural information about nucleic acids. As more small DNA fragments, ranging from 8 to 12 basepairs in length, have been crystallized and their structures solved, the relevance of such structures to DNA free in solution has been questioned. Of great concern is the extent to which crystal packing forces govern DNA structure. In an attempt to assess the features of DNA structure which are susceptible to packing forces, we have undertaken systematic molecular dynamics simulations of DNA fragments in both a crystalline environment and free in solution. Two sequences have been studied: a decamer and dodecamer, each containing a common central 5'-ATATAT site. Experimentally, these sites were observed to have different minor groove widths, depending on the length of the flanking sequences. These differences are believed largely to be a result of interhelical contacts, as the two crystals are not isomorphous, with the helices packing in a different manner. Molecular dynamics, using the AMBER forcefield, has allowed us to determine whether the two sequences converge to a common structure and to what extent this structure differs from that found within the crystal. Based on these differences, the influence of packing forces on local helix parameters has been assessed, leading to greater insight as to the nature of DNA, both crystallized and free in solution

MOLECULAR DYNAMICS SIMULATIONS ON THE 14 BASE-PAIR OLIGONUCLEOTIDE CONTAINING THE ADENOVIRUS MAJOR LATE PROMOTER TATA ELEMENT AND RELATED SEQUENCES: DOES TATA-BOX BINDING PROTEIN PREFER CURVED OR FLEXIBLE DNA SEQUENCES?

O.N. de Souza, K. Miaskiewicz, and R.L. Ornstein

Biomolecular Simulation and Modeling, Environmental Molecular Sciences Laboratory, Pacific Northwest National Laboratory, Richland, Washington

ABSTRACT

Using AMBER 4.0 force field, we investigated the static and dynamic properties of the TATA-box binding protein- binding-oligonucleotide GC**TATA**AAAGGGCA and four other related sequences where the consensus element TATA was systematically altered. The dynamic trajectory was analyzed using the CURVES algorithm; results will be discussed.

BIOLOGICAL APPLICATIONS OF NEAR-FIELD SINGLE-MOLECULE SPECTROSCOPY

R.C. Dunn,[1] E.V. Allen,[1] X. S. Xie,[1] and L. Mets[2]

[1]Chemical Sciences Department, Pacific Northwest National Laboratory, Richland, Washington

[2]Department of Molecular Genetics and Cell Biology, University of Chicago, Chicago, Illinois

ABSTRACT

Recent advances in near-field optical microscopy make it possible to conduct molecular spectroscopy at the nanometer dimension. The technique involves scanning a spot of light, of a dimension much smaller than the optical wavelength, in close proximity to the surface of a sample. For biological applications, fluorescence measurements with near-field optics are particularly informative, providing spectroscopic information not accessible with other scanning probe techniques such as SEM, STM, and AFM.

A few groups, including ours, have recently demonstrated the sensitivity of the near-field technique by imagining the fluorescence of single dye molecules with sub- diffraction resolution. Furthermore, we have completed a study on the fluorescence lifetimes of single dye molecules by incorporating in our microscope a picosecond time-correlated single photon counting ability. This study demonstrated the feasibility of lifetime imaging with single molecule sensitivity, nanometric resolution, and picosecond temporal resolution. In our efforts to study photosynthetic membranes, we have imaged intact photosynthetic membrane fragments from *Chlamydomonas reinhardtii* PSI-PSII double deficient C2 strain. Membrane fragments from this strain contain only the LHCII. The emission images reveal a homogeneous distribution of LHCII throughout the membrane fragments, in agreement with recent SEM results. Lifetimes were also measured at various points within the membrane fragments. These studies are aimed at spectroscopically mapping the photosynthetic membrane, and further progress along these directions will be presented.

INDUCTION OF BENZO(a)PYRENE METABOLISM IN HUMAN MAMMARY CELLS AS INDICATOR OF EXPOSURE TO POLYCYCLIC AROMATIC HYDROCARBONS

R. Goth-Goldstein

Lawrence Berkeley Laboratory, Berkeley, California

ABSTRACT

Human mammary epithelial cells derived from normal reduction mammoplasties can be cultured and have been shown to metabolize polycyclic aromatic hydrocarbons (PAH), such as benzo(a)pyrene (BaP), by cytochrome P450 monooxygenase. This enzyme can be induced by PAH and dioxin derivatives. In a standardized cell system, the increase of BaP metabolism can be used as a cell system, using cells from one individual at a certain passage, to evaluate the biological effects of coal tar residues from manufactured gas plants. Cells were preincubated with hexane-extractable coal tar material for 16 hr, then incubated with ^{3}H-BaP for 2 hr. The amount of BaP metabolites in cell extracts was determined by HPLC. An up to 5-fold increase over controls was seen in BaP metabolites after preincubation with 0.2, μg/ml coal tar material (a noncytotoxic dose). Induction of BaP metabolism was also seen after preincubation of cells with cigarette-smoke condensate. The results show that coal tars and other P450-inducing compounds can act as cocarcinogens by enhancing the genotoxic effects of compounds metabolized by the P450 enzyme system.

NMR INVESTIGATION OF MOLECULAR DYNAMICS IN A ZINC-FINGER DNA-BINDING DOMAIN

D.E. Hyre, M.A. Schmiedeskamp, and R.E. Klevit

Biomolecular Structure Center and Department of Biochemistry, University of Washington, Seattle, Washington

ABSTRACT

ADR1z' is a peptide representing the minimal DNA-binding domain of the yeast transcription factor ADR1, which includes two Cys_2His_2 zinc-finger domains and N-terminal residues shown to be required for DNA binding. This protein activates transcription of the gene coding for the enzyme alcohol dehydrogenase-2 by binding in tandem to its cognate sequence, the Upstream Activation Sequences (UAS). Solution structures of peptides corresponding to the individual zinc-finger domains have been reported. The N-terminal finger exists in a single conformation nearly identical to that observed in other Cys_2His_2 zinc fingers. The C-terminal finger exhibits a multiplicity of peaks whose nature is dependent on the presence of two prolines in its sequence. Mutation of the first proline to alanine significantly reduces peak doubling but it also reduces DNA binding affinity 100-fold. Mutation of the second proline to alanine slightly increases peak doubling and increases DNA binding twofold. In addition, there are four residues between the zinc-binding histidines, in contrast to the more common three-residue separation.

The correlation between the presence of the prolines and the appearance of peak doubling in the C-terminal finger, and DNA-binding affinity suggests that dynamics may play a role in the binding of ADR1. To this end, we have begun direct measurements of the heteronuclear relaxation properties of the ADR1z' peptide. Results of these measurements and their correlation to the properties of the individual fingers will be presented. In particular, similarities between the observed relaxation properties and the observed differences in conformational forms will be discussed.

REFINED SOLUTION STRUCTURE AND BACKBONE DYNAMICS OF THE PHOSPHOCARRIER PROTEIN HPr FROM *BACILLUS SUBTILIS*

B.E. Jones, R. Rajagopal, and R.E. Klevit

Biomolecular Structure Center and Department of Biochemistry, University of Washington, Seattle, Washington

ABSTRACT

A solution structure of the phosphocarrier protein histidine-containing protein (HPr) from *Bacillus subtilis*, based primarily on two-dimensional NMR data, has previously been reported. Although the structure was similar to the reported X-ray crystal structure, a detailed comparison was not possible, due in part to the moderate definition of the NMR structure. Therefore, we have refined the solution structure with data obtained primarily from three-dimensional NMR experiments. The resulting set of calculated structures, although very similar to the crystal structure, show a significant deviation in a region of the protein (residues 45-54) which contains the site of regulatory phosphorylation (Ser46). Interestingly, even though this region is α-helical (residues 49-53 in the solution structure and 47-53 in the crystal structure), previous NMR experiments have indicated that this region is significantly more flexible than the majority of the protein. We have therefore investigated the backbone dynamics of HPr based on ^{15}N relaxation measurements. The details of the structural differences between the solution state and solid state will be discussed in light of the protein's dynamical properties.

HETERONUCLEAR NMR STUDIES OF CALMODULIN FROM *SACCHAROMYCES CEREVISIAE*

S.Y. Lee, T.N. Davis, and R.E. Klevit

Bimolecular Structure Center and Department of Biochemistry, University of Washington, Seattle, Washington

ABSTRACT

Calmodulin is a 16-kD Ca^{2+} binding protein that acts as a transducer of Ca^{2+} signals in numerous biological pathways. Calmodulin from the yeast *Saccharomyces cerevisiae* (yCaM) shares only 60% sequence identity with vertebrate calmodulin, which is in contrast to the >95% sequence identity found among most other calmodulins. More interestingly, yCaM displays different biochemical properties from that of the vertebrate, including its Ca^{2+}-binding properties and its ability to activate target enzymes of vertebrate calmodulin. These biochemical differences suggest that certain aspects of the yCaM structure will be different and a structure determination is therefore warranted.

In addition, it has been shown *in vivo* and *in vitro* that yCaM can bind target enzymes in a Ca^{2+}-independent manner, suggesting that the Ca^{2+}-free form of the protein is biologically relevant. In an attempt to gain some insight into the structural ramifications of Ca^{2+} binding, we have initiated multidimensional heteronuclear NMR experiments aimed toward the elucidation of the three-dimensional structure of both the Ca^{2+}-bound and the Ca^{2+}-free form of yeast calmodulin. Having the *S. cerevisiae* structure available to compare and contrast with the existing vertebrate calmodulin structure will provide a structural framework to analyze the Ca^{2+}-binding properties of the two calmodulins. To this end, yCaM has been overexpressed in *Escherichia coli* and uniformly labeled with either ^{15}N or with ^{15}N and ^{13}C. Progress toward the assignment of the polypeptide backbone as well as spectra showing the effects of Ca^{2+} interacting with yCaM will be presented.

THE THREE-DIMENSIONAL STRUCTURE OF WATER NEAR COMPLEX SOLUTES: A NEW THEORY AND A NEW WATER MODEL

Y. Liu and T. Ichiye

Department of Biochemistry and Biophysics, Washington State University, Pullman, Washington

ABSTRACT

An integral equation theory for the three-dimensional (3-D) structure of water around complex, globular solutes is presented. A key ingredient of this single-site theory is that it uses a new potential for liquid water, referred to as the soft, sticky dipole (SSD) model. The SSD gives good structure and energetic descriptions of water, has a dielectric constant of 76, and is nearly an order of magnitude more efficient in simulations than most existing multisite models. The hypernetted-chain (HNC) type of integral equation theory utilizes the simplicity (single site) of the SSD to give the 3-D structure of water. The theory is tested by studying water around various types of solutes: methane, a NA^+ or Cl^- ion, clusters of water molecules, and formamide. The theory predicts water structure in qualitatively good agreement with results from Monte Carlo simulations.

COMBINED QUANTUM MECHANICS/MOLECULAR MECHANICS METHODS AND THEIR APPLICATIONS TO BIOMOLECULAR MODELING

X.P. Long,[1] R.L. Ornstein,[1] M.D. Paulsen,[1] G.E. Amord,[1] J.I. Manchester,[1] M.F. Guest,[2] J.B Nicholas,[2] and J. Nichols[2]

[1]Biomolecular Modeling and Simulations Group, Environmental Molecular Sciences Laboratory, Pacific Northwest National Laboratory, Richland, Washington

[2]High Performance Computational Chemistry Group, Environmental Molecular Sciences Laboratory, Pacific Northwest National Laboratory, Richland, Washington

ABSTRACT

We have developed combined quantum mechanics/molecular mechanics (QM/MM) methods in *NWchem*, Pacific Northwest Laboratory's "home-made," highly optimized computational chemistry package. In these methods, a large molecular system can be partitioned into two regions, QM and MM. The MM method is to treat by the standard MM empirical force field, e.g., Amber. The QM method is to treat by either *ab initio* or semiempirical Hartree-Fock methods, or density functional theory (DFT). By partitioning the chemically active part of a system (which is usually a small portion, such as an enzyme active site) as a QM region and the chemically inactive part as an MM region, the quantum chemistry behavior of a large molecular system, such as an enzyme, can be studied at an acceptable computation cost.

We have made extensive calculations on small model systems, such as water clusters and dipeptides, to test the accuracy and reliability of the methods in calculating structure, energetics, and electron density distributions of QM atoms. We have also started to apply the methods to biomolecular modeling, particularly to studies of cytochrome P450cam and haloalkane dehalogenase catalysis.

We will present a general description of the methods and some tests and preliminary P450cam and dehalogenase application results.

PREDICTION OF THE RATE OF REDUCTIVE DEHALOGENATION CATALYZED BY NOVEL P450CAM VARIANTS

J. Manchester and R.L. Ornstein

Environmental Molecular Sciences Laboratory, Pacific Northwest National Laboratory, Richland, Washington

ABSTRACT

Chlorinated methanes and ethanes undergo reductive dehalogenation under a variety of biotic and abiotic conditions. The one-carbon compounds react via hydrogenolysis of a carbon-halogen bond, eliminating a single chloride for every two reducing equivalents consumed. On the other hand, two chlorides and the corresponding ethylene product are detected for the two-carbon compounds, consistent with a ß-elimination mechanism. In both cases, however, the first step is the transfer of a single electron into the substrate. The observation by several investigators that the reduction potential of the substrate appears to correlate with its reactivity toward reductive dehalogenation suggests that this first electron transfer is the rate-limiting step in the reaction. Under the Marcus theory, the electron transfer rate depends on the difference in midpoint potential between the donor and acceptor and can thus be estimated from the redox properties of these groups in their ground states. Practically, this electron transfer event occurs when the donor and acceptor are associated in a precursor complex. The apparent electron transfer rate, and therefore the observed rate of dehalogenation, is mediated by both the frequency of precursor complex formation and the redox potentials of the donor and acceptor molecules.

We have used molecular dynamics (MD) simulations to estimate association constant K_A of the precursor complex consisting of the heme cofactor and a number of chlorinated hydrocarbons that are substrates in the reductive dehalogenation catalyzed by cytochrome P450cam under anaerobic conditions. The reduction potentials of these substrates have been estimated using their calculated vertical electron affinities. Using the Marcus theory, an expression relating these two calculated quantities to the apparent rate constant for the overall dehalogenation reaction is readily obtained. We will present the results of MD simulations on several mutant enzymes, and the estimated rate enhancement over the wild-type enzyme for various substrates, according to this expression. Predictions will be compared with experimental results where available.

STUDY OF INTERSUBUNIT INTERACTIONS IN METHEMOGLOBIN ASSOCIATED WITH PERTURBATION IN THE LIGAND POCKET

R. Nirmala and J.H. Rifkind

Laboratory of Cellular and Molecular Biology, Gerontology Research Center, National Institute on Aging, National Institutes of Health, Baltimore, Maryland

ABSTRACT

The tertiary and quarternary structural changes caused by ligand binding to hemoglobin has been studied by X-ray crystallography and NMR spectroscopy as well as by theoretical methods. In these studies, it is generally assumed that changes in the geometry of the ligand binding site in one subunit is not the direct consequence of structural changes in the heme pocket, directly induced by ligand binding at another subunit. Instead, it is thought to be transmitted via the proximal side of the heme through the changes in the close packing of the interface between the subunits. We have earlier shown in experimental EPR studies that distal-side perturbations can be transmitted between the subunits via the $\alpha_1\beta_1$ interface. We have now investigated the communication between the subunits when the heme pocket is altered by perturbing the position of the distal histidine. For this purpose, employing QUANTA simulation package, different structures were generated from the crystal structure of methemoglobin by moving the distal histidine closer to Fe using distance constraints and minimization for 4000 steps by the Adopted Basis Newton-Raphson method. This was followed by molecular dynamics simulations on these structures using CHARMM in Intel parallel computer for 100 ps. We observed that the structural changes in one binding site could be transmitted across the $\alpha_1\beta_1$ interface through the correlated motion between H helices. This has been accomplished by refined molecular dynamics studies, using domain correlation and distance maps. The residues lying along the interface undergo dihedral transitions in a way to favor such communication between the subunits. These interactions are expected to modulate the changes associated with the T to R transition. The potential energy surface proposed based on spectroscopic results is reproduced by these computer simulations supporting the existence of bis(Histidine) complex.

THE ROLE OF THUMB MOVEMENT AND DNA BENDING IN POLYMERASE FIDELITY

H. Pelletier,[1] M.R. Sawaya,[1] W. Wolfle,[2] S.H. Wilson,[2] and J. Kraut[1]

[1]Department of Chemistry and Biochemistry, University of California, San Diego, California

[2]Sealy Center for Molecular Science, University of Texas Medical Branch, Galveston, Texas

ABSTRACT

Crystal structures of pol β-DNA binary complexes and pol ß-DNA-nucleotide ternary complexes have revealed two structurally homologous DNA binding motifs in the pol β: the principal determinants in constructing a model for a pol β-gapped DNA complex. One motif is located in the fingers subdomain and binds to the DNA primer near the 3'-OH terminus; the other motif, which is located in the 8kD domain, is adjacent to a lysine-rich 5'-PO_4 binding pocket and most likely binds to the 5' terminus of the downstream oligonucleotide of a gapped-DNA substrate. The resulting model for productive binding of pol β to a gapped-DNA substrate suggests a possible 90° bend in the single-stranded DNA template which may enhance nucleotide selectivity significantly during processive polymerization. When structural data are combined with kinetic data, all results point to an induced-fit mechanism for polymerase fidelity. In vivo, by means of this mechanism, only the "correct" nucleotide can trigger a global, rate-limiting conformational change, where both a bent DNA template and a highly flexible thumb subdomain simultaneously close down on the polymerase active site and allow the nucleotidyl transfer reaction to occur.

PROTEIN DYNAMICS STUDIES VIA NMR RELAXATION EXPERIMENTS

J.W. Peng

Vertex Pharmaceuticals Inc., Cambridge, Massachusetts

ABSTRACT

The relationship between protein dynamics and protein function is drawing increasing scrutiny by both theoretical and experimental approaches. NMR spectroscopy offers an attractive experimental approach. In particular, current multidimensional NMR techniques allow one to obtain protein dynamics information on an essentially residue-by-residue basis, over a broad range of time scales.

Such techniques typically involve the measurement of various relaxation parameters for nuclei such as 13C, 15N, and 1H. Accordingly, this presentation will discuss both the current methodologies for the measurement of such relaxation parameters, and we will also discuss how such parameters are used to construct a comprehensive picture of protein flexibility. Examples of applications of these methodologies aimed at elucidating the backbone dynamics of various proteins, as well as ligand-binding processes, will also be presented.

REDESIGN OF COPPER SITES IN CUPREDOXINS

J. Sanders-Loehr, C.R. Andrew, and N. Nakamura

Department of Chemistry and Biochemistry, Oregon Graduate Institute of Science and Technology, Portland, Oregon

ABSTRACT

Complexes in which copper is coordinated to a cysteine sulfur atom are frequently used as electron transfer sites in proteins. The type 1 proteins (or cupredoxins), with their unusual His_2Cys trigonal coordination and short Cu-S bond, are particularly adapted to rapid electron transfer since the geometry is well-suited to both Cu^{1+} and Cu^{2+} oxidation states. We have found that resonance raman (RR) spectroscopy, a technique that detects molecular vibrations, is an excellent tool for identifying these sites; the Cu-S stretching frequency is a sensitive indicator of Cu-S bond length. In the protein, azurin, replacement of either *His* ligand by *Gly* leads to a surprising flexibility in coordination geometry. Addition of exogenous imidazole recreates the trigonal site with its short Cu-S bond (Cu-S stretch near 400 cm^{-1}); addition of a bidentate exogenous ligand such as histidine leads to a 4-coordinate square planar structure (Cu-S stretch near 300 cm^{-1}) more typical of a type 2 site. In Cu,Zn superoxide dismutase, the tetrahedral Zn site has been altered by the replacing a *His* ligand with *Cys* and replacing Zn with Cu to generate a trigonal type 1 Cu site. In this case, the protein structure is strongly influencing the coordination geometry. Finally, addition of a second *Cys* ligand (and a second Cu) to a cupredoxin produces a mixed-valence species similar to the Cu_A site in cytochrome *c* oxidase. The RR spectrum is suggestive of the presence of two type 1 Cu-cysteinates with novel Cu-Cu bond. Such a Cu(1.5)-Cu(1.5) species would be an excellent facilitator of rapid electron transfer.

EFFECT OF N-CAP MUTATIONS ON THE STRUCTURE AND DYNAMICS OF *ESCHERICHIA COLI* HPr

R. Thapar,[1] J.M. Scholtz,[2] E.B. Waygood,[3] and R.E. Klevit[1]

[1]Department of Biochemistry, University of Washington, Seattle, Washington

[2]Department of Biochemistry, Texas A&M University, College Station, Texas

[3]Department of Biochemistry, University of Saskatchewan, Saskatoon, Saskatchewan, Canada

ABSTRACT

Histidine-containing protein (HPr) is a bacterial phosphocarrier protein, and different HPr variants have been well-characterized by crystallographic and NMR techniques. The overall folding topology of HPR is that of an open-faced β-sandwich with three helices whose apolar side chains pack on one face of a four-stranded β-sheet. Helix-B of HPR forms the binding interface for its interaction with Enzyme IIA and, intriguingly, this helix is dynamically averaged in solution. Biochemical studies have shown that substitution of the helix N-cap residue Ser-46 with aspartic acid makes HPr a poor substrate in the PTS. In order to understand the mechanism of inactivation of S46D HPr, we have characterized *Escherichia coli* S46D by NMR spectroscopy and circular dichroism. NMR parameters such as chemical shift, NOEs, amide exchange rates, and coupling constants indicate that no conformational change occurs in S46D Hpr. However, amide protons in helix-B become protected from exchange in conformational stability of S46D HPr. In addition, solvent denaturation studies reveal that the global conformational stability of S46D HPr is increased by 1 kcal/mole compared to the wild-type protein. To further investigate the mechanism by which such stabilization occurs we have similarly characterized the mutants S46N and S46A. The structural, dynamic, and thermodynamic consequences of mutagenesis at the helix N-cap will be presented.

MODELING STRUCTURES OF SINGLE-STRANDED DNA MOLECULES IN THE GENE V PROTEIN DNA COMPLEXES

C.S. Tung and T.C. Terwilliger

Los Alamos National Laboratory, Los Alamos, New Mexico

ABSTRACT

The gene V protein binds to single-stranded nucleic acids with high affinity but little sequence specificity. It provides an ideal system for studying strong yet non-sequence-specific interactions between protein and DNA. A modeled structure of the gene V protein in the protein-DNA complex form was proposed, based on the crystal structure solved using the multiwavelength anomalous diffraction (MAD) technique. This modeled protein structure, which forms a left-handed helix, is used as a template for constructing the single-stranded DNA molecule in the complex. An algorithm that is capable of modeling single-stranded DNA molecules in atomic details with fixed end-to-end distances has been developed in our laboratory. Based on this algorithm and the Metropolis Monte Carlo simulation approach, we have developed a method that is capable of sampling different allowed conformations of the repeating ss-DNA unit in the protein-DNA complex. DNA molecules with different sequences and strand orientations were modeled and studied. Ensembles of DNA structures were found that could fit inside the DNA-binding cleft of the protein in the helical structure. This family of allowed structures indicates that the space accessible to the backbone of the DNA molecule is smaller than that accessible to the bases. Our results also suggest that the DNA molecule might bind equally well in both strand orientations. Specific interactions between the ss-DNA and the protein will be discussed in detail.

CRYSTALLOGRAPHIC STUDIES OF A NEW HEAT-LABILE ENTEROTOXIN, LT-IIB

F. van den Akker, S. Sarfaty, and W.G.J. Hol

Department of Biological Structure and Biochemistry, Biomolecular Structure Center and Howard Hughes Medical Research Institute, University of Washington, Seattle, Washington

ABSTRACT

Certain strains of enterotoxigenic *Escherichia coli* produce choleratoxin-like, heat-labile enterotoxins (LT-I). In recent years a second class of LT has been discovered (LT-II) which has a very similar quaternary AB_5 subunit arrangement as well as the toxic capability to activate adenylate cyclase. The catalytic A subunit possesses about 55% sequence identity with LT-I, while within the B-subunit no sequence similarity with LT-I could be detected. The structure of LT-I was solved in our laboratory a few years ago, and determining the structure of this new LT-II will give us more insight into the peculiar subunit assembly and receptor recognition of the AB_5 bacterial toxins.

The space group of LT-IIb appeared to be $P3_121$ with cell dimensions a=105.5 Å, c=171.7 Å. A 2.25-Å data set has been collected from a single crystal with an R_{merge} of 6.3%. The structure is solved by SIR with K_2PTCl_4 as the derivative. An initial 15-5-Å SIR map, without anomalous data, shows features of a 5-fold arrangement of long rods. These rods could be superimposed onto the 5 long α-helices found in the LT-IB pentamer and provided the first evidence of structural similarity between the B subunits of the two toxins. Solvent flattening, combined with phase extension and 5-fold averaging, improved the electron density dramatically. The LT-IIb structure is currently refined to 2.25-Å resolution with an R-factor of 19.6% with good geometry. The B-pentamer of LT-IIb shows the same fold (the "OB-fold") as the pentamers of LT-I and other AB_5 toxin structures reported to date; e.g., pertussis toxin and shiga toxin. This constitutes a remarkable level of structural homology, even in the absence of detectable sequence identity. A detailed B subunit comparison between LT-IIb and LT-I reveals that, of 99 amino acids, 68 are in structural equivalent positions and can be superimposed with an rmsd of 1.4 Å. The overall sequence identity between the LT-IIb and LT-I B subunit sequence is 11%.

INDEX TO AUTHORS